Processes and Applications of Plant Ecology

Processes and Applications of Plant Ecology

Edited by Jude Boucher

SYRAWOOD
PUBLISHING HOUSE

New York

Published by Syrawood Publishing House,
750 Third Avenue, 9th Floor,
New York, NY 10017, USA
www.syrawoodpublishinghouse.com

Processes and Applications of Plant Ecology
Edited by Jude Boucher

© 2019 Syrawood Publishing House

International Standard Book Number: 978-1-68286-662-7 (Hardback)

Cataloging-in-Publication Data

Processes and applications of plant ecology / edited by Jude Boucher.
    p. cm.
Includes bibliographical references and index.
ISBN 978-1-68286-662-7
1. Plant ecology. 2. Botany. 3. Ecology. I. Boucher, Jude.
QK901 .P76 2019
581.7--dc23

TABLE OF CONTENTS

PREFACE

Plant ecology is a sub-discipline of ecology that studies the relationship between plants with reference to the environment. Comparative studies of plant morphology, evolutionary history and population biology are used to create a comprehensive understanding of plant groups and sub-groups. Plant ecology also extends to soil studies and helps agriculturists study and understand the influence of biotic and abiotic factors on various components of the plant ecosystem. Some of the diverse topics covered in this book address the processes and applications of plant ecology. It is a valuable compilation of topics, ranging from the basic to the most complex advancements in this field. This book aims to equip students and experts with the advanced topics and studies in the field of plant ecology.

This book is the end result of constructive efforts and intensive research done by experts in this field. The aim of this book is to enlighten the readers with recent information in this area of research. The information provided in this profound book would serve as a valuable reference to students and researchers in this field.

At the end, I would like to thank all the authors for devoting their precious time and providing their valuable contribution to this book. I would also like to express my gratitude to my fellow colleagues who encouraged me throughout the process.

Editor

Travelling at a slug's pace: possible invertebrate vectors of *Caenorhabditis* nematodes

Carola Petersen, Ruben Joseph Hermann, Mike-Christoph Barg, Rebecca Schalkowski, Philipp Dirksen, Camilo Barbosa and Hinrich Schulenburg*

Abstract

Background: How do very small animals with limited long-distance dispersal abilities move between locations, especially if they prefer ephemeral micro-habitats that are only available for short periods of time? The free-living model nematode *Caenorhabditis elegans* and several congeneric taxa appear to be common in such short-lived environments, for example decomposing fruits or other rotting plant material. Dispersal is usually assumed to depend on animal vectors, yet all current data is based on only a limited number of studies. In our project we performed three comprehensive field surveys on possible invertebrate vectors in North German locations containing populations of *C. elegans* and two related species, especially *C. remanei*, and combined these screens with an experimental analysis of persistence in one of the vector taxa.

Results: Our field survey revealed that *Caenorhabditis* nematodes are commonly found in slugs, isopods, and chilopods, but are not present in the remaining taxonomic groups examined. Surprisingly, the nematodes were frequently isolated from the intestines of slugs, even if slugs were not collected in close association with suitable substrates for *Caenorhabditis* proliferation. This suggests that the nematodes are able to enter the slug intestines and persist for certain periods of time. Our experimental analysis confirmed the ability of *C. elegans* to invade slug intestines and subsequently be excreted alive with the slug feces, although only for short time periods under laboratory conditions.

Conclusions: We conclude that three invertebrate taxonomic groups represent potential vectors of *Caenorhabditis* nematodes. The nematodes appear to have evolved specific adaptations to enter and persist in the harsh environment of slug intestines, possibly indicating first steps towards a parasitic life-style.

Keywords: *Caenorhabditis elegans*, *Caenorhabditis remanei*, Phoresy, Commensalism, Parasitism, *Arion*, Vector-mediated migration, Ephemeral habitats

Background

The laboratory model system *Caenorhabditis elegans* is used in many biological disciplines, however, information on its natural life history is still scarce. A more natural context is needed to enhance our understanding of gene function, especially for those genes that are only relevant for worm life-history in the field [1]. *C. elegans* has been found frequently in ephemeral environments like rotting fruits or decomposing plant material [2–6]. These environments lack continuity because abiotic (e.g. temperature) and biotic factors (e.g. food availability) often fluctuate. Because of these fluctuations the worm seems to face a high level of unpredictability in nature, including periods with highly unfavorable conditions (e.g., high temperatures, absence of food microbes, or the presence of pathogenic bacteria), which it can escape in space, time, or a combination thereof. Dauer larvae formation represents a likely strategy for an escape in time and is very well studied under laboratory conditions [7–9]. In contrast, we currently have very little information about

*Correspondence: hschulenburg@zoologie.uni-kiel.de
Department of Evolutionary Ecology and Genetics, Zoological Institute
Christian-Albrechts University, Am Botanischen Garten 1-9, 24118 Kiel,
Germany

escape in space, especially as *C. elegans* seems unlikely to possess the necessary mobility itself considering its small size and sensitivity to desiccation [10]. *C. elegans* shares its habitats with various invertebrates and even seems to be closely associated with some of the species. These associations are commonly assumed to be part of a dispersal strategy to avoid harsh environmental conditions [11]. Interestingly, escape in space seems to be connected to escape in time, because dauer larvae are often found in association with diverse invertebrates, particularly snails, slugs, and isopods [2–4, 11–16]. The characteristic waving behavior of dauer larvae may therefore represent an adaptation to nematode-invertebrate association [11].

It is further conceivable that *C. elegans* exhibits other types of interactions with invertebrates, including necromeny and parasitism, as reported for other nematode species [11, 14, 15]. Particularly slugs show a large variety of associated nematodes which are found attached to the body or also proliferating internally [14, 17]. *C. elegans* and other *Caenorhabditis* species have been found occasionally inside of slugs [14, 17–19]. It is currently unknown whether this type of association is common or may represent an escape strategy in space with immediate access to a novel source of food, such as bacteria present in the slug's intestines.

Here, we present our results on quantitative analysis of a wide range of invertebrates over a time span of three years to characterize their association with *Caenorhabditis* species. An initial screen focused on slugs and isopods as they are known to associate with *Caenorhabditis*. Sampling of a broader range of invertebrates subsequently aimed at identifying novel associations. These two screens revealed that *Caenorhabditis* nematodes are commonly found in the intestine of the slugs, especially of the genus *Arion*. A third screen aimed at validating this finding through a more detailed analysis of 544 slugs, mainly of the *Arion* genus, originating from 21 sampling sites. We complemented our findings with the help of two laboratory experiments, in which we assessed the ability of different nematode stages to invade and persist in the gut of *Arion* slugs across time.

Methods
Sampling sites and sampled invertebrates
We carried out three independent screens of invertebrates to reveal their association with common *Caenorhabditis* species. The samplings were carried out between July 2011 and October 2014. During the first screen between July 2011 and October 2012 a total of 23 slugs and 93 isopods were sampled from compost and rotten apples from three North German locations (Kiel, Münster, and Roxel). Isopods and slugs were collected in parallel to substrate samples, which we analyzed

previously and found to harbor *C. elegans* and *C. remanei* and occasionally *C. briggsae* at all three sampling locations (for further details see our previous work [6]). In Kiel the invertebrates were collected in the botanical garden (54°20′N and 10°06′E) from three large compost heaps and additionally from a locally separated apple heap. In Münster, the invertebrates originated from a compost heap and apple trees in close vicinity on a meadow of the city's farming museum (51°56′N and 7°36′E). In Roxel (51°57′N and 7°32′E) the invertebrates were collected in a private garden from three small compost heaps. A second independent screen was performed in the botanical garden in Kiel between July and September 2013 to include a broader spectrum of invertebrates. A total of 373 invertebrates (93 isopods, 56 flies, 51 chilopods, 41 spiders, 41 beetles, 35 slugs, 12 locusts, 10 bugs and 34 other invertebrates) were sampled exclusively from compost. A third independent screen was carried out between July and October 2014 to examine the potential of slug intestines for *Caenorhabditis* dispersal. 544 slugs were collected from 21 locations in Kiel or the close surroundings (Table 1). Additionally, 123 substrate samples (e.g. soil, grass, straw, leaves) were sampled from the same locations to assess whether the slugs picked up the worms at the corresponding sampling sites. The substrate samples were each collected in separate 50 ml Falcon tubes directly from underneath or within 10 cm distance to a slug. The sampling sites included six parks, four private gardens, five paths, four compost heaps, a forest and a meadow (Table 1, Additional files 1, 2, 3).

Collection of invertebrates and isolation and identification of *Caenorhabditis* species
The invertebrates were collected and depending on their size placed individually in either 2 ml Eppendorf or in 50 ml Falcon tubes. Substrate samples were collected in plastic bags or 50 ml Falcon tubes. All invertebrates and substrates were processed within 24 h after sampling. The invertebrates were killed with a scalpel and placed individually on a peptone free medium (PFM) agar plate [20]. A spot of *Escherichia coli* OP50 was used to attract worms. Approximately 5 g of a substrate sample was placed around an OP50 spot on separate plates. Throughout the second sampling screen the slugs were analyzed in more detail. During our sampling we focused on the slug family Arionidae, which was the most frequent family to be found. The slugs were killed by cutting off the head with a scalpel. The intestines were extracted, and the slug body separated in four equally sized parts (from anterior to posterior end), in order to obtain a first approximate indication of the slug body region which contains the associated *Caenorhabditis* nematodes. Each part was analyzed for the presence of *Caenorhabditis*

Table 1 Description of sampling sites used for slug mass sampling between July and October 2014

Code	Loc type	Description	GPS
G	Path[a]	Path "Schwarzer Weg"	54°20'58.4"N 10°06'51.4"E
J	Path[a]	Path "Schwarzer Weg"	54°21'35.0"N 10°07'10.2"E
P	Path[a]	Small path "Russee" along a brook, between a living area and a lake	54°18'08.0"N 10°05'12.8"E
W	Path[a]	Path "Melsdorf Landstrasse" close to street	54°18'59.4"N 10°02'08.9"E
BB	Path[a]	Path "Mühlenweg"	54°20'36.4"N 10°07'03.7"E
L	Park	Surroundings of the lake "Russee" shady because of trees	54°17'55.9"N 10°05'01.4"E
N	Park	Park "Moorteichwiesen" with some smaller water areas, frequently used by humans	54°18'37.7"N 10°07'15.1"E
R	Park	Park "Schützenwallpark" next to a lake, many stinging-nettles, frequently used by humans	54°19'00.7"N 10°06'47.4"E
U	Park	Flower bed in a park "Schlossgarten"	54°19'40.9"N 10°08'40.6"E
V	Park	Meadow in the old botanical garden, meadow is surrounded by trees	54°19'50.8"N 10°08'47.3"E
Z	Park	Park "Forstbaumschule", meadows and trees	54°20'56.1"N 10°08'29.8"E
H	Garden	Private garden without compost	54°22'32.3"N 10°08'07.7"E
Q	Garden	Path "Russee" close to garden plots, close to a brook, apple and plum trees available	54°18'14.6"N 10°05'28.9"E
S	Garden	Private garden with apple trees in a village close to Kiel	54°13'13.2"N 10°03'54.1"E
Y	Garden	Natural finish gardens in a trailer park, apple trees and other fruit trees including rotting fruits	54°18'41.4"N 10°05'00.2"E
M	Compost	3 Big compost heaps in the botanical garden, different decomposing stages, partly covered by straw and pumpkins	54°20'47.0"N 10°07'03.8"E
O	Compost	Compost close to a beach volleyball cort, mainly grass, leaves and soil, approx. 15 m distance to a sports field	54°20'38.3"N 10°06'56.1"E
T	Compost	Private garden with compost, mostly grass and some kitchen garbage	54°20'30.2"N 10°05'52.0"E
X	Compost	Private garden with several compost heaps and an apple tree	54°20'25.1"N 10°07'35.7"E
I	Meadow	Meadow "Kopperpahler Au" with weed, separated by a path, partly next to a small river	54°20'46.5"N 10°05'27.9"E
K	Forest	Forest "Tiergehege Tannenberg"	54°21'53.0"N 10°07'04.0"E

loc type location type.

[a] A path either tarred or made out of sand, often some grass areas in close proximity, but without the big grass lawn found in parks (see Additional file 2A–E).

nematodes [20] on an individual plate. The third sampling of 2014 focused exclusively on the slug intestine. All plates were checked for nematodes within 5 h after placing the sample on the plate and again after approximately 24 and 48 h. Worms that showed characteristics of *Caenorhabditis* [20] were isolated and placed individually on 6 cm PFM plates with an OP50 spot. Occasionally occurring males where placed together with a female or hermaphrodite from the same sample. The isolated worms were left for 5–7 days at room temperature and DNA was isolated from worms that produced offspring [6]. During the sampling of various invertebrates we focused on the identification of the most common *Caenorhabditis* species found in Northern Germany. *Caenorhabditis* species were characterized following previously established and commonly used *Caenorhabditis* sampling approaches [4–6, 16, 20], based on three criteria: (1) morphological features characteristic for *Caenorhabditis* [20], (2) the production of offspring from single individuals, which is at least indicative of one of the self-fertilizing hermaphrodite taxa; and (3) a positive result in diagnostic species-specific PCRs. For identification of *C. elegans* the two primer pairs nlp30-F and nlp30-R, targeting a variable part of the immunity gene *nlp-30* [6], and zeel/peel-left-F and zeel/peel-left-R, targeting the *zeel-1/peel-1* compatibility locus [21], were used. *C. remanei* was identified using the primer pair Cre-ITS2-F1 and Cre-ITS2-R4, targeting the ribosomal internal transcribed spacer 2 (ITS2) region [6]. During the mass sampling of slugs, the primers Cbriggsae-F and Cbriggsae-R, targeting the *glp-1* gene, were additionally used for identification of *C. briggsae* [20]. All primer pairs have been established to be diagnostic for the indicated species [6, 17].

Experimental analysis of the ability of *C. elegans* to invade and persist in slug intestines

To test *C. elegans'* ability to enter and persist in the slug intestine, we performed two laboratory-based experiments. In the first experiment, slugs were exposed to different stages of red fluorescent *C. elegans*, followed by microscopic analysis of dissected slugs. The slugs were freshly collected from nature. *C. elegans* of the frIs7 transgenic strain containing the p*nlp-29*::GFP (GFP, green fluorescent protein) and p*col-12*::dsRed reporters were used [22]. The p*col-12*::dsRed red fluorescent reporter is expressed constitutively in the epidermis, starting from the late first larval stage (L1) onwards, thus allowing identification of experimental worms in dissected intestines and thus their distinction from worms already associated with the freshly collected slugs. Approximately 15,000 synchronized *C. elegans* at either first larval (L1), fourth larval (L4), adult, or dauer larva stage were distributed on top of 25 g flower soil which was moistened with approximately 6 ml of water, a piece of cucumber and a piece of salad in a 500 ml plastic box. Each box contained only one specific stage of synchronized *C. elegans*. One slug was added to each box and boxes were closed with small meshed net to allow aeration and moistening, however preventing the escape of the slugs. Boxes without worms were used as controls. To test for worm invasion and persistence, the slugs were dissected and their intestines assessed for the presence of fluorescent worms after 24, 48 h or 6 days. At each of these time points, the intestine and the rest of the body of the killed slugs were placed separately on PFM plates with an *E. coli* OP50 spot to attract worms. For the 48 h treatment, slugs were transferred after 24 h and again after 30 h post initial exposure to a new box with fresh food and damp paper towel instead of soil, in order to separate them from fluorescent worms in the soil environment. In particular, the paper towel limits nematode

survival and proliferation outside of the slug, thus minimizing the likelihood of repeated *C. elegans*-uptake by the slugs. For the 6 day-treatment, slugs were similarly transferred every 24 h to a new box with damp paper towel and fresh food. The dissected intestines and body remainder were analyzed for presence of worms 24–30 h after placing them on the PFM plate. Worm abundance was scored in five categories: no worms (category 0), 1–10 worms (category 1), 11–30 worms (category 2), 31–50 worms (category 3) and more than 50 worms (category 4). Scoring was performed without knowledge of the *C. elegans* stage that was initially added to the slug, in order to avoid observer bias. For this experiment, we analyzed a total of 31 slugs after 24 h (6 slugs for the dauer larvae treatment, 9 for the L1, 6 for the L4, 5 for the adult, and 5 for the no-worms control treatment); 25 slugs after 48 h (7 slugs for the dauer larvae treatment, 5 for the L1, 6 for the L4, 5 for the adult, and 2 for the no-worms control treatment); and 28 slugs after 6 days (9 slugs for the dauer larvae treatment, 6 for the L1, 6 for the L4, 3 for the adult, and 4 for the no-worms control treatment).

In a separate second experiment, we analyzed slug feces to assess whether *C. elegans* is able to survive the entire passage through the digestive system. A total of nine slugs were included in this experiment. Of these, two served as negative controls, which were not exposed to the fluorescent *C. elegans*. The remaining seven slugs were each exposed to approximately 15,000 fluorescent adult worms for 3 h. The worms were added first in 2 ml M9 buffer (42 mM Na2HPO$_4$, 22 mM KH$_2$PO$_4$, 86 mM NaCl, and 1 mM MgSO$_4$·7H$_2$O) to 25 g wet flower soil, a piece of cucumber and salad in 500 ml plastic boxes. Slugs were placed in the box immediately after transferring the worms. The slugs were transferred to 1,000 ml boxes with fresh food but without soil in regular intervals to reduce the likelihood of worm survival outside of slugs. Within the first 12 h the feces were collected hourly to avoid secondary colonization by nematodes. Slugs were left unobserved overnight (9 h; total of 21 h after start) and feces collected every 3 h in the following 15 h (total of 36 h after start). After another 12 h (total of 48 h after start) the last feces were collected and slugs killed, resulting in a total of seven time points, for which feces were analyzed. The slug intestine was analyzed for the presence of remaining *C. elegans*. Each of the droppings was transferred to a separate 6 cm PFM plate with an OP50 spot and analyzed for the presence of fluorescent worms after 24 h.

DNA barcoding analysis of *Arion* slug species

Species identity of a representative subset of the slugs collected in the third field screen and of the slugs used in the first experiment was characterized using an established DNA barcoding approach, based on DNA sequencing of a polymerase chain reaction (PCR)-amplified fragment of the mitochondrial cytochrome c oxidase subunit I gene (COI) [23]. DNA was extracted from intestinal slug tissue frozen at −20°C directly after the slug was killed. Approximately 25 mg of tissue was processed following the standard protocol of a DNeasy Blood & Tissue Kit (QIAGEN, Hilden, Germany). A 710 bp COI fragment was amplified using the universal primer pair LCO1490 and HCO2198 [23]. PCR was performed in 30 μl reaction volume, containing 1 μl isolated DNA, 1 unit of Taq polymerase and otherwise following polymerase manufacturer's instructions (Promega, Madison, USA). PCR cycling consisted of 95°C for 2 min, followed by 35 cycles of 95°C for 1 min, 40°C for 1 min and 72°C for 1.5 min, and a final extension step at 72°C for 7 min. The PCR product was directly subjected to Sanger sequencing in both directions with the PCR primers at the Sequencing facility of the Institute of Clinical Molecular Biology, Kiel University, Germany. For each sample, the resulting corresponding two sequences were aligned and a consensus sequence was produced for the overlapping part, yielding a mean fragment length of 594.8 ($\pm$2.52 standard error of the mean, SEM). This fragment was subjected to a BLAST comparison with the public NCBI Nucleotide collection (nr/nt) database [24]. We recorded species designations of the first three most similar sequences, which in all cases showed a similarity of more than 99.1% (average BLAST bitscore of 1095 $\pm$ 4.31 SEM).

Statistics

The current study explores the association of *Caenorhabditis* with various invertebrates. Three types of statistical tests were applied with caution to each invertebrate group or body part separately, in order to assess the overall variation in species prevalence or nematode occurrence in a certain group or part. We used Fisher's exact test for pairwise comparisons of the number of independent invertebrate individuals containing either of the different *Caenorhabditis* species. The comparison was performed across the entire sampling period and for each invertebrate group separately. Fisher's exact test was also used for comparison of nematode abundance in different slug parts. The first experiment on the abundance of nematode stages in slug intestines and remainder was compared using ANOVA and posthoc pairwise comparisons with Tukey's HSD test. The second experiment on the amount of worms in feces across time was also compared with an ANOVA. All statistical tests were performed with the program R version 3.1.1. For each of the analyses, multiple testing was accounted for by adjusting the significance level using the false discovery rate (FDR;

[25]). Graphs were produced with R version 3.1.1 and Inkscape version 0.48.

Results

First screen: both *Caenorhabditis* species are associated with isopods and slugs

93 isopods (69 from compost, 24 from rotten apples) and 23 slugs (21 from compost, two from rotten apples) were collected from three North German locations between July 2011 and October 2012 and analyzed for the presence of *Caenorhabditis* species. Our aim was to obtain a first approximate idea of the invertebrate groups, which harbor the *Caenorhabditis* nematodes in Northern Germany. Therefore, we did not determine exact species identities of all collected isopods and slugs. We nevertheless noted that isopods mainly included three of the species that are abundant in Northern Germany, namely *Porcellio scaber*, *Oniscus asellus*, and *Armadillidium vulgare*, while almost all slugs belonged to the genus *Arion*. In addition, our previous work showed that substrate samples (i.e., compost material and/or rotten apples) from all three sites can harbor *C. elegans*, *C. remanei* and occasionally *C. briggsae* [6]. Since *C. elegans* and *C. remanei* were the dominant species in the substrates we focused on these two species. *C. elegans* and *C. remanei* were both found in association with isopods and *C. elegans* additionally with slugs (Additional file 4). *C. elegans* was isolated from eight isopods (8.6% of all isopods) and eight slugs (34.8% of all slugs). One isopod from compost carried both *Caenorhabditis* species simultaneously. *C. remanei* was found on 14 isopods (15.1% of all isopods), but was not associated with slugs during this screen. Compost isopods carried *C. elegans* (11.6%; 8 of 69 isopods) and *C. remanei* (10.1%; 7 of 69 isopods), whereas apple isopods never carried *C. elegans,* but only *C. remanei* (29.2%; 7 of 24 isopods). Consistent with this finding the apples from which the isopods were collected were not found to contain any *C. elegans* [6].

Second screen: analysis of associations with a large variety of invertebrate taxa

373 invertebrate specimens from various taxonomic groups were analyzed for the presence of the two *Caenorhabditis* species between July and September 2013. *C. elegans* and *C. remanei* were found in association with 15 or 3.8% of the sampled invertebrates, respectively, all of them from three invertebrate groups: slugs, isopods, and chilopods (Figure 1; Additional file 5). *C. elegans* and *C. remanei* differed significantly in abundance on the three invertebrate groups (Fisher's exact test for a r x c contingency table, total n = 179, $P < 0.001$; FDR-adjusted for multiple testing at $\alpha < 0.05$): 13 of 35 slugs (37.1%), 30 of 93 isopods (32.3%) and 13 of 51 chilopods (25.5%) carried

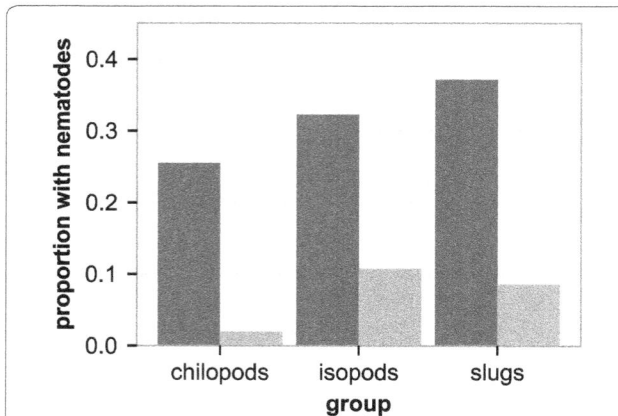

Figure 1 Proportion of chilopods, isopods and slugs found in association with *C. elegans* and *C. remanei* during the second screen between July and September 2013. Neither *C. elegans* (*dark grey*) nor *C. remanei* (*light grey*) have been found associated with other invertebrates. The overall occurrence of *C. elegans* differed from that of *C. remanei*, yet each species was isolated in similar relative frequencies from the three invertebrate groups (*C. elegans* was found in 13 out of 35 assayed slugs, 30 out of 93 isopods, and 13 out of 51 chilopods; *C. remanei* was isolated from 3 out of 35 slugs, 10 out of 93 isopods, and 1 out of 51 chilopods; Additional file 5).

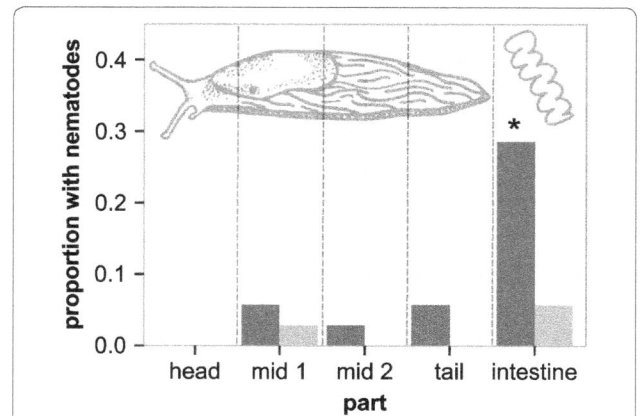

Figure 2 Proportion of different slug body sections and intestine associated with either *C. elegans* or *C. remanei* during the second screen between July and September 2013. *C. elegans* (*dark grey*) and *C. remanei* (*light grey*) proportions were calculated in relation to the total number of 35 slugs analyzed for this screen. The only value with significant variation to all others is indicated by an *asterisk*.

C. elegans, whereas *C. remanei* was isolated from three of 35 slugs (8.6%), 10 of 93 isopods (10.8%) and one of 51 chilopods (2%). As for the first screen, we focused on only the broad taxonomic invertebrate groups and did not characterize species identities. We noted again that most isopods belonged to *P. scaber*, *O. asellus*, and *A. vulgare*, while almost all slugs were from the genus *Arion*. Separate analysis of the invertebrate groups showed no significant difference in *C. elegans* and *C. remanei* occurrence on either slugs, isopods, or chilopods (in all cases, Fisher exact test for 2×2 tables, $P > 0.1$). *C. elegans* and *C. remanei* co-occurred in one slug, three isopods and one chilopod. 56 flies, 41 spiders, 41 beetles, 12 locusts, 10 bugs and 34 other invertebrates did not carry any *C. elegans* or *C. remanei*.

The collected slugs were additionally dissected to assess exact localization of the nematodes. Head, two middle parts (mid 1 and mid 2), tail and intestine were analyzed separately and variation in worm prevalence among slug parts was tested with Fisher's exact test for r x c contingency tables (Figure 2, including the total of 35 slugs for both *Caenorhabditis* species). Significantly more *C. elegans* were found in the intestine compared to the head region (Fisher's exact test, n = 35, $P = 0.008$; FDR-adjusted for multiple testing at $\alpha < 0.05$) and the second middle part ($P = 0.027$) and numbers in the intestine tended to be different to the first middle part ($P = 0.052$) and the tail ($P = 0.052$). The abundance of *C. elegans* was not significantly different between head, the two middle parts and tail (in all cases, $P > 0.1$). There were no significant differences in *C. remanei* occurrence between slug parts (in all cases, $P > 0.1$).

Third screen: comprehensive analysis of *Caenorhabditis* species prevalence in slug intestines

We characterized the intestines of a total of 544 slugs (almost all of the genus *Arion* and one *Limax maximus*; see also below) between July and October 2014. We found nematodes of three *Caenorhabditis* species in 54 of these (9.9%), originating from 16 of 21 sampling sites (76.2%; Figure 3; Table 2; Additional file 6). The sampling sites were grouped in several broad categories of location types (i.e., park, garden, compost, forest, etc.; see Table 1 and Additional files 1, 2) with general differences in structure and habitat properties, in order to assess to what extent such location differences may influence the occurrence of *Caenorhabditis*-containing slugs. *C. remanei* was found in 45 (8.5% of 544 slugs; Table 2), *C. elegans* in 15 (2.8%) and *C. briggsae* in 6 slugs (1.1%). *C. remanei* thus occurred significantly more often than *C. elegans* (Fisher's exact test for a 2×2 table, n = 544, $P < 0.001$; FDR-adjusted for multiple testing at $\alpha < 0.05$) and *C. briggsae* ($P < 0.001$), while *C. elegans* tended to occur more often than *C. briggsae* ($P = 0.05$). Variation in *C. remanei* prevalence could be explained by the sampling site (ANOVA, $P = 0.002$) but not the different location types (see Table 1) or the interaction between these two (in both cases, $P > 0.1$). Sampling sites (ANOVA, $P < 0.001$) but not location types or the

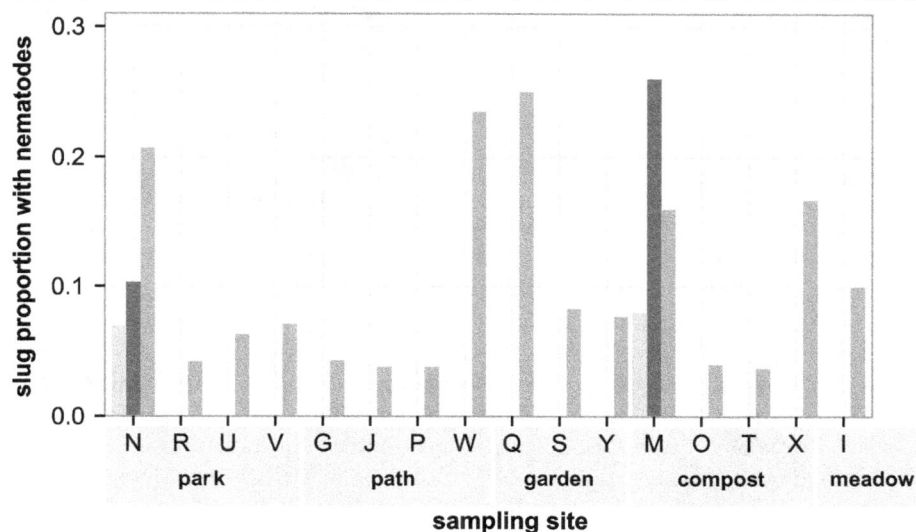

Figure 3 Occurrence of *C. elegans* (*dark grey*), *C. remanei* (*grey*) and *C. briggsae* (*light grey*) in slug intestines during the third screen between July and October 2014. *Letters* indicate the different sampling sites and are grouped by location type (see also Table 1). Slug intestines from five sampling sites did not harbor any *Caenorhabditis* (sites H, K, L, Z and BB; data not shown). The indicated proportions are always calculated in relation to the total number of slug intestines assayed at the corresponding sampling site (Table 2; Additional file 6).

Table 2 Slugs and substrates associated with *Caenorhabditis* nematodes during the third screen in 2014

Site	Loc type	Total[a]		With CR[b]		With CE[b]		With CB[b]	
		Slugs	Sub	Slugs	Sub	Slugs	Sub	Slugs	Sub
L	Park	30	4	0 (0)	0 (0)	0 (0)	0 (0)	0 (0)	0 (0)
N	Park	29	8	6 (0.21)	0 (0)	3 (0.1)	0 (0)	2 (0.07)	1 (0.13)
R	Park	24	5	1 (0.04)	0 (0)	0 (0)	0 (0)	0 (0)	0 (0)
U	Park	32	6	2 (0.06)	0 (0)	0 (0)	0 (0)	0 (0)	0 (0)
V	Park	28	6	2 (0.07)	0 (0)	0 (0)	0 (0)	0 (0)	0 (0)
Z	Park	23	5	0 (0)	0 (0)	0 (0)	0 (0)	0 (0)	0 (0)
BB	Path	13	2	0 (0)	0 (0)	0 (0)	0 (0)	0 (0)	0 (0)
G	Path	23	5	1 (0.04)	0 (0)	0 (0)	0 (0)	0 (0)	0 (0)
J	Path	26	7	1 (0.04)	0 (0)	0 (0)	0 (0)	0 (0)	0 (0)
P	Path	26	4	1 (0.04)	0 (0)	0 (0)	0 (0)	0 (0)	0 (0)
W	Path	17	3	4 (0.24)	0 (0)	0 (0)	0 (0)	0 (0)	0 (0)
H	Garden	28	8	0 (0)	0 (0)	0 (0)	0 (0)	0 (0)	0 (0)
S	Garden	24	5	2 (0.08)	0 (0)	0 (0)	0 (0)	0 (0)	0 (0)
Q	Garden	24	4	6 (0.25)	0 (0)	0 (0)	0 (0)	0 (0)	0 (0)
Y	Garden	26	11	3 (0.12)	2 (0.18)	0 (0)	0 (0)	0 (0)	0 (0)
M	Compost	50	10	8 (0.16)	0 (0)	12 (0.24)	3 (0.3)	4 (0.08)	1 (0.02)
O	Compost	25	7	1 (0.04)	0 (0)	0 (0)	0 (0)	0 (0)	0 (0)
T	Compost	27	6	1 (0.04)	1 (0.17)	0 (0)	0 (0)	0 (0)	0 (0)
X	Compost	18	6	3 (0.17)	1 (0.17)	0 (0)	0 (0)	0 (0)	0 (0)
I	Meadow	30	6	3 (0.1)	0 (0)	0 (0)	0 (0)	0 (0)	0 (0)
K	Forest	21	5	0 (0)	0 (0)	0 (0)	0 (0)	0 (0)	0 (0)

CR C. remanei, CE C. elegans, CB C. briggsae, loc type location type, *sub* substrate sample.

[a] Total number of independent slugs or substrate samples.

[b] Number and proportion (in brackets) of independent slugs or substrate samples that contained the respective *Caenorhabditis* species per site.

interaction between both ($P > 0.1$) accounts for variation in *C. elegans* occurrence. Neither the location nor the location type nor the interaction between the two factors explained the variation in *C. briggsae* occurrence (in all cases, $P > 0.1$).

C. elegans and *C. remanei* co-occurred in five slugs, four from compost (sampling site M) and one from a park (sampling site N). *C. elegans* and *C. briggsae* were isolated together from one slug and one substrate sample from compost (sampling site M). *C. elegans, C. remanei* and *C. briggsae* co-occurred in three slug intestines, one sampled from compost and two in a park (sampling site M or N, respectively). We also scored the stage of the *Caenorhabditis* that were isolated within 5 h after placing the intestine on plate. Three out of twelve isolated *C. remanei* were adults originating from sampling site G and Q (2×), one was a third instar larvae (L3; placed together with a male; sampling site G) and all other eight isolates were dauer larvae. We only isolated a single *C. elegans* and a single *C. briggsae* at this early time point and both were dauer larvae. Of the 123 additionally collected substrate samples (Additional file 6), eight (6.5%) contained *Caenorhabditis* nematodes. *C. remanei* was found in four substrates, two originating from compost (sampling site T and X) and two from a garden with rotten apples (sampling site Y). *C. elegans* was found in three compost samples (sampling site M) and *C. briggsae* occurred in two substrates (compost M and park N).

We used a DNA barcoding approach on a subset of the collected slugs to obtain a better understanding of the exact species, which harbored the different *Caenorhabditis* species. For this analysis, a total of 252 slugs was characterized (Additional files 6, 7). 194 of these belong to *Arion lusitanicus* (77.0% of the total of 252), 55 to uncharacterized *Arion* species (21.8%), and one each to *A. rufus* and *A. subfuscus* (0.4% in each case). We also confirmed identity of *L. maximus* with this approach. 23 individuals of *A. lusitanicus* harbored *Caenorhabditis* nematodes (11.9% of the total number of analyzed *A. lusitanicus*). Of these, 15 contained *C. remanei* (7.7% of all 194 tested *A. lusitanicus*), 4 *C. elegans* (2.1%), and 4 *C. briggsae* (2.1%). 16 individuals of uncharacterized *Arion* species had *Caenorhabditis* nematodes (29.1% of the 55 examined individuals of this group), twelve with *C. remanei* (21.8%) and four with *C. elegans* (7.3%). The other two identified *Arion* species did not harbor any *Caenorhabditis* worms, whereas *L. maximus* was associated with both *C. elegans* and *C. remanei* (Additional file 7). Taken together, the two most common *Arion* taxa (*A. lusitanicus* and uncharacterized *Arion* species) were most often associated with *Caenorhabditis* nematodes, especially with the most frequent nematode species of this screen, *C. remanei*.

Slug experiment: various *C. elegans* stages can enter and survive the slug intestine

We tested the ability of different *C. elegans* stages to enter and persist the intestines of slugs in two experiments. We used red fluorescent *C. elegans* to distinguish the experimental worms from nematodes that may have already been associated with the slugs, which were originally collected from nature. DNA barcoding analysis of a representative subset of slugs from the first experiment revealed that 66.7% (22 out of 33 tested; Additional file 8) belong to *A. lusitanicus* and the rest to uncharacterized *Arion* species (33.3%; 11 out of 33; Additional file 8). In this first experiment, *C. elegans* was found in the intestine and on the remainder of the body after 24 h post initial exposure, but rarely after 48 h or 6 days (Figure 4). The statistical analysis was thus focused on the 24 h exposure time point (original data available from the dryad repository, doi:10.5061/dryad.9j850). At this time point, the overall number of worms was significantly higher in the intestine compared to the remainder (ANOVA, $P < 0.001$; analysis based on a total of 26 slugs). In general, all stages were found in the intestine and on the rest of the body. Moreover, the frequencies of the various *C. elegans* stages in the intestines varied significantly (ANOVA, $P = 0.003$; analysis based on the 26 slugs; Figure 4). Dauer larvae tended to be more frequent in the intestine than the remainder (Tukey HSD, $P = 0.089$), whereas no such difference was significant for the other stages (in all cases, Tukey HSD, $P > 0.1$). A pairwise comparison of the frequencies of the various *C. elegans* stages in the slug intestines at the 24 h time point additionally revealed that L1 stages were significantly less frequent than L4 (Tukey HSD, $P = 0.019$; analysis based on the 26 slugs; Figure 4), adults (Tukey HSD, $P = 0.006$) and dauer larvae (Tukey HSD, $P = 0.044$). The number of adults, L4 s, and dauer larvae from slug intestines did not differ significantly between each other (Tukey HSD, $P > 0.1$), while in the slug remainder pairwise comparison of the frequencies of the various nematode stages did not yield any significant difference (Tukey HSD, $P > 0.1$).

Experimental analysis of feces

To test whether and for how long the nematodes are able to survive the entire passage through the digestive system of the slug we analyzed slug feces for the presence of living fluorescent nematodes (Figure 5). We found that *C. elegans* adult stages are able to enter and survive the passage, but the number of worms decreased significantly over time (ANOVA, $P < 0.001$; analysis based on seven slugs studied across seven time points; original data available from the dryad repository, doi:10.5061/dryad.9j850). Worms could no longer be recovered from the feces after 30 h. Additionally, the intestines of all slugs were

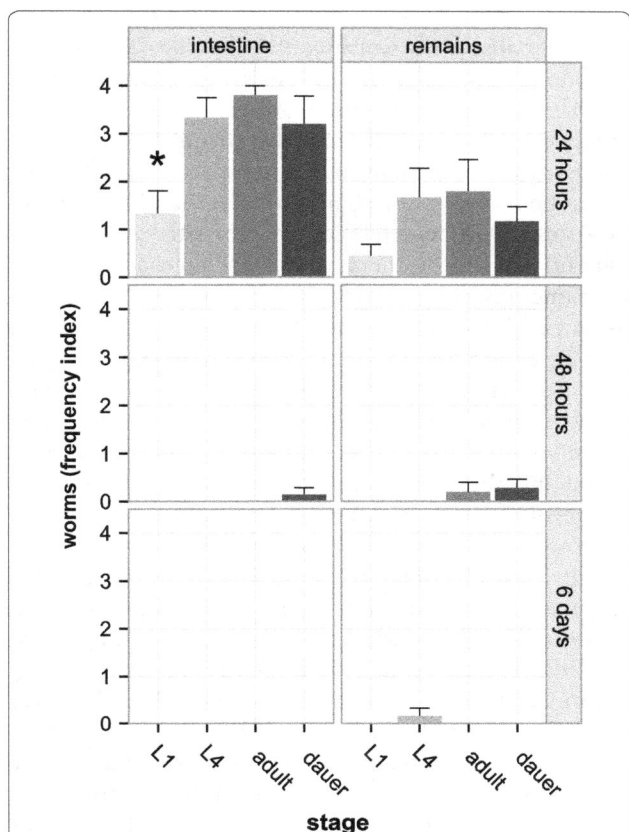

Figure 4 Different *C. elegans* stages were found in slug intestines and body remainder. The experiment was based on characterization of a total of 84 slugs. Of these, 31 were analyzed after 24 h (*top row*; 6 slugs for the dauer larvae treatment, 9 for the L1, 6 for the L4, 5 for the adult, and 5 for the no-worms control treatment); 25 slugs after 48 h (*middle row*; 7 slugs for the dauer larvae treatment, 5 for the L1, 6 for the L4, 5 for the adult, and 2 for the no-worms control treatment); and 28 slugs after 6 days (*bottom row*; 9 slugs for the dauer larvae treatment, 6 for the L1, 6 for the L4, 3 for the adult, and 4 for the no-worms control treatment). The graph does not show results for the control treatment, because these did not contain any of the labeled *C. elegans*. For the other treatments (given along the X axis), the presence of worms was separately analyzed for slug intestines and the remainder of the body (*left* and *right columns*, respectively). Worms were counted in categories (category 0 = no worms, category 1 = 1–10 worms, category 2 = 11–30 worms, category 3 = 31–50 worms, category 4 = more than 50 worms). For illustration, we calculated a frequency index by taking the average of the ordered categories per worm stage, slug body part, and time point. The Y axes show the worm frequency indices (±standard error). *C. elegans* L1, L4, adult and dauer larva stages were able to enter the intestine of slugs within 24 h. Worms associated also with the outside of the slugs and could be found on the remains. After 48 h (24 h after separating the slugs from the worms) almost no worms were found in the intestine or on the remainder. The only value that differed significantly from all others from the same body part and time point is indicated by an *asterisk*.

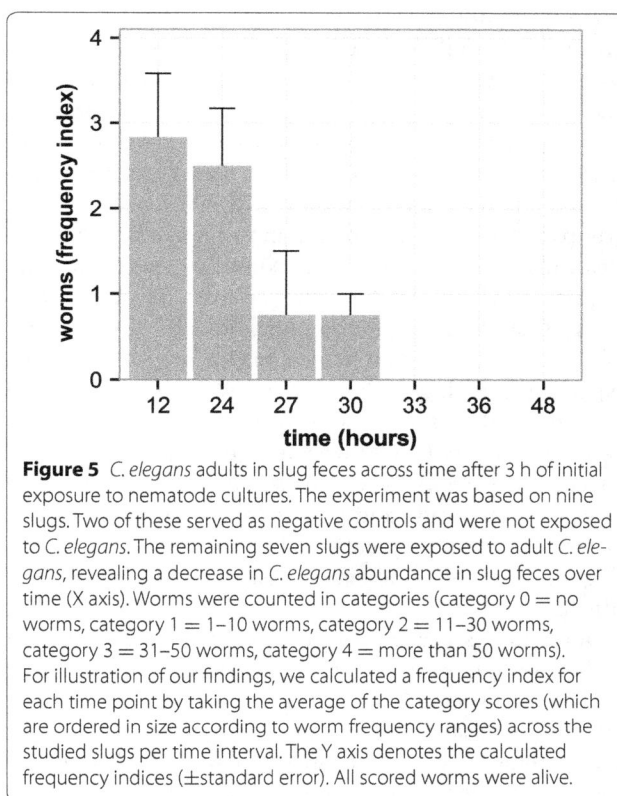

Figure 5 *C. elegans* adults in slug feces across time after 3 h of initial exposure to nematode cultures. The experiment was based on nine slugs. Two of these served as negative controls and were not exposed to *C. elegans*. The remaining seven slugs were exposed to adult *C. elegans*, revealing a decrease in *C. elegans* abundance in slug feces over time (X axis). Worms were counted in categories (category 0 = no worms, category 1 = 1–10 worms, category 2 = 11–30 worms, category 3 = 31–50 worms, category 4 = more than 50 worms). For illustration of our findings, we calculated a frequency index for each time point by taking the average of the category scores (which are ordered in size according to worm frequency ranges) across the studied slugs per time interval. The Y axis denotes the calculated frequency indices (±standard error). All scored worms were alive.

dissected at the end of the experiment (48 h) and did not contain any nematode. *C. elegans* recovered from feces were generally fertile, because we repeatedly observed eggs and L1 larvae on the assay plates, onto which feces had been transferred.

Discussion

Caenorhabditis association with possible invertebrate vectors

In this study we explore the importance of possible vectors for small-sized animals that live in ephemeral habitats using the model nematode *C. elegans*. *C. elegans* is suited for such studies for three main reasons. Firstly, *C. elegans* and several congeneric taxa are common inhabitants of short-lived environments [3, 6, 11]. Secondly, field studies can be efficiently combined with laboratory experiments in these taxa, because of the ease with which this nematode can be controlled and manipulated under laboratory conditions. Thirdly and most importantly, the available toolkit for *C. elegans* functional genetic analysis can in the future be used to dissect the genes involved in interactions with invertebrate vectors. This information is currently not available for other taxa with similar life history and would similarly help to enhance our understanding of the biology of this intensively studied model taxon. Interactions with invertebrates are assumed to be an escape strategy used by the worms under unfavorable environmental conditions. In our field survey we analyzed a total of 1034 invertebrates of different taxonomic groups for the presence of common *Caenorhabditis* species and revealed that these are commonly found in slugs, isopods, and chilopods, however are absent in other invertebrate taxa. We found indications for the presence of *C. elegans* in slug intestines, in the large third screen especially the slug *A. lusitanicus*. We then exposed 93 slugs to a total of 1200000 worms of different stages in two experiments (15000 worms to each of 80 slugs plus 13 control slugs). We confirmed hereby the ability of *C. elegans* to invade and persist for a short time in slug intestines and subsequently to be excreted alive with the slug feces, possibly indicating first steps towards a parasitic life-style.

Our findings can be placed in context of our current understanding of *Caenorhabditis* ecology. Most *Caenorhabditis* species have been found in microbe-rich organic material [3, 4, 6], an environment which they often share with various invertebrates. Several invertebrate taxa including isopods, millipedes, snails, and slugs were previously reported to harbor *Caenorhabditis* nematodes [2–4, 11–17, 26]. Our comprehensive screens revealed that *C. elegans* and *C. remanei* are commonly associated with slugs, isopods and chilopods, but with none of the other invertebrate taxa studied. Humidity may be of key importance for the ability of nematodes to attach to invertebrates, consistent with our previous observation of the influence of humidity for the general occurrence of *Caenorhabditis* species in rotting plant material [6]. Most nematode stages suffer severely from dehydration if exposed to dry environments [10, 11]. Even *C. remanei* dauer larvae were previously observed to stay for up to 5 days attached to their isopod host in dry environments but abandoned the isopod immediately in a damp environment [12]. *C. elegans* and related species may take advantage of the moist micro-environment encountered in some invertebrates, especially snails and slugs, which have previously been found to harbor *Caenorhabditis* nematodes [14, 26]. These gastropod groups constantly produce mucus, for example to aid locomotion [27, 28] or to attach to substrates [29], thus providing a generally humid environment favorable for the nematodes. As discussed in more detail below, slug intestines may be even more advantageous because they provide humidity as well as potential food microbes.

Neither *C. elegans* nor *C. remanei* occurred on flies, beetles, spiders, locusts and bugs or other invertebrates although these invertebrates were collected from the same compost as the worm-containing slugs, isopods and chilopods. This suggests that initiation or maintenance of infestation in the former groups is somehow constrained. One reason may be lack of sufficient humidity. Alternative explanations may be a consequence of nematode chemosensation, choice behavior, and/or possible defenses of the invertebrate taxa. In particular, worms may be repelled by chemical defense mechanisms, used by numerous invertebrates, such as beetles [30] or harvestmen [31]. Such chemical defense compounds may repulse or prevent *C. elegans* from attaching. Recognition of a carrier invertebrate may also be species-specific. Host preference has been observed for *C. remanei* within Porcellionidae, a family of isopods [12]. It has also been found in other *Caenorhabditis* species, such as dauer larvae of *C. japonica* and their attraction towards the burrower bug *Parastrachia japonensis* [15]. Chemotactic attraction of nematodes towards invertebrate hosts is currently best described for *Pristionchus pacificus*, which is associated with scarabaeid beetles [32]. These nematodes are able to specifically detect chemical signatures of their host beetle species and then use these to navigate towards their hosts [32]. It is as yet unclear how *C. elegans* is able to specifically detect and respond to cues of invertebrate taxa in its natural environment. Analysis of such interactions may be of particular value for our understanding of the worm's biology and could be comparatively easily achieved in the future using the established toolkit of *C. elegans* behavioral assays and functional genetic analysis methods.

Moving at slug's pace

Our analysis highlighted that *C. elegans* is particularly common in slug intestines when compared to the rest of the slug body, especially in the most abundant slug species *A. lusitanicus*. This finding confirms the previous reports of *C. elegans* inside of slugs collected in Africa [14] and Germany [17] and of *C. briggsae* inside of slugs from the US [18]. In these three previous studies a wide spectrum of slug-associated nematodes was explored and Ross et al. examined mainly parasitic nematodes, thus, *C. elegans* ecology was not the primary focus. In our study we analyzed 544 slug intestines from 21 sampling sites in Kiel, Germany, and were able to recover all three species, *C. remanei*, *C. elegans*, and *C. briggsae*. The prevalence of *C. elegans* and *C. remanei* in slug intestines varied among the two relevant screens, possibly due to random differences among years or other factors, which were not controlled. It would be interesting in the future to assess which exact factors may account for such variation. Although *Caenorhabditis* was found in slugs from 76.2% of the sampling sites, most of the directly associated substrate samples (e.g., substrate collected below or directly adjacent to the studied slugs) did not harbor any *Caenorhabditis*. These results are a possible indication that worms were picked up earlier and already transported by the slugs for some time. This strongly indicates that these worms are able to survive at least the time required to migrate from nematode containing substrates to the sampling location in the intestinal environment. In addition, to our knowledge, our study is the first to have isolated the three species, *C. elegans*, *C. remanei*, and *C. briggsae*, from the same substrate sample. In this case, the individual worm-mixture-containing slugs were collected from the compost in the botanical garden and a park in Kiel. This finding implies that all three species coexist, two or more *Caenorhabditis* species can directly interact in nature, and thus they may indeed compete for the same or at least a related ecological niche [3, 6]. Our screen was specifically focused on the nematode genus *Caenorhabditis*. We regularly noticed presence of other nematode taxa, which could represent additional competitors for the *Caenorhabditis* species and should thus be considered in future studies.

Various *C. elegans* stages can enter and leave the slug intestine

The dauer larva stage is predominantly found in association with invertebrates [11, 12]. This stage has thus been suggested to be specifically adapted for attachment and subsequent transport by vectors, especially if worms attach to the outside of other invertebrates [11]. Interestingly, even though dauer larvae were most frequent, we occasionally also observed other stages in the slug intestines in our field survey. Thus, different stages seem to be able to enter and persist in slug intestines, possibly because slugs unintentionally take up any of the stages while feeding on rotting plant material. Alternatively, slugs ingest dauer larvae which subsequently form proliferating populations. With our experiments, we specifically tested which *C. elegans* stages are able to enter and persist in the intestines. Our results highlight that all stages are able to enter, but not with similar efficiency. L1 worms were less proficient in establishing themselves in the slug gut than the other tested stages. One possible explanation is that the L1 stage is less likely to pass the slug radula unharmed or to resist the digestive system. Nevertheless, even though various stages are able to enter slugs, our finding from the field survey of a high abundance of dauer larvae suggests that this stage is particularly favored under natural conditions to be taken up by slugs, possibly because of specific behavioral adaptations (e.g., nictation behavior [9, 11]) or because dauer larvae are common on substrates preferred by the slugs.

During the experiments, the tested *C. elegans* stages were not able to persist for much more than 1 day in the intestines, indicating an only short-term interaction with the slug. These results contrasted with the field results, where worm-containing slugs were often found in no close association with substrates suitable for *Caenorhabditis* proliferation (e.g., on sidewalks close to streets or on large grass areas without rotting plant material), which may imply that worms can travel with the slugs for more than 1 day. Nevertheless, the same outcome may also be achieved by repeated re-invasion of the slugs. Alternatively our experimental conditions differ too much from field conditions, especially as to the maintenance of the slugs. It would be of particular value in the future to quantify the exact dynamics of *C. elegans*-slug interactions under natural conditions, for example by assessing slug feces collected from distinct field locations, which either offer or lack suitable nematode substrates.

Similarly, based on our results, it is not entirely clear which type of association is formed between *C. elegans* and slugs. A purely phoretical interaction with invertebrates with weak or no effects on host fitness has previously been proposed for *C. elegans* [11]. In our study we found that the worms enter and leave the slug intestine without any obvious harm and generally being fertile, suggesting that slugs represent suitable means of transport. At the same time, the slugs survived infestation with large worm numbers without any obvious damage. Both findings support the idea of a phoretic association, but are also consistent with a commensal or even mutualistic interaction. In addition, the slug's bacterial community may be exploited by *C. elegans* as food during the short-term inhabitation. This may be supported by our finding of occasional non-dauer larva stages in the slug intestines, possibly suggesting that the nematodes proliferate and reproduce inside the slug, Moreover, although rarely found for *Caenorhabditis* [11], a parasitic association between *C. elegans* and slugs may not yet be excluded. In fact, parasitic relationships are known for other nematodes that use slugs as intermediate and final hosts [14, 17, 18, 33, 34]. One prominent example is the commercially available strain *Phasmarhabditis hermaphrodita* [35], which actively searches for slugs and kills them through infection with its gut bacteria [36]. A distinction of these alternative interaction types requires a more detailed, long-term analysis of the *C. elegans*-slug associations under controlled conditions.

Conclusions

C. elegans and *C. remanei* can be regularly found in association with diverse invertebrates in Northern Germany being more prevalent on slugs, isopods and chilopods than on other taxa including beetles, flies and spiders possibly as a consequence of carrier specificity or chemical host defense mechanisms. *Caenorhabditis* nematodes can especially be found in slug intestines in higher frequencies. The exact nature of this association is hitherto unknown. Our analysis indicates that slugs are a suitable means of transport for diverse *Caenorhabditis* species in different developmental stages and that slug intestines may provide more advantages during dispersal hinting on possible mutualistic, commensal or possibly even parasitic interactions. Therefore, detailed long-term analysis of the *C. elegans*-slug associations under controlled conditions may provide a better understanding on the nature of the interaction and the underlying dynamics.

Additional files

Additional file 1: Overview of sampling sites. The pictures show some of the 21 sampling sites, from which slugs were collected, including six parks (A-F) and four compost heaps (G, H, J, K). Picture I shows a slug on compost. The other sampling sites are shown in Additional file 3. Letters in brackets refer to the code used for the individual sampling sites (Table 1).

Additional file 2: Overview of sampling sites. The pictures show some of the 21 sampling sites, from which slugs were collected, including five paths (A-E), four private gardens (F-I), one meadow (J) and one forest (K). The other sampling sites are shown in Additional file 2. Letters in brackets refer to the code used for the individual sampling sites (Table 1).

Additional file 3: Map of sampling locations in Kiel. Slugs were sampled from 21 locations in Kiel and the surroundings. Different letters indicate different locations. For details see Table 1 and Additional Files 2 and 3. Location S is not shown as it is outside of Kiel.

Additional file 4: Original data for the first field screen in 2011 and 2012. *C. elegans* and *C. remanei* were found in association with isopods and slugs during the first screen between July 2011 and October 2012.

Additional file 5: Original data for the second field screen in Kiel in 2013. 373 invertebrates from various taxonomic groups were sampled between July and September 2013. *C. elegans* and *C. remanei* were found in association with chilopods, isopods and slugs.

Additional file 6: Original data for the third field screen in Kiel, Northern Germany, in 2014. 544 slugs and 123 substrate samples were collected between July and October 2014 from 21 locations in Kiel and the surroundings.

Additional file 7: Association of different slug species from the third field screen with *C. elegans*, *C. remanei* or *C. briggsae*. Slug species identity was characterized for a representative subset of the slugs from the third screen with a DNA barcoding approach using a fragment of the mitochondrial COI gene.

Additional file 8: Overview of slug species used in the first laboratory experiment. Species identity was determined for a subset of slugs with a DNA barcoding approach using a fragment of the mitochondrial COI gene.

Abbreviations

ANOVA: analysis of variance; COI: cytochrome oxidase subunit I; FDR: false discovery rate; GFP: green fluorescent protein; ITS2: internal transcribed spacer 2 of the ribosomal cistron; L1: first instar nematode larvae; L3: third instar nematode larvae; L4: fourth instar nematode larvae; M9: nematode medium (42 mM Na2HPO$_4$, 22 mM KH$_2$PO$_4$, 86 mM NaCl, 1 mM MgSO$_4$·7H$_2$O); NGM: nematode growth medium; PCR: polymerase chain reaction; PFM: peptone free nematode growth medium; SEM: standard error of the mean.

Authors' contributions

CP conceived the project, carried out the first sampling and part of the third sampling, processed the samples, isolated and processed worms, characterized slugs using the DNA barcoding approach, supervised RJH and MCB, carried out part of the statistical analysis and drafted the manuscript. RJH performed part of the second and third sampling, processed the samples, performed the slug experiments, and contributed to the DNA barcoding sequence analysis. MCB performed part of the second sampling, processed and analyzed the samples. RS helped with the third sampling. CB carried out part of the statistical analysis. PD performed part of the DNA barcoding sequence analysis. HS conceived and supervised the project and helped to draft the manuscript. All authors read and approved the final manuscript.

Acknowledgements

We are grateful to all collaborating owners of private gardens for support while sampling of slugs, especially Antje Thomas and Christina Griebner. We thank Eike Strathmann for help with the sampling and the members of the Schulenburg group for general advice and support. CP and HS were funded by the ESF Eurocores EEFG project NEMADAPT (DFG grant SCHU 1415/11-1) and Kiel University for the sequencing analysis. CP, PD, and CB were additionally supported by the International Max Planck Research School (IMPRS) for Evolutionary Biology. We are grateful for support from the Kiel ICMB sequencing team headed by Philip Rosenstiel (especially Markus Schilhabel, Melanie Friskovec, Melanie Schlapkohl). The funders had no role in study design, data collection and analysis, decision to publish, or preparation of the manuscript.

Competing interests

The authors declare that they have no competing interests.

References

1. Petersen C, Dirksen P, Schulenburg H (2015) Why we need more ecology for genetic models such as *C. elegans*. Trends Genet 31:120–127
2. Caswell-Chen EP (2005) Revising the standard wisdom of *C. elegans* natural history: ecology of longevity. Sci ging Knowl Environ 2005:pe30
3. Félix M-A, Braendle C (2010) The natural history of *Caenorhabditis elegans*. Curr Biol 20:R965–R969
4. Félix M-A, Duveau F (2012) Population dynamics and habitat sharing of natural populations of *Caenorhabditis elegans* and *C. briggsae*. BMC Biol 10:59
5. Haber M, Schüngel M, Putz A, Müller S, Hasert B, Schulenburg H (2005) Evolutionary history of *Caenorhabditis elegans* inferred from microsatellites: evidence for spatial and temporal genetic differentiation and the occurrence of outbreeding. Mol Biol Evol 22:160–173
6. Petersen C, Dirksen P, Prahl S, Strathmann EA, Schulenburg H (2014) The prevalence of *Caenorhabditis elegans* across 1.5 years in selected North German locations: the importance of substrate type, abiotic parameters, and *Caenorhabditis* competitors. BMC Ecol 14:4
7. Viney ME, Gardner MP, Jackson JA (2003) Variation in *Caenorhabditis elegans* dauer larva formation. Dev Growth Differ 45:389–396
8. Green JWM, Snoek LB, Kammenga JE, Harvey SC (2013) Genetic mapping of variation in dauer larvae development in growing populations of *Caenorhabditis elegans*. Heredity 111:306–313

9. Lee H, Choi M, Lee D, Kim H, Hwang H, Kim H et al (2012) Nictation, a dispersal behavior of the nematode *Caenorhabditis elegans*, is regulated by IL2 neurons. Nat Neurosci 15:107–112

10. Erkut C, Penkov S, Khesbak H, Vorkel D, Verbavatz J-M, Fahmy K et al (2011) Trehalose renders the dauer larva of *Caenorhabditis elegans* resistant to extreme desiccation. Curr Biol 21:1331–1336

11. Kiontke K, Sudhaus W (2006) Ecology of *Caenorhabditis* species. In: WormBook (ed) The *C. elegans* research community. WormBook. doi:10.1895/wormbook.1.37.1. http://www.wormbook.org

12. Baird SE (1999) Natural and experimental associations of *Caenorhabditis remanei* with *Trachelipus rathkii* and other terrestrial isopods. Nematology 1(5):471–475

13. Barrière A, Félix M-A (2005) High local genetic diversity and low outcrossing rate in *Caenorhabditis elegans* natural populations. Curr Biol 15:1176–1184

14. Ross JL, Ivanova ES, Sirgel WF, Malan AP, Wilson MJ (2012) Diversity and distribution of nematodes associated with terrestrial slugs in the Western Cape Province of South Africa. J Helminthol 86:215–221

15. Okumura E, Tanaka R, Yoshiga T (2013) Species-specific recognition of the carrier insect by dauer larvae of the nematode *Caenorhabditis japonica*. J Exp Biol 216:568–572

16. Barrière A, Félix M-A (2007) Temporal dynamics and linkage disequilibrium in natural *Caenorhabditis elegans* populations. Genetics 176:999–1011

17. Mengert DH (1953) Nematoden und Schnecken. Z Für Morphol Ökol Tiere 41:311–349

18. Ross JL, Ivanova ES, Severns PM, Wilson MJ (2010) The role of parasite release in invasion of the USA by European slugs. Biol Invasions 12:603–610

19. Ross JL, Ivanova ES, Spiridonov SE, Waeyenberge L, Moens M, Nicol GW et al (2010) Molecular phylogeny of slug-parasitic nematodes inferred from 18S rRNA gene sequences. Mol Phylogenet Evol 55:738–743

20. Barrière A, Félix M-A (2006) Isolation of *C. elegans* and related nematodes. In: WormBook (ed) The *C. elegans* research community. WormBook. doi:10.1895/wormbook.1.43.1. http://www.wormbook.org

21. Seidel HS, Rockman MV, Kruglyak L (2008) Widespread genetic incompatibility in *C. elegans* maintained by balancing selection. Science 319:589–594

22. Pujol N, Cypowyj S, Ziegler K, Millet A, Astrain A, Goncharov A et al (2008) Distinct innate immune responses to infection and wounding in the *C. elegans* epidermis. Curr Biol 18:481–489

23. Folmer O, Black M, Hoeh W, Lutz R, Vrijenhoek R (1994) DNA primers for amplification of mitochondrial cytochrome c oxidase subunit I from diverse metazoan invertebrates. Mol Mar Biol Biotechnol 3:294–299

24. Altschul SF, Gish W, Miller W, Myers EW, Lipman DJ (1990) Basic local alignment search tool. J Mol Biol 215:403–410

25. Benjamini Y, Hochberg Y (1995) Controlling the false discovery rate: a practical and powerful approach to multiple testing. J R Stat Soc 57(1):289–300

26. Chen J, Lewis EE, Carey JR, Caswell H, Caswell-Chen EP (2006) The ecology and biodemography of *Caenorhabditis elegans*. Exp Gerontol 41:1059–1065

27. Denny M (1980) Locomotion: the cost of gastropod crawling. Science 208:1288–1290

28. McDonnell R (2009) Slugs: a Guide to the Invasive and Native Fauna of California. UCANR Publications. http://anrcatalog.ucdavis.edu/pdf/8336.pdf

29. Denny MW, Gosline JM (1980) The physical properties of the pedal mucus of the terrestrial slug *Ariolimax columbianus*. J Exp Biol 88:375–394

30. Poinar GO, Marshall CJ, Buckley R (2007) One hundred million years of chemical warfare by insects. J Chem Ecol 33:1663–1669

31. Machado G, Carrera PC, Pomini AM, Marsaioli AJ (2005) Chemical defense in harvestmen (arachnida, opiliones): do benzoquinone secretions deter invertebrate and vertebrate predators? J Chem Ecol 31:2519–2539

32. McGaughran A, Morgan K, Sommer RJ (2013) Natural variation in chemosensation: lessons from an island nematode. Ecol Evol 3:5209–5224

33. Grewal PS, Grewal SK, Tan L, Adams BJ (2003) Parasitism of molluscs by nematodes: types of associations and evolutionary trends. J Nematol 35:146–156

34. Blaxter M, Koutsovoulos G (2014) The evolution of parasitism in Nematoda. Parasitology 142:S26–S39

35. Rae R, Verdun C, Grewal PS, Robertson JF, Wilson MJ (2007) Biological control of terrestrial molluscs using *Phasmarhabditis hermaphrodita*—progress and prospects. Pest Manag Sci 63:1153–1164

36. Tan L, Grewal PS (2002) Endotoxin activity of *Moraxella osloensis* against the grey garden slug *Deroceras reticulatum*. Appl Environ Microbiol 68:3943–3947

2

Wild pollinators enhance oilseed rape yield in small-holder farming systems in China

Yi Zou[1,2] (iD), Haijun Xiao[3*], Felix J. J. A. Bianchi[4], Frank Jauker[5], Shudong Luo[6] and Wopke van der Werf[1]

Abstract

Background: Insect pollinators play an important role in crop pollination, but the relative contribution of wild pollinators and honey bees to pollination is currently under debate. There is virtually no information available on the strength of pollination services and the identity of pollination service providers from Asian smallholder farming systems, where fields are small, and variation among fields is high. We established 18 winter oilseed rape (*Brassica napus* L.) fields along a large geographical gradient in Jiangxi province in China. In each field, oilseed rape plants were grown in closed cages that excluded pollinators and open cages that allowed pollinator access. The pollinator community was sampled by pan traps for the entire oilseed rape blooming period.

Results: Oilseed rape plants from which insect pollinators were excluded had on average 38% lower seed set, 17% lower fruit set and 12% lower yield per plant, but the seeds were 17% heavier, and the caged plants had 28% more flowers and 18% higher aboveground vegetative biomass than plants with pollinator access. Oilseed rape plants thus compensate for pollination deficit by producing heavier seeds and more flowers. Regression analysis indicated that local abundance and diversity of wild pollinators were positively associated with seed set and yield/straw ratio, while honey bee abundance was not related to yield parameters.

Conclusions: Wild pollinator abundance and diversity contribute to oilseed rape yield by enhancing plant resource allocation to seeds rather than to above-ground biomass. This study highlights the importance of the conservation of wild pollinators to support oilseed rape production in small-holder farming systems in China.

Keywords: Ecosystem services, Canola, Compensation, Honey bee, Pollination, Pollinator diversity, Wild bee

Background

A wide range of agricultural crops depend on pollination by insects [1]. The decline of pollinators in terms of abundance and species richness has caused great concern about the risk of a deterioration of crop pollination and the associated crop production [2–6]. Potential drivers for the loss of wild pollinators include habitat loss and fragmentation, insecticides, pathogens, invasive species, climate change and the interactions between them [5]. The consequences of the decline of wild pollinators for pollination services may partially be offset by managed honey bees, compensating for the loss of wild pollinators [3, 7, 8]. This view, however, has recently been challenged

after assessing the contribution of wild bees, hoverflies, butterflies, moths, wasps and beetles [9, 10].

The vast majority of the studies focussing on the interplay between wild and managed pollinators in providing agricultural pollination services originates from Europe and North America, where industrialization of agriculture has resulted in agroecosystems dominated by monocultures in large fields. In contrast, Chinese agroecosystems, particular in South China, are characterised by relatively small fields, leading to a high heterogeneity in terms of crop species, field management and field edges [11]. This high heterogeneity may favour wild pollinators by providing nesting sites and floral resources [12, 13]. We therefore expect that the small-holder agroecosystems in China support a high abundance and diversity of wild pollinators contributing to pollination services

*Correspondence: hjxiao@jxau.edu.cn
[3] Institute of Entomology, Jiangxi Agricultural University, Nanchang 330045, China
Full list of author information is available at the end of the article

that significantly exceed the contribution of managed pollinators.

A globally important crop benefitting from pollination services is oilseed rape (*Brassica napus* L.), of which China is one of the world's largest producers with more than 7.5 million ha cultivated area for the production of cooking oil, feed and biofuel [14]. Although oilseed rape is considered a self-pollinating plant species [15], insect pollination can further increase yield and quality [3, 7, 8, 16]. Seed yield of individual oilseed rape plants is determined by the number of seeds per pod (seed set), the number of pods per plant (fruit set), and the individual seed weight. Seed set is mainly determined by the amount of pollen grains deposited on the stigma of flowers during the receptive period [17], which can be increased by pollinator-mediated pollen transfer [18]. Similarly, pollination usually enhances fruit set, i.e. the proportion of flowers developing into pods [19, 20]. Oilseed rape plants show variation in their ability to compensate for a pollination deficit, which may depend on the cultivar [21–23] and the pollination efficiency of flower visiting insects [8, 24–26]. However, the potential to compensate for pollination deficit by allocating resources into heavier seeds or increased flowering has received little attention [but see 23, 27], but may have important consequences for the oilseed rape production potential in situations of pollinator declines.

The aim of this study is twofold. First, to assess the relationship between pollinator communities and oilseed rape yield parameters. We hypothesise that yield parameters will be positively influenced by more abundant and more diverse pollinator communities. Second, to assess the relative contribution of wild pollinators versus honey bees to oilseed rape pollination. Here, we expect that yield of oilseed rape is positively associated with the abundance of both wild pollinators and honey bees.

Methods
Study area
We selected 18 oilseed rape fields across a large geographical area in Jiangxi Province, China (N28.35°–N28.99°, E115.26°–E115.82°). The mean distance between fields was 36.9 km (range: 5.8–75.2 km). As the maximum foraging range of most pollinator species is less than 2 km [28, 29], individual pollinators are unlikely to visit more than one field, and hence the pollinator communities in the study fields can be considered independent. The mean size of study fields was 845 ± 86 m^2 (range: 400–1400 m^2) and all fields were sown between the middle and the end of October 2014 with the same traditional open-pollinated winter oilseed rape cultivar YangGuang-2009.

Experimental design and plant yield parameters
In the centre of each field, eight oilseed rape plants at a similar growth stage were selected, spaced 4 m apart. Each plant was covered by an individual cage (alternating open and closed). Closed cages had a base of 0.6×0.6 m^2, a height of 2.0 m and were entirely covered with 1×1 mm^2 mesh to exclude pollinators. This mesh size has only a limited influence on the microclimate in the cage [30]. The open cage was set as a control treatment and consisted of a similar frame as the closed cage, but only contained mesh at the roof and the top 0.3 m such that pollinators had access to the plants. This resulted in a similar shading of plants in closed and open cages. Neighbouring plants were removed to provide space for setting up cages, and the cages were established about one week before blooming and were removed during harvest.

After harvest, the number of pods and total number of flower stalks were counted for each plant. Seeds were removed from pods, weighed and counted using an automatic seed counter (SLY-C, Zhejiang Top Instrument, China). The following yield parameters were measured and calculated per plant: seed set (number of seeds per pod), number of pods (siliques) per plant, number of flowers per plant, fruit set (pod/flower ratio), seed weight (total seed weight divided by the total number of seeds) and yield (total weight of all seeds). Plants were dried for 30 days in the greenhouse and the total aboveground dry vegetative biomass excluding seeds and pods (referred to as straw) was assessed.

Insect sampling
The pollinator community in each experimental oil seed rape field was sampled by pan traps, which is a suitable method for sampling pollinators such as bees [31], hoverflies [32] and butterflies [33]. Each pan trap station consisted of three cups (8.3 cm diameter, 13.5 cm height and a volume of 450 ml) fixed on a wooden stick at a distance of 1.5 m above the ground. The cups were white from the outside, and painted ultraviolet (UV) yellow, UV blue and UV white from the inside, respectively [34]. We used water saturated with kitchen salt (NaCl) as a killing agent with several drops of detergent to break water surface tension. At 3 cm from the brim, two 3 mm diameter holes were drilled in order to drain off rainwater and sufficient water was added to prevent the traps from drying out. In each field, four stations were installed at the corners of a 20 m × 20 m square in the centre of the field. Traps were set up before the onset of bloom, at the same time as the cages, and were monitored until harvest for a period ranging from 49 to 52 days. This difference in the sampling period was mainly caused by different trap establishment dates,

but since the traps were established before the activity period of most pollinators, there was a negligible effect on the catch. Samples were collected five times at approximately 10-day intervals.

Insect pollinator specimens were collected and stored at -20 °C, and then sorted and pinned. All specimens were identified to species level when possible. Pollinator specimens were separated into wild pollinators (including wild bees, hoverflies, butterflies and moths, social and solitary wasps) and honey bees (*Apis mellifera* and *Apis cerana*). Asynchronous flowering times of the oilseed rape crops prevented the separation of insect communities that were visiting the oilseed fields during and after flowering (see flower cover data in Additional file 1). Therefore, all specimens collected from the same field were pooled in the analysis.

Data analysis

We conducted two main analyses. The first analysis focused on the effect of pollinator exclusion on plant yield parameters using linear mixed effect models. Response variables were calculated as the difference in plant yield parameters between plants with pollinator access (open cages) and plants without pollinator access (closed cages), and included (1) seed set, (2) number of pods per plant, (3) number of flowers per plant, (4) fruit set (pod/flower ratio), (5) thousand seed weight, (6) plant yield, (7) straw biomass, and (8) yield/straw ratio. Plant yield/straw ratio is the ratio between total seed weight and total dry vegetative biomass. It expresses the ratio of assimilates to seeds or vegetative growth, and provides a useful indicator for limitation in active seed sinks on the plant [35, 36] as a result of pollination deficit of oilseed rape. Treatment (closed cage versus open cage) was used as an explanatory variable (fixed factor) and study field as a random factor. Transformations were applied for response variables to meet normal distribution requirements.

The second analysis focused on the effect of pollinator abundance and diversity on yield parameters. Generalized linear mixed effect models were used with study field as a random factor. Data from the open and closed cage treatments were analysed separately. The purpose of the analysis on open cages was to assess the role of different pollinator taxa in pollination, while the analysis on closed cages was conducted to verify that pollinator abundance and community composition did not affect plant yield parameters in closed cages. Response variables included (1) seed set (Gaussian error distribution with identity-link function), (2) fruit set (gamma error distribution with log-link function), (3) thousand seed weight (gamma error distribution with log-link function),

and (4) yield/straw ratio (gamma error distribution with log-link function). Explanatory variables included (a) wild pollinator abundance, (b) honey bee abundance, (c) wild pollinator diversity, and (d) plant straw biomass. Study field was included as a random factor. Wild pollinator diversity was characterised in terms of the back-transformed Shannon entropy index [37], the rarefied number of species (n = 54) [38] and the Fisher's alpha index [39]. As the back-transformed Shannon entropy index was strongly correlated with both the rarefied number of species (Pearson r = 0.94, $P < 0.001$) and Fisher's alpha (r = 0.91, $P < 0.001$), and is a robust indicator for mobile insects and uneven sample sizes [40–42], we selected it as an indicator for pollinator diversity (with a focus on species richness) in the statistical analysis. Plant straw biomass was included as a control variable to account for variation in plant size, but was excluded for the analysis of yield/straw ratio.

All models were validated by checking residuals according to the protocol of Zuur et al. [43] and deviance residuals met normality and homoscedasticity assumptions. In addition, model residuals were checked for spatial autocorrelation using Moran's I coefficient [44]. No significant spatial autocorrelation was found in the fitted models ($P > 0.05$). All calculations and analyses were conducted in R (v3.1.2) [45] using the "nlme" package for linear mixed effect models [46], the "lme4" package for generalized linear mixed effect models [47], and the "ape" package for spatial autocorrelation [48]. Means and standard errors are reported throughout the paper.

Results
Pollinator community

A total of 5148 specimens representing 60 pollinator species were collected from the pan traps. These included 3931 Hymenoptera comprising 44 species, 52 hoverflies (7 species), and 1165 Lepidoptera (9 species). The top five most abundant species were cabbage butterfly (*Pieris rapae*), two wild bee species *Eucera chinensis* and *Lasioglossum proximatum*, and two honey bee species *A. mellifera* and *A. cerana*, accounting for 21.6, 20.8, 16.5, 9.1 and 8.1% of the overall specimens, respectively (see complete species list in Additional file 1). The overall abundance of collected wild pollinators across the 18 fields was 237 (±40) individuals, ranging from 54 to 720 individuals per field, while the abundance of honey bees averaged 49 (±12) individuals, ranging from 1 to 195 individuals per field, highlighting substantial between-field variation. This variation allows a meaningful analysis of the relationship between pollinator abundance and diversity on the one hand and plant yield parameters on the other.

Plant yield parameters

Results indicated that pollinator exclusion significantly influenced plant yield parameters (Table 1). Oilseed rape plants in closed cages had 38% (±4%) lower seed set, 17% (±4%) lower fruit set, 12% (±14%) lower yield and 35% (±7%) lower yield/straw ratios than plants in open cages (Fig. 1). However, plants in the closed cage treatment had 22% (±7%) higher seed weight, 28% (±9%) more flowers, and 39% (±11%) more straw biomass than plants in open cages. The number of pods per plant was not significantly different between treatments.

Table 1 Results of linear mixed effect models showing the effects of pollinator exclusion on oilseed rape yield parameters

Response variable	Data transformation	df	t	P
Seed set	None	113	7.86	<0.001
Number of pods	Square root	115	−1.06	0.3
Number of flowers	Square root	115	−2.99	0.003
Fruit set	Arcsine	115	4.26	<0.001
Thousand seed weight	Logarithm	111	−3.25	0.002
Plant yield	Logarithm	111	2.31	0.02
Plant straw biomass	Square root	115	−3.72	<0.001
Yield/straw ratio	Square root	111	5.21	<0.001

Negative t values indicate a higher value for the plants with pollinator exclusion than for the plants with pollinator access

Influence of pollinators on plant yield parameters

When pollinators had access to oilseed rape plants, seed set was positively associated with wild pollinator abundance and diversity (Table 2). In addition, a strong positive association was observed between the yield/straw ratio and the abundance and diversity of wild pollinators. In contrast, the abundance of honey bees was not significantly associated with any of the plant yield parameters (Table 2). The control analysis using the data from the closed cage treatment did not reveal any significant effects of the abundance or diversity of pollinators on yield components, indicating that the exclusion treatment functioned well and confirming that associations between wild pollinators and yield components of oilseed rape in the open cages can indeed be attributed to insect pollination.

Discussion

The heterogeneous landscape mosaic in Southern China, which is characterized by small field sizes, harboured a rich pollinator community. Accordingly, wild pollinators contributed substantially to oilseed rape yield, confirming our expectations. The fact that we did not find statistical support for the contribution of honey bees to oilseed rape yield parameters, despite their well-documented contribution crop pollination in other parts of the world, substantiates the relevance of natural service providers to small-holder farming.

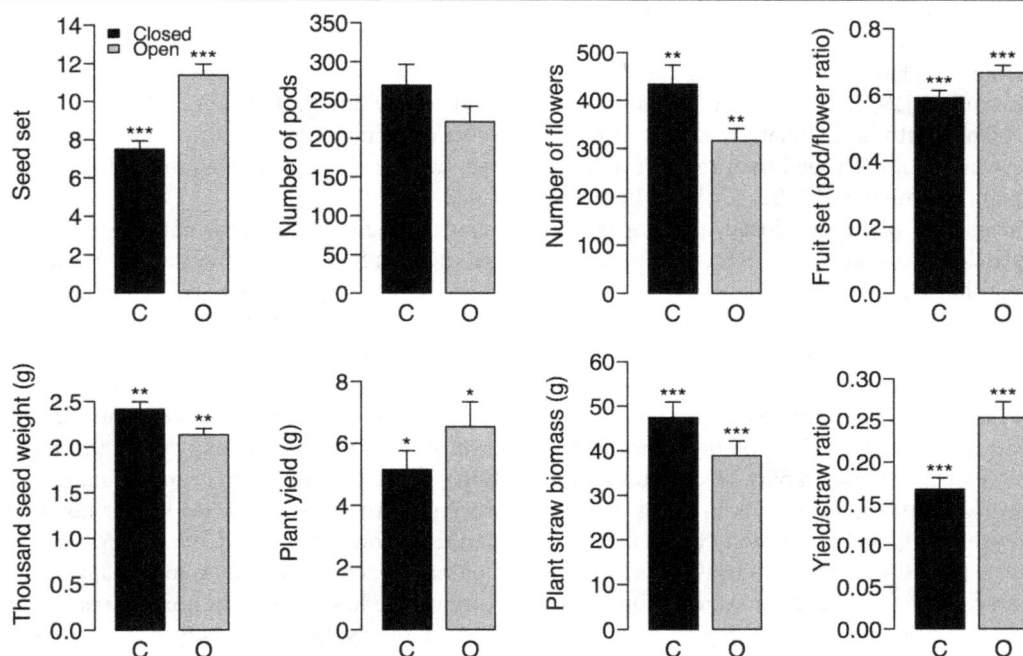

Fig. 1 Plant yield parameters of oilseed rape plants in closed (*C*) and open (*O*) cages. *Bars* represent SEM. *Asterisks* show the significance level based on an analysis with mixed models (see Table 2; * ≤0.05; ** ≤0.01; *** <0.001)

Table 2 Results of generalized linear mixed effect models showing the relationship between plant yield parameters and pollinator variables for oilseed rape plants with pollinator access (open cage)

Response variable	Error distribution	Straw biomass	Wild pollinator abundance	Wild pollinator diversity	Honey bee abundance
Seed set	Gaussian	–	0.015 ± 0.007*	0.801 ± 0.352*	–
Fruit set	Gamma	–	–	–	–
Seed weight	Gamma	–	–	–	–
Yield/straw ratio	Gamma	/	0.003 ± 0.001***	0.159 ± 0.053**	–

Values indicate estimates and standard errors, a dash (–) indicates that the variable was not significant, a slash (/) indicates that the variable was not entered in the model because of dependency on the response variable, and asterisks show significance levels (*$\leq$0.05; **$\leq$0.01; ***<0.001)

We identified 60 insect pollinator species, 44 of which were Hymenoptera species. This represents a high number of pollinator species in comparison with other landscape-scale studies in oilseed rape. For example, 20 species (honeybees, bumblebees and solitary bees) from 1181 individuals were reported in Wiltshire, UK [49], 36 flower-visiting species from 1866 individuals Uppsala, Sweden [3], and 26 bee and hoverfly species in Ireland (number of individuals was not mentioned) [8]. Our findings were in line with our expectation that the study region contains a high diversity of wild pollinators, which may partly due to the high heterogeneity in field size, crop species and crop management, and high diversity of wild plant species in field edges in small-holder agroecosystems [11].

Insect pollinated plants showed higher seed set and overall higher yields than plants deprived from pollinators. Therefore, our study contributes to a body of evidence that insect pollination matters for oilseed production despite its capacity for self-pollination [3, 21, 50–52]. At the same time, insect pollination lowered some yield parameters such as seed weight, suggesting compensation mechanisms of the plants also in line with previous studies [19, 21]; but see Bommarco et al. [53]. Often, plants with a pollination deficit produce fewer seeds per pod, but each seed then receives a higher share of the plant assimilates [19, 21, 52, 53]. The higher number of flowers on plants in closed as compared to open cages provides further support for compensatory responses to a pollination deficit [23].

The higher straw biomass of oilseed rape plants in closed cages points to an increased allocation of assimilates to the above ground vegetative plant parts, which supports a lack of sink strength resulting from a pollination deficit. The positive effect of pollinator exclusion on straw weight may also in part be due to the high energetic cost of producing fatty acids in seeds as compared to the lower energetic conversion costs to leaf and stem dry matter [54]. Overall, compensation effects did not fully counterbalance the yield loss due to the lack of pollination as exemplified by the 12% higher yield when comparing plants with pollinator access to plants without pollinator access.

Wild pollinator abundance and diversity were positively associated with oilseed rape seed set and yield/straw ratio, but not with fruit set and seed weight, suggesting that their benefits to oilseed rape yield per plant mostly result from an increased number of seeds per pod. The control analysis, which showed no relationship between plant yield parameters and pollinator abundance and diversity for closed cages, confirmed the effectiveness of the exclusion treatment and the overall consistency of the experimental setup. This also suggested that pollinator collections from pan traps can be used as a proxy in reflecting the pollinator communities and pollination service at the landscape scale [55, 56].

We assessed the contribution of insect pollination on isolated plants where neighbouring plants were removed. The focus on isolated plants may have also resulted in an reduction of plant-to-plant pollen transfer [35] and a reduction in plant competition for water, nutrients and light. Therefore, we may underestimate the potential of closed oil seed rape stands to compensate for pollinator limitation by mechanical and wind pollination [35, 57] and refrain from estimating agronomic benefits at the field level [22]. Furthermore, the mesh tents may have reduced wind pollination, even though the same mesh size has been widely applied in pollinator exclusion studies [3, 16, 19, 53].

Surprisingly, our analysis gave no support for the contribution of honey bees to oilseed rape yield, even though there were large differences in honey bee abundance between fields. While the contribution of honey bees to crop pollination is widely documented [see review in 58], this result is in line with a current global meta-analysis that highlighted the importance of wild pollinators in crop pollination [9]. In our study, the higher contribution of wild pollinators to crop pollination can in part be attributed to their five times higher abundance as compared to honey bees. Indeed, wild pollinators dominate pollinator communities in many agroecosystems [59]. The relative low number of honey bees may have resulted from a relatively low density of bee hives in the study areas. Also, some wild pollinator species may be as efficient or even more efficient than honey bees [24, 26]. In our case, the abundant cabbage butterfly (*Pieris rapae*) might be an important pollinator [60]. As their larvae are considered a pest, however, this species may have both a positive and negative effect on oilseed rape production.

Conclusion

Our study demonstrates that wild pollinators play an important role in the pollination of oilseed crops in small-holder farming systems in China. Wild pollinator abundance and diversity contribute to oilseed rape yield by mediating increased allocation to seeds rather than above-ground straw biomass, but oilseed rape plants suffering from a pollination deficit can compensate to some extent by generating heavier seeds, more flowers and higher straw biomass. This study highlights the importance of conserving wild pollinators in order to maximise oilseed rape production, especially in heterogeneous landscapes where their pollination service is exceeding the service provided by managed pollinators.

Authors' contributions
YZ, HX, FJJAB, FJ and WW designed the experiments. YZ, HX and SL performed the experiments. YZ analysed the data. YZ, FJJAB and WW wrote the manuscript; other authors provided editorial advice. All authors read and approved the final manuscript.

Author details
[1] Centre for Crop Systems Analysis, Wageningen University, P.O. Box 430, 6700 AK Wageningen, The Netherlands. [2] Present Address: Department of Environmental Science, Xi'an Jiaotong-Liverpool University, Suzhou 215123, China. [3] Institute of Entomology, Jiangxi Agricultural University, Nanchang 330045, China. [4] Farming Systems Ecology, Wageningen University, P.O. Box 430, 6700 AK Wageningen, The Netherlands. [5] Department of Animal Ecology, Justus Liebig University, Heinrich-Buff-Ring 26-32, 35932 Giessen, Germany. [6] Institute of Apicultural Research, Chinese Academy of Agricultural Sciences, Beijing 100093, China.

Acknowledgements
We thank Mario van Telgen, Junhui Chen, Chao Zou, Yuekun Wu and Weizhao Sun for help in the fieldwork. We also thank Huanli Xu for helping with pollinator identification.

Competing interests
The authors declare that they have no competing interests.

Funding
This study was funded by the Division for Earth and Life Sciences of the Netherlands Organization for Scientific Research (Grant 833.13.004), the National Natural Science Foundation of P.R. China (31360461), the Agricultural Science and Technology Innovation Program (CAAS-ASTIP-2015-IAR) and the Cultivation Plan for Young Scientists of Jiangxi Province (20153BCB23014).

References
1. Klein A-M, Vaissiere BE, Cane JH, Steffan-Dewenter I, Cunningham SA, Kremen C, Tscharntke T. Importance of pollinators in changing landscapes for world crops. Proc R Soc B. 2007;274(1608):303–13.
2. Allen-Wardell G, Bernhardt P, Bitner R, Burquez A, Buchmann S, Cane J, Cox PA, Dalton V, Feinsinger P, Ingram M, et al. The potential consequences of pollinator declines on the conservation of biodiversity and stability of food crop yields. Conserv Biol. 1998;12(1):8–17.
3. Bommarco R, Marini L, Vaissiere BE. Insect pollination enhances seed yield, quality, and market value in oilseed rape. Oecologia. 2012;169(4):1025–32.
4. Biesmeijer JC, Roberts SPM, Reemer M, Ohlemüller R, Edwards M, Peeters T, Schaffers AP, Potts SG, Kleukers R, Thomas CD, et al. Parallel declines in pollinators and insect-pollinated plants in Britain and the Netherlands. Science. 2006;313(5785):351–4.
5. Potts SG, Biesmeijer JC, Kremen C, Neumann P, Schweiger O, Kunin WE. Global pollinator declines: trends, impacts and drivers. Trends Ecol Evol. 2010;25(6):345–53.
6. Gill RJ, Baldock KCR, Brown MJF, Cresswell JE, Dicks LV, Fountain MT, Garratt MPD, Gough LA, Heard MS, Holland JM, et al. Protecting an ecosystem service: approaches to understanding and mitigating threats to wild insect pollinators. Adv Ecol Res. 2016;54:135–206.
7. Sabbahi R, De Oliveira D, Marceau J. Influence of honey bee (Hymenoptera: Apidae) density on the production of canola (Crucifera: Brassicacae). J Econ Entomol. 2005;98(2):367–72.
8. Stanley DA, Gunning D, Stout JC. Pollinators and pollination of oilseed rape crops (*Brassica napus* L.) in Ireland: ecological and economic incentives for pollinator conservation. J Insect Conserv. 2013;17(6):1181–9.
9. Garibaldi LA, Steffan-Dewenter I, Winfree R, Aizen MA, Bommarco R, Cunningham SA, Kremen C, Carvalheiro LG, Harder LD, Afik O, et al. Wild pollinators enhance fruit set of crops regardless of honey bee abundance. Science. 2013;339(6127):1608–11.
10. Rader R, Bartomeus I, Garibaldi LA, Garratt MPD, Howlett BG, Winfree R, Cunningham SA, Mayfield MM, Arthur AD, Andersson GKS, et al. Non-bee insects are important contributors to global crop pollination. PNAS. 2016;113(1):146–51.
11. Liu Y, Duan M, Yu Z. Agricultural landscapes and biodiversity in China. Agric Ecosyst Environ. 2013;166:46–54.
12. Benton TG, Vickery JA, Wilson JD. Farmland biodiversity: is habitat heterogeneity the key? Trends Ecol Evol. 2003;18(4):182–8.
13. Fahrig L, Baudry J, Brotons L, Burel FG, Crist TO, Fuller RJ, Sirami C, Siriwardena GM, Martin JL. Functional landscape heterogeneity and animal biodiversity in agricultural landscapes. Ecol Lett. 2011;14(2):101–12.
14. FAO. FAOSTAT; 2013. http://faostat3.fao.org. Accessed 1 Jan 2017.
15. Williams IH, Martin AP, White RP. The pollination requirements of oil-seed rape (*Brassica napus* L.). J Agric Sci. 1986;106(01):27–30.
16. Jauker F, Wolters V. Hover flies are efficient pollinators of oilseed rape. Oecologia. 2008;156(4):819–23.
17. Pechan PM. Ovule fertilization and seed number per pod determination in oil seed rape (*Brassica napus*). Ann Bot. 1988;61(2):201–7.
18. Abrol DP. Honeybees and rapeseed: a pollinator-plant interaction. Adv Bot Res. 2007;45:337–67.
19. Steffan-Dewenter I. Seed set of male-sterile and male-fertile oilseed rape (*Brassica napus*) in relation to pollinator density. Apidologie. 2003;34(3):227–35.
20. Jauker F, Bondarenko B, Becker HC, Steffan-Dewenter I. Pollination efficiency of wild bees and hoverflies provided to oilseed rape. Agric For Entomol. 2012;14(1):81–7.
21. Hudewenz A, Pufal G, Bögeholz A-L, Klein A-M. Cross-pollination benefits differ among oilseed rape varieties. J Agric Sci. 2014;152(5):770–8.
22. Lindström SAM, Herbertsson L, Rundlöf M, Smith HG, Bommarco R. Large-scale pollination experiment demonstrates the importance of insect pollination in winter oilseed rape. Oecologia. 2016;180(3):759–69.
23. Marini L, Tamburini G, Petrucco-Toffolo E, Lindström SAM, Zanetti F, Mosca G, Bommarco R. Crop management modifies the benefits of insect pollination in oilseed rape. Agric Ecosyst Environ. 2015;207:61–6.
24. Holzschuh A, Dudenhöffer J-H, Tscharntke T. Landscapes with wild bee habitats enhance pollination, fruit set and yield of sweet cherry. Biol Conserv. 2012;153:101–7.

25. Hoyle M, Cresswell JE. The effect of wind direction on cross-pollination in wind-pollinated GM crops. Ecol Appl. 2007;17(4):1234–43.

26. Rader R, Howlett BG, Cunningham SA, Westcott DA, Newstrom-Lloyd LE, Walker MK, Teulon DA, Edwards W. Alternative pollinator taxa are equally efficient but not as effective as the honeybee in a mass flowering crop. J Appl Ecol. 2009;46(5):1080–7.

27. Sutter L, Albrecht M. Synergistic interactions of ecosystem services: florivorous pest control boosts crop yield increase through insect pollination. Proc R Soc B. 1824;2016:283.

28. Chifflet R, Klein EK, Lavigne C, Le Feon V, Ricroch AE, Lecomte J, Vaissiere BE. Spatial scale of insect-mediated pollen dispersal in oilseed rape in an open agricultural landscape. J Appl Ecol. 2011;48(3):689–96.

29. Steffan-Dewenter I, Munzenberg U, Burger C, Thies C, Tscharntke T. Scale-dependent effects of landscape context on three pollinator guilds. Ecology. 2002;83(5):1421–32.

30. Martin EA, Reineking B, Seo B, Steffan-Dewenter I. Natural enemy interactions constrain pest control in complex agricultural landscapes. PNAS. 2013;110(14):5534–9.

31. Cane JH, Minckley RL, Kervin LJ. Sampling bees (Hymenoptera: Apiformes) for pollinator community studies: pitfalls of pan-trapping. J Kans Entomol Soc. 2000;73(4):225–31.

32. Irvin NA, Wratten SD, Frampton CM, Bowie MH, Evans AM, Moar NT. The phenology and pollen feeding of three hover fly (Diptera: Syrphidae) species in Canterbury, New Zealand. NZ J Zool. 1999;26(2):105–15.

33. Cizek O, Bakesová A, Kuras T, Benes J, Konvicka M. Vacant niche in alpine habitat: the case of an introduced population of the butterfly *Erebia epiphron* in the Krkonoše Mountains. Acta Oecol. 2003;24(1):15–23.

34. Fortel L, Henry M, Guilbaud L, Guirao AL, Kuhlmann M, Mouret H, Rollin O, Vaissiere BE. Decreasing Abundance, Increasing Diversity and changing structure of the wild bee community (Hymenoptera: Anthophila) along an urbanization gradient. PLoS ONE. 2014;9(8):e104679.

35. Diepenbrock W. Yield analysis of winter oilseed rape (*Brassica napus* L.): a review. Field Crops Res. 2000;67(1):35–49.

36. Iglesias FM, Miralles DJ. Changes in seed weight in response to different sources: sink ratio in oilseed rape. IJARIT. 2014;4(1):44–52.

37. Jost L. Entropy and diversity. Oikos. 2006;113(2):363–75.

38. Hurlbert SH. The nonconcept of species diversity: a critique and alternative parameters. Ecology. 1971;52(4):577–86.

39. Fisher RA, Corbet AS, Williams CB. The relation between the number of species and the number of individuals in a random sample of an animal population. J Anim Ecol. 1943;12(1):42–58.

40. Fiedler K, Truxa C. Species richness measures fail in resolving diversity patterns of specious forest moth assemblages. Biodivers Conserv. 2012;21(10):2499–508.

41. Beck J, Schwanghart W. Comparing measures of species diversity from incomplete inventories: an update. Methods Ecol Evol. 2010;1(1):38–44.

42. Jost L. Partitioning diversity into independent alpha and beta components. Ecology. 2007;88(10):2427–39.

43. Zuur A, Ieno EN, Walker N, Saveliev AA, Smith GM. Mixed effects models and extensions in ecology with R. Stanford: Springer Science & Business Media; 2009.

44. Gittleman JL, Kot M. Adaptation: statistics and a null model for estimating phylogenetic effects. Syst Zool. 1990;39:227–41.

45. R Core Team. R: A language and environment for statistical computing. Vienna: R Foundation for Statistical Computing; 2015. p. 2014.

46. Pinheiro J, Bates D, DebRoy S, Sarkar D, R Core Team: nlme: linear and nonlinear mixed effects models. R package version 31-118. 2014.

47. Bates D, Maechler M, Bolker B, Walker S. lme4: Linear mixed-effects models using Eigen and S4. R package version 11-7. 2014.

48. Paradis E, Claude J, Strimmer K. APE: analyses of phylogenetics and evolution in R language. Bioinformatics. 2004;20(2):289–90.

49. Woodcock BA, Edwards M, Redhead J, Meek WR, Nuttall P, Falk S, Nowakowski M, Pywell RF. Crop flower visitation by honeybees, bumblebees and solitary bees: behavioural differences and diversity responses to landscape. Agric Ecosyst Environ. 2013;171:1–8.

50. Bartomeus I, Potts SG, Steffan-Dewenter I, Vaissiere BE, Woyciechowski M, Krewenka KM, Tscheulin T, Roberts SP, Szentgyörgyi H, Westphal C. Contribution of insect pollinators to crop yield and quality varies with agricultural intensification. PeerJ. 2014;2:e328.

51. Mishra RC, Kumar J, Gupta JK. The effect of mode of pollination on yield and oil potential of *brassica campestris* L. var. *Sarson* with observations on insect pollinators. J Apic Res. 1988;27(3):186–9.

52. Free JB, Nuttall PM. The pollination of oilseed rape (*Brassica napus*) and the behaviour of bees on the crop. J Agric Sci. 1968;71(01):91–4.

53. Durán XA, Ulloa RB, Carrillo JA, Contreras JL, Bastidas MT. Evaluation of yield component traits of honeybee pollinated (*Apis mellifera* L.) Rapeseed canola (*Brassica napus* L.). Chilean. J Agric Res. 2010;70(2):309–14.

54. Penning de Vries FWT, van Laar HH. Simulation of growth processes and the model BACROS. In: Penning de Vries FWT, van Laar HH, editors. Simulation of plant growth and crop production. Wageningen: Pudoc; 1982. p. 114–35.

55. Westphal C, Bommarco R, Carré G, Lamborn E, Morison N, Petanidou T, Potts SG, Roberts SPM, Szentgyörgyi H, Tscheulin T. Measuring bee diversity in different European habitats and biogeographical regions. Ecol Monogr. 2008;78(4):653–71.

56. Kovacs-Hostyanszki A, Haenke S, Batary P, Jauker B, Baldi A, Tscharntke T, Holzschuh A. Contrasting effects of mass-flowering crops on bee pollination of hedge plants at different spatial and temporal scales. Ecol Appl. 2013;23(8):1938–46.

57. Angadi SV, Cutforth HW, McConkey BG, Gan Y. Yield adjustment by Canola grown at different plant populations under semiarid conditions. Crop Sci. 2003;43(4):1358–66.

58. Thapa R. Honeybees and other insect pollinators of cultivated plants: a review. J Inst Agri Anim Sci. 2006;27:1–23.

59. Kleijn D, Winfree R, Bartomeus I, Carvalheiro LG, Henry M, Isaacs R, Klein A-M, Kremen C, M'Gonigle LK, Rader R, et al. Delivery of crop pollination services is an insufficient argument for wild pollinator conservation. Nat Commun. 2015;6:7414.

60. Conner JK, Davis R, Rush S. The effect of wild radish floral morphology on pollination efficiency by four taxa of pollinators. Oecologia. 1995;104(2):234–45.

3

Host choice in a bivoltine bee: how sensory constraints shape innate foraging behaviors

Paulo Milet-Pinheiro[1,3]*, Kerstin Herz[1], Stefan Dötterl[2] and Manfred Ayasse[1]

Abstract

Background: Many insects have multiple generations per year and cohorts emerging in different seasons may evolve their own phenotypes if they are subjected to different selection regimes. The bivoltine bee *Andrena bicolor* is reported to be polylectic and oligolectic (on *Campanula*) in the spring and summer generations, respectively. Neurological constraints are assumed to govern pollen diet in bees. However, evidence comes predominantly from studies with oligolectic bees. We have investigated how sensory constraints influence the innate foraging behavior of *A. bicolor* and have tested whether bees of different generations evolved behavioral and sensory polyphenism to cope better with the host flowers available in nature when they are active.

Results: Behavioral and sensory polyphenisms were tested in choice assays and electroantennographic analyses, respectively. In the bioassays, we found that females of both generations (1) displayed a similar innate relative reliance on visual and olfactory floral cues irrespective of the host plants tested; (2) did not prefer floral cues of *Campanula* to those of *Taraxacum* (or vice versa) and (3) did not display an innate preference for yellow and lilac colors. In the electroantennographic analyses, we found that bees of both generations responded to the same set of compounds.

Conclusion: Overall, we did not detect seasonal polyphenism in any trait examined. The finding that bees of both generations are not sensory constrained to visit a specific host flower, which is in strict contrast to results from studies with oligolectic bees, suggest that also bees of the second generation have a flexibility in innate foraging behavior and that this is an adaptive trait in *A. bicolor*. We discuss the significance of our findings in context of the natural history of *A. bicolor* and in the broader context of host-range evolution in bees.

Keywords: *Andrena bicolor*, *Campanula*, Olfactory and visual cues, Oligolecty, Polylecty, Seasonal polyphenism

Background

Bees visit flowers mainly to collect nectar and pollen. These floral rewards are essential for both their own nutritional requirements and brood provision [1]. The spectrum of plants visited for pollen collection varies greatly from one species to another. Some bees restrict pollen gathering to a few (single) species within a genus or family (oligolecty), whereas others collect pollen from various species of distinct families (polylecty) [2, 3].

The evolution of pollen preference in bees has long puzzled scientists but has only recently received special attention. In contrast to previous long-held assumptions (see for example, [4–6]), recent studies show that many polylectic lineages are derived from oligolectic ancestors [7–10]. Whereas the basal state of oligolecty is now well acknowledged, the ecological and physiological aspects governing host choice in bees remain poorly investigated. In bee-plant interactions, pollen plays a paradoxical role; it is, on the one hand, the male gametophyte of plants but, on the other hand, the food of future pollinators (i.e. bee larvae). Consequently, in addition to being pollinators, bees can be seen as herbivores and plants might be under selective pressure to reduce pollen harvesting

*Correspondence: miletpinheiro@hotmail.com
[3] Present Address: Departamento de Química Fundamental, Universidade Federal de Pernambuco, Av. Prof. Moraes Rego, s/n, Recife 50670-901, Brazil
Full list of author information is available at the end of the article

by animals [11]. Recently, plants have been suggested to have evolved chemical defensive properties (i.e. toxic pollen) to avoid excessive pollen consumption [12, 13]. These defensive properties have to be overcome by bees and are assumed to influence pollen diet. Furthermore, nutritional content (e.g. protein, amino acids, lipids, etc.) and the digestibility of pollen vary immensely among the different taxa [14] and, thus, the digestion of various types of pollen might be challenging for bees [15]. Accordingly, evidence is available that the ability of bee larvae to digest and develop on different pollen types varies considerably among species and this holds true for both oligolectic [12, 16, 17] and polylectic [13, 18, 19] representatives. Together, these findings show that pollen is not an easy-to-use resource and that physiological adaptations are necessary for efficient pollen digestion [10, 12, 13, 20].

In addition to physiological constraints, neurological adaptations are assumed to govern host range in bees [12, 13]. However, the exact mechanisms by which neurological adaptations govern host-flower preference remain elusive. Linsley [21] was the first to speculate that the tendency of newly-emerged oligolectic bees to collect pollen only on host flowers used by the previous generation arises from their experience with pollen aroma during the larval stage (i.e. imprinting or conditioning). Although intensively discussed in the literature, this hypothesis has been tested only in three solitary species whose pollen preferences vary considerably (Tepedino, cited in [22–24]). The results of these studies suggest that host-flower preference is genetically controlled. Dobson et al. [24], for example, have found that adults of the broadly polylectic *Osmia bicornis* (Megachilidae), which had been reared during the larval stage on pollen loads of either *Brassica napus* (Brassicaceae) or *Onobrychis viciifolia* (Fabaceae), showed no clear preference for these two hosts when each was offered them together with seven alternative host plants. In *Megachile rotundata* (Megachilidae), a polylectic bee with a more restricted pollen diet, adults selected their preferred host, *Medicago sativa* (Fabaceae), even if they had been reared on a pure pollen diet of *Daucus* (Apiaceae) (Tepedino, cited in [22]). Finally, Praz et al. [23] have found that females of *Heriades truncorum* (Megachilidae) restrict pollen gathering to their hosts (Asteraceae), irrespective of the pollen diet on which they had been reared as larvae, and have suggested that the innate foraging behavior of oligolectic bees is constrained by genetically based neurological adaptations (e.g. vision and olfaction) that drive their flight towards host flowers. Obviously, the extent to which these findings can be generalized to other bee species remains to be established.

Studies investigating the role of visual and olfactory floral cues in the innate foraging behavior of solitary bees are scarce and have focused mainly on oligolectic species. Nevertheless, they provide interesting insights into the way that neurological adaptations can restrict host range in bees. The general tendency emerging from these studies suggests that oligolectic bees innately prefer visual and olfactory floral cues of host plants over those of non-host plants [25–28], thereby implying that oligolectic bees are neurologically adapted to detect some cues that are characteristic for host flowers. In terms of visual cues, for example, some oligolectic species have been shown to display an innate preference for the color of host flowers [26, 28]. According to these authors, color might act as a filter that drives foraging flights of bees towards potential host flowers but, given its unspecific nature (i.e. flowers of species belonging to very distinct taxa might share the same color), color alone would not be a reliable cue for host recognition. Instead, they suggest that floral scents, which are assumed to have an infinite diversity [29, 30], provide a reliable signature for host flowers. Indeed, several oligolectic bees have been shown to rely innately on a single or a few host-typifying floral scent compounds in order to recognize host flowers [25, 31–35]. This might prevent oligolectic bees foraging for pollen on non-host plants that might not be digestible by their larvae. Further studies focusing on the innate reliance on visual and olfactory cues by bees with a different degree of pollen preference would help in understanding the way that sensory adaptations shape the evolution of host range in bees as a whole.

Unlike oligolectic bees, which normally have a very short flight period that is synchronized with the flowering of their host plants, some polylectic bees exhibit long flight activity and present two (bivoltinism) or more (multivoltinism) generations per year [1, 36]. Multivoltine insects (including bees) are assumed to be under distinct selective pressures depending on the biotic (e.g. food availability, predators) and abiotic (e.g. temperature, day length) conditions of the environment in which each generation appears (reviewed by [37–39]). Thus, individuals of distinct generations might evolve their own phenotypes to cope better with specific conditions typical to their surrounding environment, a phenomenon known as polyphenism (i.e. distinct phenotypes produced by the same genotype) [39]. Indeed, many insects show seasonal polyphenism that can influence a variety of traits, morphology and color being the most well investigated [38, 40]. Naturally, the same differences in selection pressures that cause seasonal polyphenism in morphology and color are also likely to cause behavioral and neurological polyphenisms. Evidence of seasonal behavioral and neurological polyphenisms exists for insects such as butterflies (measured as mating propensity; [41, 42]) and locusts [measured as brain size; [43], respectively, but have never been investigated in any respect (e.g. host preference, color preference and olfactory receptors) in bivoltine bees.

The ability of bees to detect a given floral scent compound is often assessed by gas chromatography coupled to electroantenographic detection (GC-EAD), an analytical technique in which an insect antenna is used as a parallel detector for compounds separated on a GC column to identify subsets of complex odor blends that are physiologically active and likely show biological activity [44]. Assuming that first- and second-generation individuals of bivoltine bees are under distinct selective pressure exerted by the different host flowers, and thus different scents, available in each season, we might expect that individuals of each generation would sensory evolve to detect better their respective host plants. Under this perspective, it seems reasonable to speculate that the antennae of first-generation bees respond to a higher number of compounds released by plants flowering in the spring than antennae of first-generation bees and vice versa.

Andrena bicolor (Fabricus 1775) (Andrenidae) is a European solitary bivoltine species. Adults of the first generation are active between March and May and those of the second generation between June and August. Whereas the first generation is assumed to be polylectic, the second has a strong preference for flowers of *Campanula* [45, 46], (Westrich, pers. comm.). In the present study, we have investigated the innate responses of *A. bicolor* females to visual and olfactory floral cues of two common host plants, namely *Taraxacum officinale* (Fig. 1a) and *Campanula trachelium* (Fig. 1b), whose blooming peaks coincide with the flight activity of the first and second generation, respectively. We have hypothesized that bees of the second generation have evolved behavioral and sensory adaptations to cope better with flowers of *Campanula*, whereas bees of the first generation have evolved adaptations to cope with a broader spectrum of host flowers. In order to test this possible seasonal polyphenism, we have integrated behavioral and electrophysiological methods and addressed the following questions:

1. Does the relative reliance on visual and olfactory floral cues of host plants differ between bees of the first and second generation?
2. Do bees of the first and second generation prefer the floral cues (i.e. visual and olfactory) of their common pollen hosts, *Taraxacum* and *Campanula*, respectively?
3. Do bees of the two generations have different color preferences?
4. Are there differences in antennal responses to volatiles of *Taraxacum* and *Campanula* between bees of the two generations?

Fig. 1 *Andrena bicolor* and host plants used in the present study. Females of the first and second generation on flowers of *Taraxacum officinale* (**a**) and *Campanula trachelium* (**b**), respectively (Photographs: P. Milet-Pinheiro)

5. Does the preference of adults for floral cues of either *Taraxacum* or *Campanula* (if any) reflect the pollen diet provided during the larval stage?

Methods

The host plants

Taraxacum officinale F.H Wigg (Asteraceae) is a widely distributed herb capable of growing under diverse environmental conditions. The yellow flowers provide nectar and pollen and are attractive to several floral visitors, mainly bees [47]. The plant blooms from April to October (Schmeil-Fitschen [48]) and is reported as a pollen source for the first generation bees of *A. bicolor* [36, 45].

Campanula trachelium L. (Campanulaceae) has hermaphroditic violet flowers and blooms between July and September (Schmeil-Fitschen [48]). It is protandrous and pollen is secondarily presented on the style [49]. For experiments with bees of the first generation (between late March and early April), *C. trachelium* plants were cultivated in pots in the plant beds of the Botanical Garden of the University of Ulm. During late January, plants were placed in the greenhouses where they were subjected to controlled light (16 h day length) and temperature (10–12 °C) conditions to stimulate flowering during late March. Plants were vigorous and produced several flowers.

Establishment of the bee population in a flight cage

For bioassays, *A. bicolor* bees were reared in a flight cage at the Botanical Garden of the University of Ulm. The flight cage consisted of a steel frame (7 × 3.5 × 2.2 m) that was covered with a fine mesh (stitch density of 1 mm × 0.5 mm) and whose base was buried into the soil to a depth of 0.5 m.

For the establishment of the population, females of the second generation were caught in the Botanical Garden of Ulm and at wild vegetations in the "Schwäbische Alb" at the vicinities of Ulm while foraging on flowers in the summer of 2012. Individuals were exclusively found on *Campanula* flowers, even if we have searched for bees on flowers of other species. The bees were then released into the cage in which flowering plants of *C. trachelium* had been placed. After a few days, bees collected pollen and built nests in the ground.

Flower-naive bees

To test the innate responses of *A. bicolor* to floral cues of *T. officinale* and *C. trachelium*, we performed a series of two-choice bioassays (see details below) with flower-naive female bees of the first and second generation that had hatched from nests inside the cage. Flower-naive bees were defined as those that had had no previous contact with any kind of flower as adults. During the phase in which the bioassays were performed, bees were provided only with sugar water (30 % fructose and glucose 1:1) presented in black sponge feeders. In the flight cage, bees of the first generation (97 males and 76 females) emerged in early April 2013, whereas those of the second generation (100 males and 87 females) emerged in early June 2013. After completion of the bioassays with flower-naive bees of the first and second generations, we introduced plants of *T. officinale* and *C. trachelium* into the flight cage, respectively. The bees promptly started to gather pollen and nectar on flowers and to build nests in the ground. By doing this, we forced bees of the first and second generation to provide offspring exclusively with pollen of *T. officinale* and *C. trachelium*, respectively.

General design of bioassays
Experimental cylinders

The attractiveness of decoupled and combined olfactory and visual floral cues of *T. officinale* and *C. trachelium* to *A. bicolor* females was tested with three kinds of cylinders, namely (1) decoupled olfactory cues: grey cylinders with 60 small holes (diameter 0.2 cm) (Fig. 2a); (2) decoupled visual cues: transparent solid cylinders, without

holes (Fig. 2b); (3) coupled olfactory and visual cues: transparent cylinders with 60 small holes (Fig. 2c). Air containing floral scents from the enclosed inflorescences in cylinders (1) and (3) was blown out of the holes by a membrane pump (G12/01 EB; Rietschle Thomas, Puchheim, Germany) at a flow rate of 1 l min^{-1}. All cylinders had dimensions of 39 cm height and 10 cm diameter. Grey cylinders were made from PVC and transparent cylinders from Plexiglas®. Plexiglas was used because of its ultraviolet (UV) transparency. Plant samples, which consisted of 5–8 inflorescences with pedicel lengths of 15 cm for *T. officinale* and one inflorescence with 5–8 flowers and 20 cm pedicel length for *C. trachelium*, were covered with either grey or transparent cylinders (see details below). For all bioassays, the same number of *Campanula* and *Taraxacum* flowers was used.

Relative importance of visual and olfactory cues for host location

To test whether the relative reliance on visual and olfactory floral cues differed in females of the two generations, we conducted six dual-choice bioassays for each host plant. The bioassays were carried out in the following order: (1) olfactory cues vs. empty control (Fig. 2a), (2) visual cues vs. empty control (Fig. 2b), (3) olfactory + visual cues vs. empty control (Fig. 2c), (4) olfactory vs. visual cues (Fig. 2d), (5) olfactory + visual vs. olfactory cues (Fig. 2e) and (6) olfactory + visual vs. visual cues (Fig. 2f).

Attractiveness of T. officinale vs. C. trachelium

To establish whether (1) *A. bicolor* preferred the floral cues of *T. officinale* to those of *C. trachelium* (or vice versa) and (2) whether an eventual preference differed between both generations, three dual-choice bioassays were performed. In the bioassays, females of both generations were offered a choice of floral cues of *T. officinale* vs. *C. trachelium* in the following order: (1) olfactory cues (Fig. 2a), (2) visual cues (Fig. 2b) and (3) olfactory + visual cues (Fig. 2c).

Bioassay protocol

The bioassays were conducted on sunny days between 10:00 a.m and 2:00 p.m (when bees were most active). Each bioassay lasted 30 min; the position of the paired cylinders, which were placed 1 m apart, was exchanged after 15 min. Responses of the bees were characterized as either (1) approaches: flights toward the cylinder, to a distance closer than 10 cm, without landing or (2) landings: approaches followed by landing on the cylinders. All responding bees were collected, by using nets, after they had responded in order to prevent them from interfering in the attraction of other individuals; we never observed a responding bee being "followed" by another

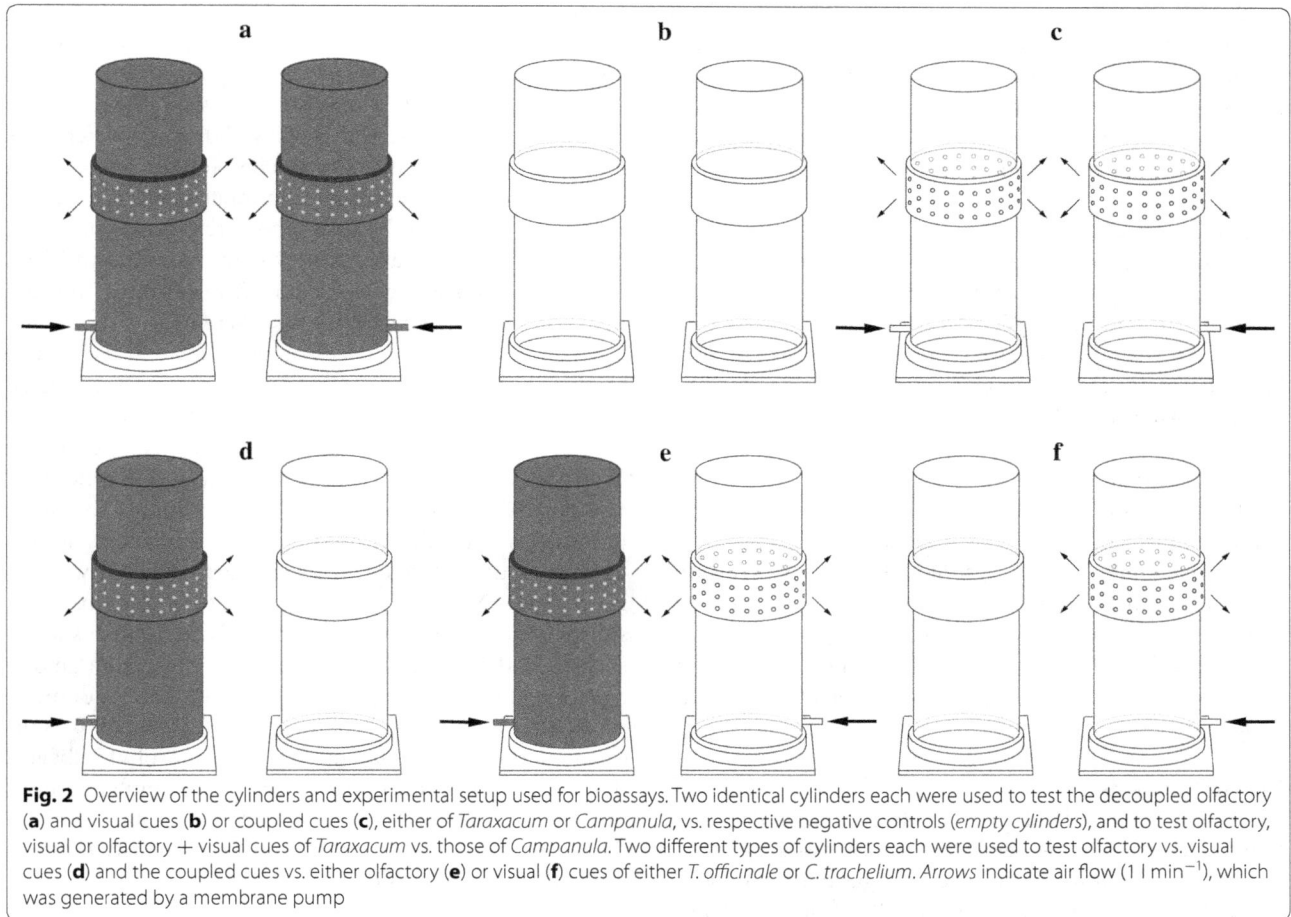

Fig. 2 Overview of the cylinders and experimental setup used for bioassays. Two identical cylinders each were used to test the decoupled olfactory (**a**) and visual cues (**b**) or coupled cues (**c**), either of *Taraxacum* or *Campanula*, vs. respective negative controls (*empty cylinders*), and to test olfactory, visual or olfactory + visual cues of *Taraxacum* vs. those of *Campanula*. Two different types of cylinders each were used to test olfactory vs. visual cues (**d**) and the coupled cues vs. either olfactory (**e**) or visual (**f**) cues of either *T. officinale* or *C. trachelium*. *Arrows* indicate air flow (1 l min^{-1}), which was generated by a membrane pump

bee. Approaching bees were collected when they flew away from the cylinders and landing bees after they had landed. All responding bees were stored in an icebox until the end of the experiment, at which time they were released back into the flight cage and could participate in subsequent tests. Thus, an individual bee could respond only once in each specific bioassay.

Color preference
To test whether *A. bicolor* females displayed an innate preference for a given color and whether color preference differed according to bee generation, dual-choice experiments with artificial flowers were performed. For these experiments, we selected human-lilac and human-yellow colors, because of both their representativeness in nature and their high attractiveness to insects in general [50–52]. Although the colors of the artificial flowers used here did not resemble exactly the colors of flowers of *T. officinale* and *C. trachelium* (see results), they did resemble those of several Asteraceae (Milet-Pinheiro, unpublished data) and Campanulaceae [28]. Lilac and yellow bell-shaped artificial flowers (length 4.5 cm; diameter at

the top 2 cm), were fashioned out of construction paper and each artificial flower was fixed at the base on a thin wooden stick (length: 23 cm; diameter: 2 mm). Two groups of artificial flowers (choices) were presented to the bees simultaneously. Each group was composed of three artificial lilac flowers or three yellow flowers. Artificial flowers within a group were arranged in a triangle and 5 cm apart from each other. The two groups of artificial flowers were placed 1 m apart. The general testing procedures were the same as those for the cylinder tests (see bioassay protocol).

Evaluation of visual and olfactory cues
Color measurements and bee color hexagon
To improve our understanding of the significance of color in the innate behavioral responses of *A. bicolor*, we measured reflectance properties of both natural ($n = 3$ for each species) and artificial ($n = 3$ measurements per paper) flowers by using a Varian Cary 5 spectrophotometer equipped with a Praying Mantis accessory (Varian, Inc, Palo Alto, California). The mean reflectance profiles were then converted into color loci of the color hexagon

space [53]. The hexagon space is a model of bee color vision applicable to a large number of hymenopteran species and allows an interpretation of the way in which colors are perceived and discriminated by bees (for full details concerning reflectance measurements and bee color hexagon, see Additional file 1).

Sampling of floral scent

To obtain samples for electrophysiological investigations ($n = 1$ per species from 50 flowers each), floral volatiles from the flowers of *T. officinale* and *C. trachelium* were collected by using a standard dynamic headspace method (for full details of the procedures applied and material used, see Additional file 1).

Electrophysiology

Analyses of gas chromatography coupled with electroantennographic detection (GC/EAD) were performed with *A. bicolor* females ($n = 5$ antennae for each generation) to determine the antennal perception of the bee for compounds in the floral scent bouquet of *T. officinale* and *C. trachelium* [for full details of equipment specifications and configurations and of antennal preparation, Additional file 1).

Chemical analyses

To identify the floral volatiles eliciting antennal depolarization in *A. bicolor*, the headspace samples of *T. officinale* and *C. trachelium* were analyzed on a gas chromatograph coupled to a mass spectrometer (GC/MS) (for full details of equipment specifications and configurations and of the elucidation and quantification of compounds, see Additional file 1).

Statistical analyses

To test for differences in total bee responses (pooled approaches and landings) between the paired treatments in each bioassay, exact binominal tests were performed. The two types of behaviors were pooled because visual cues of flowers of *T. officinale* and *C. trachelium* and a combination of visual and olfactory cues triggered approach and landing responses in similar proportions (Fisher's exact tests: $0.49 < P < 1$, see Additional file 2). In the bioassays testing the relative importance of floral cues within a plant species, one-tailed exact binomial tests were used to test the null hypothesis that both visual and olfactory cues attract ≤number of bees than the negative controls (empty cylinders) and that combined cues attract ≤number of bees than the decoupled cues. The one-tailed design was used because it is highly unlikely that visual and olfactory cues of *T. officinale* and *C. trachelium* have repellent properties for *A. bicolor*. For the bioassays testing cues of *T. officinale* vs. those of *C.*

trachelium, two-tailed exact binominal tests were used to test the hypothesis that floral cues (alone or in combination) of both species are equally attractive to bees. Fisher's exact tests were used to test whether bees of the first and second generation displayed distinct preferences for visual and olfactory floral cues (either alone or in combination) of a given host. Binomial and Fisher's exact tests were calculated by using the spreadsheets provided by http://www.biostathandbook.com/exactgof.html and http://www.biostathandbook.com/fishers.html, respectively (Accessed 10 Nov 2014; see also [54]).

Differences in antennal responses between females of the first and second generation of *A. bicolor* were tested by using an analysis of similarity (ANOSIM). For this purpose, we prepared a table with the EAD responses (presence/absence) of bees of the first and second generation. We then calculated the Sørensen similarity index. This index determines pairwise similarities among the individuals. Based on this similarity matrix, we performed an ANOSIM analysis (10,000 permutations) considering the generation as a factor. ANOSIM yields a test statistic R that is a relative measure of separation among a priori defined groups. It is based on differences of mean ranks among and within groups. An R value of '0' indicates random grouping, whereas a value of '1'indicates that all samples within groups are more similar to each other than to any sample from a different group. Software Primer 6.1.6 was used to calculate the similarity index of Sørensen and the ANOSIM analysis [55].

Results
Relative importance of visual and olfactory floral cues

When testing the relative importance of floral cues, we found that bees of the first and second generation responded similarly to the floral cues, irrespective of the host plants tested. In all cases, visual cues were significantly more attractive to bees when offered together with an empty cylinder control, whereas olfactory cues were not (Fig. 3a, b). When offered together with olfactory cues, visual cues were significantly more attractive to the bees. Finally, a combination of visual and olfactory cues was more attractive to bees than either cue alone.

Comparison of attractiveness of floral cues of the different host plants

In the dual-choice bioassays testing floral cues of *T. officinale* vs. those of *C. trachelium*, we found that bees of the first and second generations did not display any preference for the cues of one host plant over the other. In all cases, decoupled visual cues of *T. officinale* and a combination of visual and olfactory cues were equally attractive to *A. bicolor* females of both generations as those equivalent cues of *C. trachelium* (Fig. 4). Responses to olfactory

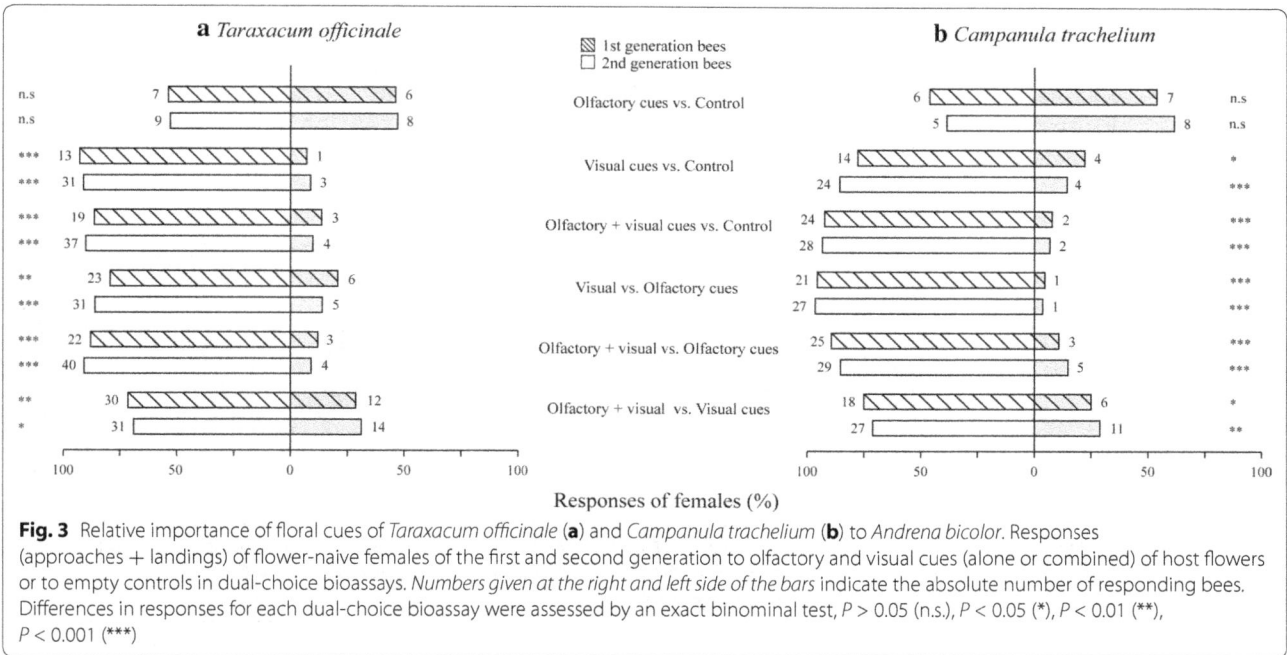

Fig. 3 Relative importance of floral cues of *Taraxacum officinale* (**a**) and *Campanula trachelium* (**b**) to *Andrena bicolor*. Responses (approaches + landings) of flower-naive females of the first and second generation to olfactory and visual cues (alone or combined) of host flowers or to empty controls in dual-choice bioassays. *Numbers given at the right and left side of the bars* indicate the absolute number of responding bees. Differences in responses for each dual-choice bioassay were assessed by an exact binominal test, $P > 0.05$ (n.s.), $P < 0.05$ (*), $P < 0.01$ (**), $P < 0.001$ (***)

cues are not shown, since these were similar to responses to an empty cylinder (see above).

Artificial flowers

The dual-choice bioassays with artificial flowers showed that yellow and lilac flowers were similarly attractive and

female bees of both generations did not display a preference for a particular color (Fig. 5).

Color analysis

The color measurements revealed that the yellow and lilac artificial flowers reflected the light in the ultra-violet (UV) color range (Fig. 6), whereas the flowers of *C. trachelium* and *T. officinale* did not. The flowers of *T. officinale* and the yellow artificial flowers reflected the light predominantly in the yellow, orange and red range (530–700 nm) and the flowers of *C. trachelium* and the lilac artificial flowers predominantly in the blue

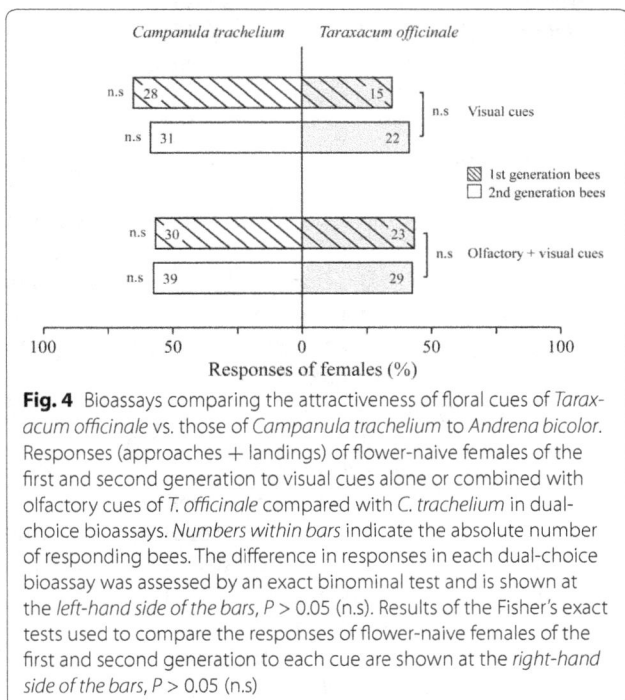

Fig. 4 Bioassays comparing the attractiveness of floral cues of *Taraxacum officinale* vs. those of *Campanula trachelium* to *Andrena bicolor*. Responses (approaches + landings) of flower-naive females of the first and second generation to visual cues alone or combined with olfactory cues of *T. officinale* compared with *C. trachelium* in dual-choice bioassays. *Numbers within bars* indicate the absolute number of responding bees. The difference in responses in each dual-choice bioassay was assessed by an exact binominal test and is shown at the *left-hand side of the bars*, $P > 0.05$ (n.s.). Results of the Fisher's exact tests used to compare the responses of flower-naive females of the first and second generation to each cue are shown at the *right-hand side of the bars*, $P > 0.05$ (n.s)

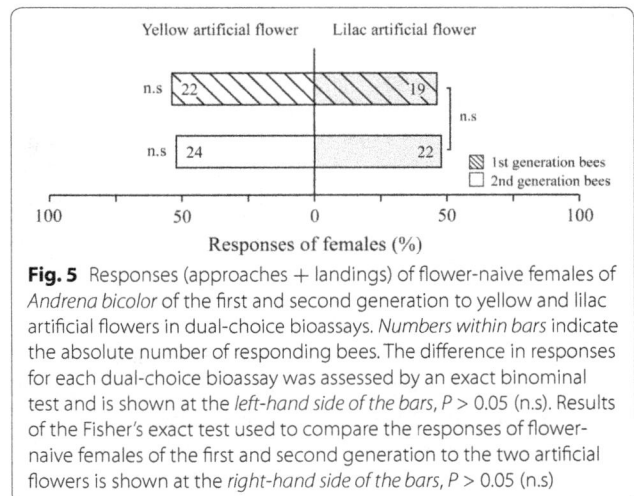

Fig. 5 Responses (approaches + landings) of flower-naive females of *Andrena bicolor* of the first and second generation to yellow and lilac artificial flowers in dual-choice bioassays. *Numbers within bars* indicate the absolute number of responding bees. The difference in responses for each dual-choice bioassay was assessed by an exact binominal test and is shown at the *left-hand side of the bars*, $P > 0.05$ (n.s). Results of the Fisher's exact test used to compare the responses of flower-naive females of the first and second generation to the two artificial flowers is shown at the *right-hand side of the bars*, $P > 0.05$ (n.s)

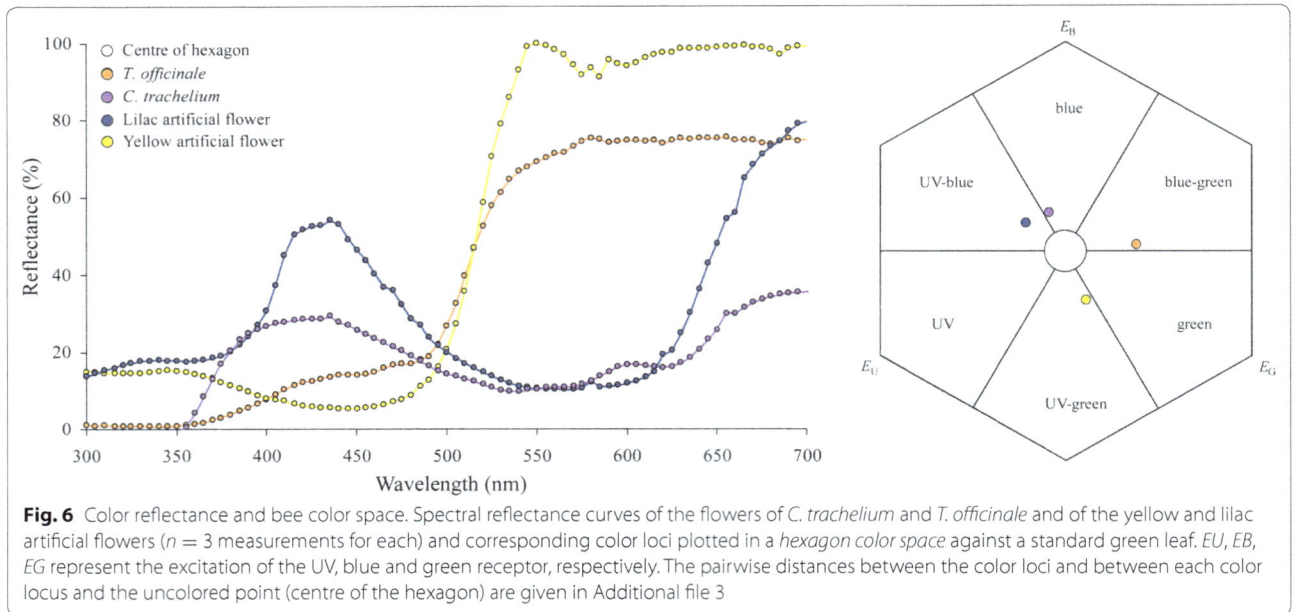

Fig. 6 Color reflectance and bee color space. Spectral reflectance curves of the flowers of *C. trachelium* and *T. officinale* and of the yellow and lilac artificial flowers (*n* = 3 measurements for each) and corresponding color loci plotted in a *hexagon color space* against a standard green leaf. *EU, EB, EG* represent the excitation of the UV, blue and green receptor, respectively. The pairwise distances between the color loci and between each color locus and the uncolored point (centre of the hexagon) are given in Additional file 3

(380–500 nm) and red (640–700 nm) range. The colors of the host flowers and artificial flowers seemed to be distinct enough to allow discrimination by the bees. When plotting the color reflectance functions into the bee color hexagon, the color loci of flowers of *C. trachelium* and *T. officinale* and those of the yellow and lilac artificial flowers were plotted into the blue, blue-green, UV-blue and UV-green bee color spaces, respectively (Fig. 6). In the color hexagon, the pairwise distances between color loci ranged from 0.12 hexagon units (comparison between the color loci of *C. trachelium* and lilac artificial flower) to 0.53 (comparison between the color loci of *T. officinale* and lilac artificial flower). The distances between the color loci of all samples to the uncolored point (centre of the hexagon) ranged from 0.2 to 0.33 units (Additional file 3).

Electrophysiology

The GC-EAD analyses with flower scent samples of *T. officinale* revealed nine electrophysiologically active compounds by using the antennae of *A. bicolor* (Fig. 7a). The compounds triggering antennal depolarization belonged to three substance classes: aromatics (benzaldehyde, acetophenone and benzoid acid), monoterpenes [(*E*)-β-ocimene and linalool] and sesquiterpenes (β-copaene).

The GC-EAD analyses with flower scent samples of *C. trachelium* revealed seven electrophysiologically active compounds by using the antennae of *A. bicolor* (Fig. 7b). These compounds belonged to three substance classes, i.e. aromatics (phenylethyl acetate), monoterpenes [(*E*)-β-ocimene, terpinolene, linalool] and sesquiterpenes

(β-elemene and (*E*)-β-caryophyllene). Only linalool and (*E*)-β-ocimene were detected in floral scent samples of both *T. officinale* and *C. trachelium*.

Antennal responses to floral scents of either *Taraxacum* (Additional file 4) or *Campanula* (Additional file 5) varied slightly among individuals. However, no significant difference was seen between the antennal responses of bees of the first and second generation. Bees of both generations responded to the same set of compounds of *T. officinale* (ANOSIM: Global $R = -0.13$; $P = 0.93$) and *C. trachelium* (ANOSIM: Global $R = -0.15$; $P = 1$).

Discussion

The findings of our study demonstrate that *A. bicolor* females of the first and second generation have the same innate search-image, in spite of the different spectrum of host flowers that they use. In the bioassays, we found that bees of the two generations (1) have the same innate relative reliance on visual and olfactory floral cues of host plants, (2) do not prefer floral cues of *Taraxacum* to those of *Campanula* (or vice versa) and (3) do not display an innate preference for the tested colors yellow and lilac. Furthermore, in the electrophysiological analyses, bees of both generations were shown to respond within a plant species to the same set of components, all of which are ubiquitous volatiles commonly reported in plants. Together, our results suggest that the visual and olfactory constraints of *A. bicolor* loosen their innate foraging behavior, in contrast to oligolectic bees in which sensory adaptations are assumed to restrict foraging flights to specific host flowers (see below).

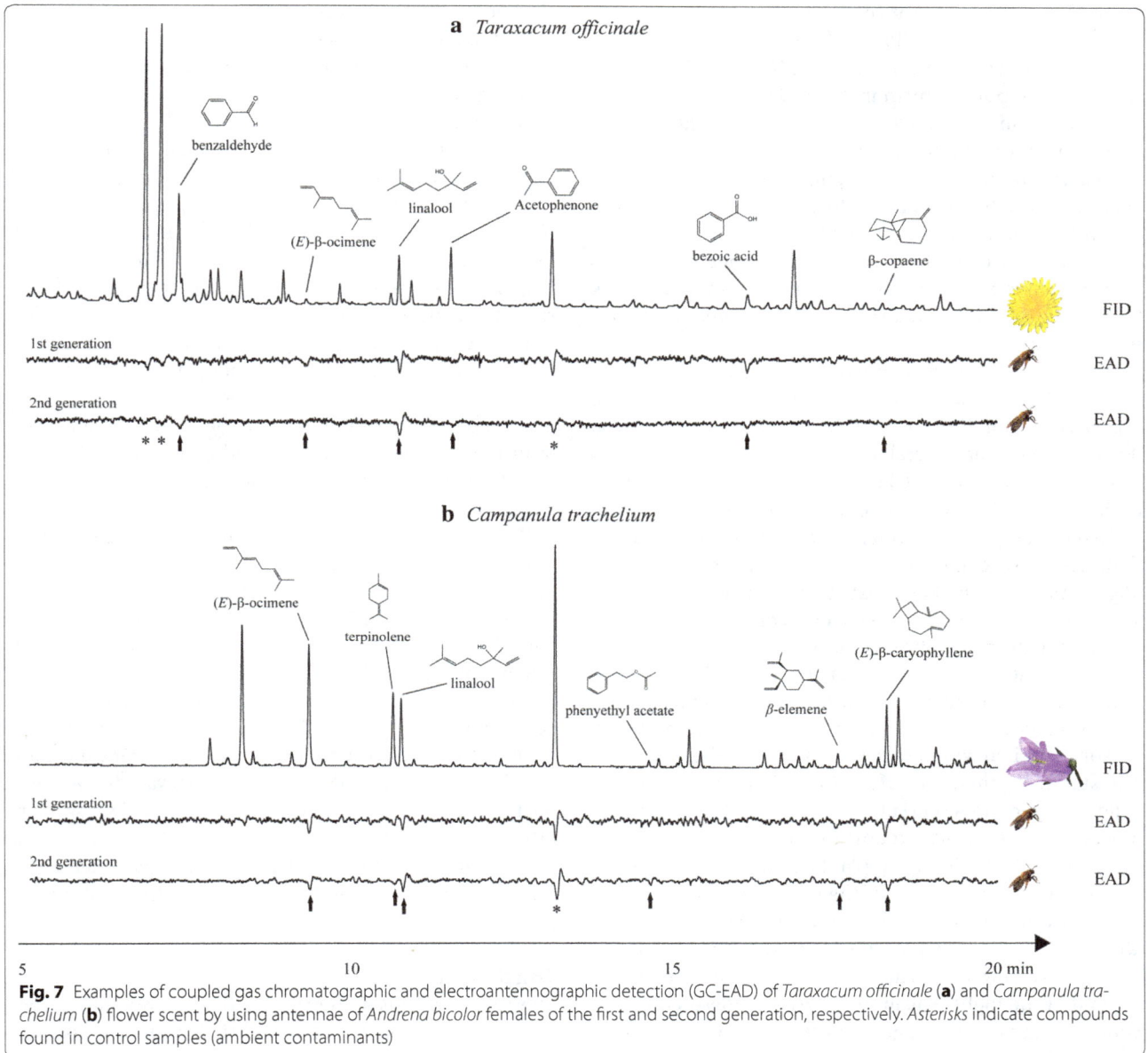

Fig. 7 Examples of coupled gas chromatographic and electroantennographic detection (GC-EAD) of *Taraxacum officinale* (**a**) and *Campanula trachelium* (**b**) flower scent by using antennae of *Andrena bicolor* females of the first and second generation, respectively. *Asterisks* indicate compounds found in control samples (ambient contaminants)

Andrena bicolor is a bivoltine solitary species that is assumed to be oligolectic on flowers of *Campanula* in the second generation but that has a broader pollen diet in the first generation [45, 46]. Based on this scenario, we have hypothesized that bees of the second generation have evolved specific sensory traits that direct their innate foraging flights towards their preferred host flowers, whereas bees of the first generation have evolved more generalized sensory traits that do not restrict their innate foraging flights to any specific host flowers. In the present study, however, various findings indicate that bees of both generations are not constrained to visit

flowers of a particular host. This is in strict contrast to our hypothesis and to reports for oligolectic bees [23, 26–28, 32, 35]. First, we have found that flower-naive females of both generations do not display a remarkable preference for the floral cues of *C. trachelium* over those of *T. officinale* and vice versa. Second, in the bioassays with lilac and yellow artificial flowers, females of both generations do not show any innate preference for one color over the other. Third, in the GC-EAD analyses, bees of both generations have been determined to respond only to ubiquitous compounds that are reported as floral scent constituents of various plant species [30], in contrast to some oligolectic bees that respond additionally to highly specific volatiles of host flowers (see, for example, [32, 33, 35]). Overall, we have not detected behavioral and sensory seasonal polyphenism in *A. bicolor* and this suggests that the distinct selective forces acting on the bees of each generation are not strong enough or are acting in a similar way to keep innate foraging behavior and sensory adaptations as broad as possible in order to reduce their dependence upon a few hosts plants. Thus, irrespective of the way in which selective forces are acting on the bees of each generation, a great flexibility in innate foraging behavior seems to be an adaptive trait in *Andrena bicolor* (see below).

The absence of a clear innate preference for the floral cues of *C. trachelium* over those of *T. officinalle* and of a preference for the UV-blue bee-color of the artificial flowers typical of several *Campanula* species; [28], as well as the capability of larvae to develop on pollen of both host plants, all point to second-generation bees of *A. bicolor* being sensory and behaviorally less constrained than the oligolectic bees [26–28, 35]. The dependence of oligolectic bees on their host flowers vary considerably among species. While some species completely refuse to collect pollen in the absence of host plants [26, 27, 35], others are more flexible and may collect pollen on alternative plants if their preferred hosts are absent [12, 16, 17]. Accordingly, there might be also a great variability in sensory and behavioral constraints among these species. The summer generation of *Andrena bicolor* is known to collect pollen preferentially on flowers of *Campanula*, however, there is also report of these bees collecting pollen on alternative host plants [36] (Westrich, pers. comm]. In Germany, bees of the second generation of *A. bicolor* are active from mid June to late May–August [36, 45], whereas *Campanula* species flower predominantly between late and early August [48, 56]. Consequently, at the end of the flight season of *A. bicolor*, flowers of *Campanula* might be no longer available and a flexibility in foraging behavior by second-generation females might allow them to resume nest provisioning using alternative pollen sources.

The findings that *Andrena bicolor* does not have a particular innate preference for the floral cues of one host over the other also provide strong evidence that host-choice by adult females is not governed by conditioning with the pollen aroma during the larval stage, as hypothesized by Linsley [21]. If conditioning to pollen aroma governed the initial foraging behavior of *A. bicolor* bees, we would have expected that flower-naive adults of the first (reared on pollen of *Taraxacum*) and second (reared on pollen of *Campanula*) generation would prefer the floral cues (in this case, floral scents) of *Taraxacum* and *Campanula*, respectively, which was not the case.

Consequently, in *A. bicolor*, the innate preference for a given host or the absence of it might be governed by sensory, genetically-based constraints that shape the innate foraging behavior of bees see also [10, 23].

In this study, we did not experimentally test the development of larvae of *A. bicolor* on pollen of *Taraxacum* and *Campanula* as, for example, carried out by Praz et al. [12] and Sedivy et al. [13] with other solitary bee species. Nevertheless, the observation that 170 individuals of the first generation (larvae reared mainly on *Campanula* pollen) and 190 individuals of the second generation (larvae reared mainly on *Taraxacum* pollen only) emerged in the flight cage strongly suggests that larvae of *A. bicolor* have a good capability for digesting pollen from these two hosts and probably also from other plants. Pollen of Campanulaceae and Asteraceae species vary enormously in nutritional content and digestibility [15]. Pollen of *Campanula* is protein-rich [15, 57] and is known to support the larval development of various bee species [12]. In contrast, several reports have appeared of insufficient or inappropriate larval development on the pollen of Asteraceae, including *Taraxacum officinale* (reviewed by [15]). The inability of bee larvae to develop on the pollen of Asteraceae is assumed to be related to either its low nutritional quality, e.g. low protein and amino acid contents [57–60], a difficulty in the extraction of essential compounds from the pollen grains [61], or an interference of toxic pollen compounds with nutrient digestion [12, 13]. The finding that *A. bicolor* larvae develop well on the pollen of Asteraceae, which is avoided by many other polylectic species [3], suggests that this species has physiologically evolved to deal with various pollen types, a finding with a strong implication for the evolution of pollen preference in bees (see below).

Conclusion

The results of this study shed light not only into several aspects of the natural history of *A. bicolor*, but also into the evolution of pollen preference in bees. In a recent study, Sedivy et al. [10] proposed the persuasive constraint hypothesis of host-range evolution in bees, based mainly on findings from studies performed with oligolectic bees. According to these authors, pollen diet breadth in oligolectic bees is evolutionary constrained by sensory or neurological adaptations that might restrict their foraging flights to host flowers and by physiological adaptations that might restrict their capability to digest different pollen types. However, evidence that polylectic bees are adapted sensory and physiologically to exploit a broad spectrum of host flowers is scarce. In this study, we have found, for the first time, evidence that polylectic solitary bees are visually and olfactory less constrained than the oligolectic bees investigated so far and that this might allow them to visit a broad spectrum of flowers. Furthermore, we have revealed that *A. bicolor* has a broad digestive capability that allows its larvae to develop even on Asteraceae pollen, which is difficult to digest for several other polylectic species [3]. The different levels of sensory and physiological constraints observed in bees might be directly related to their ability to exploit pollen from either a few or several host plants and might help to explain the continuum of pollen preference (from monolecty to broad polylecty) observed in bees. Although our understanding about the mechanisms governing the evolution of host-choice in bees has improved considerably in the last decade, the conclusions traced so far are based on a very few species. Thus, more effort is still necessary to establish to what extent sensory, neurological and physiological adaptations shape the evolution of host preference in bees.

Additional files

Additional file 1. Detailed procedures applied for color measurements, bee color hexagon, sampling of floral scents, electrophysiology and chemical analyses.

Additional file 2. Proportion of approach (AP) and landing (LA) responses of *Andrena bicolor* females to floral cues of *Taraxacum officinale* and *Campanula trachelium*.

Additional file 3. Euclidean distances (in hexagon units) among color loci of flowers of *C. trachelium* and *T. officinale* and of lilac and yellow artificial flowers.

Additional file 4. Electrophysiological responses of bees of the first and second generation of *Andrena bicolor* to floral scents of *Taraxacum*.

Additional file 5. Electrophysiological responses of bees of the first and second generation of *Andrena bicolor* to floral scents of *Campanula*.

Abbreviations
GC/EAD: gas chromatography coupled with electroantennographic detection; GC/MS: gas chromatography coupled with mass spectrometry.

Authors' contributions
PMP, SD and MA originally formulated the idea, PMP, MA and SD designed experiments, PMP and KH performed experiments, PMP and KH performed statistical analyses, PMP, KH, SD and MA wrote the manuscript. All authors read and approved the final manuscript.

Author details
[1] Institute of Evolutionary Ecology and Conservation Genomics, University of Ulm, Helmholtzstraße 10-1, 89081 Ulm, Germany. [2] Department of Ecology and Evolution, University of Salzburg, Hellbrunnerstrasse 34, 5020 Salzburg, Austria. [3] Present Address: Departamento de Química Fundamental, Universidade Federal de Pernambuco, Av. Prof. Moraes Rego, s/n, Recife 50670-901, Brazil.

Acknowledgements
We thank Petra Jentschke, Iris Albrecht and Katharina Dering for rearing bees in the flight cage, the team from the Botanical Garden of the University of Ulm for cultivating the *Campanula* and *Taraxacum* plants used in the experiments and for the rearing of bees and two anonymous referees for the valuable comments on an earlier version of this manuscript.

Competing interests
The authors declare that they have no competing interests.

Funding
This research was supported by the Deutsche Forschungsgemeinschaft (AY 12/5-1; DO 1250/6-1).

References
1. Michener CD. The bees of the world. 2nd ed. Baltimore: The Johns Hopkins University Press; 2007.
2. Cane JH, Sipes S. Characterizing floral specialization by bees: analytical methods and revised lexicon for oligolecty. In: Waser NM, Ollerton J, editors. Plant-pollinator interactions: from specialization to generalization. Chicago: The University of Chicago Press; 2006. p. 99–121.
3. Müller A, Kuhlmann M. Pollen hosts of western palaearctic bees of the genus *Colletes* (Hymenoptera: Colletidae): the Asteraceae paradox. Biol J Linn Soc. 2008;95:719–33.
4. Michener CD. Bees of Panama. Bull Am Mus Nat Hist. 1954;104:1–176.
5. Moldenke AR. Host-plant coevolution and the diversity of bees in relation to the flora of North America. Phytologia. 1979;43:357–419.
6. Hurd PD, LaBerge WE, Linsley EG. Principal sunflower bees of North America with emphasis on the southwestern United States (Hymenoptera: Apoidea). Smithson Contrib Zool. 1980;310:1–158.
7. Müller A. Host-plant specialization in western palearctic anthidiine bees (Hymenoptera: Apoidea: Megachilidae). Ecol Monogr. 1996;66:235–57.
8. Danforth BN, Sipes SD, Fang J, Brady SG. The history of early bee diversification based on five genes plus morphology. Proc Natl Acad Sci. 2006;103:15118–23.
9. Larkin LL, Neff JL, Simpson BB. The evolution of a pollen diet: host choice and diet breadth of *Andrena* bees (Hymenoptera: Andrenidae). Apidologie. 2008;39:133–45.
10. Sedivy C, Praz CJ, Müller A, Widmer A, Dorn S. Patterns of host-plant choice in bees of the genus *Chelostoma*: the constraint hypothesis of host-range evolution in bees. Evolution. 2008;62:2487–507.
11. Westerkamp C. Pollen in bee-flower relations, some considerations on melittophily. Bot Acta. 1996;109:325–32.
12. Praz CJ, Müller A, Dorn S. Specialized bees fail to develop on non-host pollen: do plants chemically protect their pollen. Ecology. 2008;89:795–804.
13. Sedivy C, Müller A, Dorn S. Closely related pollen generalist bees differ in their ability to develop on the same pollen diet: evidence for physiological adaptations to digest pollen. Funct Ecol. 2011;25:718–25.
14. Roulston TH, Cane JH, Buchmann SL. What governs protein content of pollen: pollinator preferences, pollen-pistil interactions, or phylogeny? Ecol Monogr. 2000;70:617–43.
15. Roulston TH, Cane JH. Pollen nutritional content and digestibility for animals. Plant Syst Evol. 2000;222:187–209.
16. Rozen JG. Notes on the biology of *Nomadopsis*, with descriptions of four new species (Apoidea, Andrenidae). Am Mus Novit. 1963;2142:1–17.
17. Williams NM. Use of novel pollen species by specialist and generalist solitary bees (Hymenoptera: Megachilidae). Oecologia. 2003;134:228–37.
18. Levin MD, Haydak MH. Comparative value of different pollens in the nutrition of *Osmia lignaria*. Bee World. 1957;38:221–6.
19. Guirguis GN, Brindley WA. Insecticide susceptibility and response to selected pollens of larval alfalfa leafcutting bees, *Megachile pacifica* (Panzer) (Hymenoptera: Megachilidae). Environ Entomol. 1974;3:691–4.
20. Dobson HEM, Peng YS. Digestion of pollen components by larvae of the flower-specialist bee *Chelostoma florisomne* (Hymenoptera: Megachilidae). J Insect Physiol. 1997;43:89–100.
21. Linsley EG. The ecology of solitary bees. Hilgardia. 1958;27:543–99.
22. Wcislo WT, Cane JH. Floral resource utilization by solitary bees (Hymenoptera: Apoidea) and exploitation of their stored foods by natural enemies. Annu Rev Entomol. 1996;41:257–86.

23. Praz CJ, Müller A, Dorn S. Host recognition in a pollen-specialist bee: evidence for a genetic basis. Apidologie. 2008;39:547–57.

24. Dobson HEM, Ayasse M, O'Neal KA, Jacka JA. Is flower selection influenced by chemical imprinting to larval food provisions in the generalist bee Osmia bicornis (Megachilidae)? Apidologie. 2012;43:698–714.

25. Dobson HEM, Bergström G. The ecology of pollen odors. Plant Syst Evol. 2000;222:63–87.

26. Burger H, Dötterl S, Ayasse M. Host-plant finding and recognition by visual and olfactory floral cues in an oligolectic bee. Funct Ecol. 2010;24:1234–40.

27. Milet-Pinheiro P, Ayasse M, Schlindwein C, Dobson HEM, Dötterl S. Host location by visual and olfactory floral cues in an oligolectic bee: innate and learned behavior. Behav Ecol. 2012;23:531–8.

28. Milet-Pinheiro P, Ayasse M, Dötterl S. Visual and olfactory floral cues of Campanula (Campanulaceae) and their significance for host recognition by an oligolectic bee pollinator. PLoS One. 2015;10:e0128577.

29. Williams NH. Floral fragrances as cues in animal behavior. In: Jones EC, Little RJ, editors. Handbook of experimental pollination biology. New York: Van Nostrand Reinhold; 1983. p. 50–72.

30. Knudsen JT, Eriksson R, Gershenzon J, Stahl B. Diversity and distribution of floral scent. Bot Rev. 2006;72:1–120.

31. Dötterl S, Füssel U, Jürgens A, Aas G. 1,4-Dimethoxybenzene, a floral scent compound in willows that attracts an oligolectic bee. J Chem Ecol. 2005;31:2993–8.

32. Burger H, Dötterl S, Häberlein C, Schulz S, Ayasse M. An arthropod deterrent attracts specialised bees to their host plants. Oecologia. 2012;168:727–36.

33. Milet-Pinheiro P, Ayasse M, Dobson HEM, Schlindwein C, Francke W, Dötterl S. The chemical basis of host-plant recognition in a specialized bee pollinator. J Chem Ecol. 2013;39:1347–60.

34. Carvalho AT, Dötterl S, Schlindwein C. An aromatic volatile attracts oligolectic bee pollinators in an interdependent bee-plant relationship. J Chem Ecol. 2014;40:1126–34.

35. Schäffler I, Steiner KE, Haid M, van Berkel SS, Gerlach G, Johnson SD, et al. Diacetin, a reliable cue and private communication channel in a specialized pollination system. Sci Rep. 2015;5:12779.

36. Westrich P. Die wildbienen baden-württembergs: spezieller teil: Die gattungen und arten. Stuttgart: Eugen Ulmer; 1989.

37. Moran NA. The evolutionary maintenance of alternative phenotypes. Am Nat. 1992;139:971–89.

38. Gotthard K, Nylin S. Adaptive plasticity and plasticity as an adaptation: a selective review of plasticity in animal morphology and life history. Oikos. 1995;74:3–17.

39. Simpson SJ, Sword GA, Lo N. Polyphenism in Insects. Curr Biol. 2011;21:738–49.

40. Nylin S, Gotthard K. Plasticity in life-history traits. Annu Rev Entomol. 1998;43:63–83.

41. Friberg M, Wiklund C. Generation-dependent female choice: behavioral polyphenism in a bivoltine butterfly. Behav Ecol. 2007;18:758–63.

42. Mellström HL, Friberg M, Borg-Karlson A-K, Murtazina R, Palm M, Wiklund C. Seasonal polyphenism in life history traits: time costs of direct development in a butterfly. Behav Ecol Sociobiol. 2010;64:1377–83.

43. Ott SR, Rogers SM. Gregarious desert locusts have substantially larger brains with altered proportions compared with the solitarious phase. Proc Royal Soc B Biol Sci. 2010;277:3087–96.

44. Schiestl FP, Poll FM. Detection of physiologically active flower volatiles using gas chromatography coupled with electroantennography. In: Jackson JF, Linskens HF, Inman R, editors. Molecular methods of plant analysis, volume 21: analysis of taste and aroma. Berlin: Springer; 2002. p. 173–98.

45. Müller A, Krebs A, Amiet F. Bienen: Mitteleuropäische Gattungen, Lebensweise, Beobahctung. München: Naturbuch-Verlag; 1997.

46. Amiet F, Krebs A. Bienen mitteleuropas—Gattungen, lebensweise, beobachtung. Bern, Stuttgart, Wien: Haupt Verlag; 2012.

47. Loughnan D, Thomson JD, Ogilvie JE, Gilbert B. Taraxacum officinale pollen depresses seed set of montane wildflowers through pollen allelopathy. J Pollinat Ecol. 2014;13:146–50.

48. Schmeil O, Fitschen J. Flora von Deutschland und angrenzender länder. Ulm: Quelle Meyer Verlag; 2003.

49. Shetler SG. Pollen-collecting hairs of Campanula (Campanulaceae), I: historical review. Taxon. 1979;28:205–15.

50. Lunau K, Maier EJ. Innate colour preferences of flower visitors. J Comp Physiol A. 1995;177:1–19.

51. Arnold SEJ, Savolainen V, Chittka L. Flower colours along an alpine altitude gradient, seen through the eyes of fly and bee pollinators. Arthropod Plant Interact. 2009;3:27–43.

52. Chittka L, Le Comber SC, Arnold SEJ. Flower color phenology in European grassland and woodland habitats, through the eyes of pollinators. Isr J Plant Sci. 2009;57:211–30.

53. Chittka L. The colour hexagon: a chromaticity diagram based on photoreceptor excitations as a generalized representation of colour opponency. J Comp Physiol A. 1992;170:533–43.

54. McDonald JH. Handbook of biological statistics. 2nd ed. Baltimore: Sparky House Publishing; 2009.

55. Clarke KR, Gorley RN. Primer v6: User Manual/Tutorial. Plymouth: Primer-E; 2006.

56. Campanulaceae Rosenbauer A. In: Sebald O, Seybold S, Philippi G, Wörz A, editors. Die Farn-und Blütenpflanzen Baden-Württembergs. Band 5: Spezieller Teil. (Spermatophyta, Unterklasse Asteridae) Buddlejaceae bis Caprifoliaceae. Stuttgart: Verlag Eugen Ulmer; 1996. p. 417–49.

57. Weiner CN, Hilpert A, Werner M, Linsenmair KE, Blüthgen N. Pollen amino acids and flower specialisation in solitary bees. Apidologie. 2010;41:476–87.

58. Auclair JL, Jamieson CA. A qualitative analysis of amino acids in pollen collected by bees. Science. 1948;108:357–8.

59. Loper GM, Cohen AC. Amino acid content of dandelion pollen, a honey bee (Hymenoptera: Apidae) nutritional evaluation. J Econ Entomol. 1987;80:14–7.

60. Herbert EWJ. Honey bee nutrition. In: Graham JM, editor. The hive and the honey bee. Hamilton: Dadant & Sons; 1992. p. 197–233.

61. Peng Y-S, Nasr ME, Marston JM, Fang Y. The digestion of dandelion pollen by adult worker honeybees. Physiol Entomol. 1985;10:75–82.

4

A temporal assessment of nematode community structure and diversity in the rhizosphere of cisgenic *Phytophthora infestans*-resistant potatoes

Vilma Ortiz[1,2], Sinead Phelan[1] and Ewen Mullins[1*] 🔟

Abstract

Background: Nematodes play a key role in soil processes with alterations in the nematode community structure having the potential to considerably influence ecosystem functioning. As a result fluctuations in nematode diversity and/or community structure can be gauged as a 'barometer' of a soil's functional biodiversity. However, a deficit exists in regards to baseline knowledge and on the impact of specific GM crops on soil nematode populations and in particular in regard to the impact of GM potatoes on the diversity of nematode populations in the rhizosphere. The goal of this project was to begin to address this knowledge gap in regards to a GM potato line, cisgenically engineered for resistance to *Phytophthora infestans* (responsible organism of the Irish potato famine causing late blight disease). For this, a 3 year (2013, 2014, 2015) field experimental study was completed, containing two conventional genotypes (cvs. Desiree and Sarpo Mira) and a cisgenic genotype (cv. Desiree + *Rpi-vnt1*). Each potato genotype was treated with different disease management strategies (weekly chemical applications and corresponding no spray control). Hence affording the opportunity to investigate the temporal impact of potato genotype, disease management strategy (and their interaction) on the potato rhizosphere nematode community.

Results: Nematode structure and diversity were measured through established indices, accounts and taxonomy with factors recording a significant effect limited to the climatic conditions across the three seasons of the study and chemical applications associated with the selected disease management strategy. Based on the metrics studied, the cultivation of the cisgenic potato genotype exerted no significant effect ($P > 0.05$) on nematode community diversity or structure. The disease management treatments led to a reduction of specific trophic groups (e.g. Predacious $c-p = 4$), which of interest appeared to be counteracted by a potato genotype with vigorous growth phenotype e.g. cv. Sarpo Mira. The fluctuating climates led to disparate conditions, with enrichment conditions (bacterial feeding $c-p = 1$) dominating during the wet seasons of 2014 and 2015 versus the dry season of 2013 which induced an environmental stress (functional guild $c-p = 2$) on nematode communities.

Conclusions: Overall the functional guild indices in comparison to other indices or absolutes values, delivered the most accurate quantitative measurement with which to determine the occurrence of a specific disturbance relative to the cultivation of the studied cisgenic *P. infestans*-resistant potatoes.

Keywords: GM cisgenic potato, Nematode, Diversity, Community structure

---

*Correspondence: ewen.mullins@teagasc.ie
[1] Dept. Crop Science, Teagasc, Oak Park, Carlow, Ireland
Full list of author information is available at the end of the article

Background

In terms of global production potato (*Solanum tubero-sum* L.) is the fourth most important global food crop after, maize, wheat and rice [1]. Yet, the same crop that sustains human dietary requirements across the world is susceptible to a myriad of diseases; the most economically significant [2] being potato late blight disease (causative organism *Phytophthora infestans*), which continues to 'emerge' with devastating affect [3]. While chemical control measures have maintained yields, European regulations on the use of plant protection products [4] present an additional challenge to commercial potato growers at a time when novel, more aggressive strains of *P. infestans* are dominating native populations [5]. Looking ahead, the deployment of genetic resistance into commercial varieties is the only logical solution [6, 7], in light of the legislative and environmental challenges facing the crop [8].

However, the introgression of resistance (R) genes from wild potato species into breeding populations via conventional practise is a time consuming and logistically challenging process [9], which is further complicated by the evolving potential of *P. infestans* to adapt and overcome R genes [10, 11]. However, as the characterisation of R genes has rapidly increased [7, 9, 12] in parallel to the mainstream adoption of sequencing technologies, the concept of stacking R genes via cisgenic genetic modification to deliver durable resistance is now a reality [13]. This theory is more sustainable if merged with an appropriate Integrated Pest Management (IPM) strategy [2], with the agronomic potential of a suite of R genes having been recently demonstrated in field evaluations in Belgium [14], the Netherlands [15] and separately in the UK [16].

From the perspective of the European legal framework [17], genetically modified (GM) crops must undergo a comprehensive risk assessment prior to market release; the goal of which is to determine the level of substantial equivalence between the engineered material and its conventional comparator in regards to human and animal health and the environment. Coordinated by the European Food Safety Authority (EFSA), these assessments are supported by the EFSA GMO panel, which in 2010 proposed a novel risk assessment approach for European environments based on the selection of functional groups and/or individual species within a tiered approach, such that the focus is on the analysis of functional biodiversity in receiving environments and the possible interference GM varieties could cause to the functioning of this habitat [18]. To accomplish this though, risk assessment investigations require scientific data about the possible environmental impact of cultivating a GM variety and to achieve this, a higher level of practical research

is required that relates directly to the field environment [19]. An important component of this is the overall impact cultivation may have on soil biodiversity, which supports a diversity of microbes (fungi, bacteria and algae), microfauna (protozoa) and mesofauna organisms such as arthropods and nematodes, all of which are critical to soil functionality.

Nematodes are key agents in important soil processes such as decomposition, mineralisation and nutrient cycling, with alterations in the nematode community structure having the potential to considerably influence ecosystem functioning [20]. Widespread and highly diverse, nematodes form part of the food web of soil by occupying primary, secondary and tertiary positions in at least five trophic groups: bacterial feeding (BF), fungal feeding (FF), predators (PR), omnivorous (OM) and plant feeding (PF) [21], making them excellent indicators of fluctuations in soil composition arising from for example, plant genotype and/or type of soil management and environmental conditions in the rhizosphere. To date, multiple studies have been carried out using soil nematodes as indicators in different ecosystems evaluating for example the impact of crop management [22, 23], fertilizers [24], water availability [25], seasonal fluctuations [26] as well as the application of crop protectants [27]. From the perspective of monitoring nematode diversity in response to the cultivation of GM crops, several reports have detailed interactions in regards to GM maize, carrying Cry-type insecticidal proteins [28, 29]. However, to date no study has detailed the impact on nematode diversity of cultivating GM potato. This issue is compounded by the fact that a knowledge deficit also exists in regards to describing nematode community diversity within the rhizosphere of cultivated potatoes as a whole.

The process of characterizing nematode populations can be achieved morphologically or via the sequencing of nuclear (LSU rDNA, SSU rDNA and ITS) and/or mitochondrial genes (Cytochrome c oxidase subunit). Of the targets listed the SSU rDNA has proven to be most informative for investigating nematode populations considering the semi-conserved and variable regions within the sequence which provides opportunity to identify down to the species level [30]. From this, taxonomic conclusions along with absolute values and respective indices, that integrate the responses of different nematode taxa and trophic groups to soil perturbations, can be calculated as a means to measure environmental impact on the soil ecosystem [31, 32]. In light of the application of high-throughput sequencing for characterising nematode communities, balancing the desire to achieve adequate coverage of samples taken versus the cost of detecting sequences within same samples is an important consideration. While Neher and Campbell [33] examined the issue of optimal sampling strategies via the variability of ecological indices, richness and evenness indices could also be alternative parameters with which to determine the level of inter-replicate variability.

In providing a framework for quantifying the environmental impact of a GM crop, the EFSA Guidance Document [18] details a number of areas that require focus, including; impacts of GM crops on soil biodiversity and biology. The goal of this study was to begin the process of generating a baseline, from which the temporal impact (2013–2015) of cultivating modified cisgenic potatoes (equipped with a single R gene derived from *Solanum venturii*) on soil nematode community structure and diversity could begin to be quantified. As a comparative study, the work also included the opportunity to calculate nematode diversity relative to conventional potato practises that rely on weekly chemical fungicide applications and the cultivation of an additional potato cultivar Sarpo Mira, generated through conventional breeding but which possesses five sources of genetic resistance [9]. Combined, this work provides insight into the overall impact of this specific cisgenic potato crop on soil nematode populations and begins to address the current knowledge deficit that exists in the literature on this subject. Completed as part of the EU funded AMIGA project (http://www.amigaproject.eu), the output of this study contributes to the overall AMIGA goal of supporting policymakers and society in developing an in depth understanding of the potential impacts associated with the field cultivation of GM crops in the EU [19].

Methods
Weather measurements
Information on the rainfall (mm), environmental temperature (°C), % relative humidity and soil temperature (°C) at a 30 cm depth were recorded daily at the Oak Park automated weather station, situated ~400 m from the field site but which is linked with a national network of weather stations (http://www.met.ie). Rainfall and soil temperature were considered as direct parameters affecting the nematode community, while environmental temperature and relative humidity were considered indirect parameters involved in the cultivation of the potatoes.

Experimental design, plant material and crop husbandry
The study was completed on the Oak Park campus of the Teagasc Crop Research Centre, Carlow, Ireland (GPS coordinates; 52.8560667, −6.9121167), where a fixed field (~1 ha) was split into two equal sites (No. 1 and No. 2). Previously a low-managed grass pasture for >10 years, for 2013, 2014 and 2015 each site was cultivated with plots of three potato genotypes; the conventional cultivar Desiree, the modified cisgenic Desiree line and the

conventionally bred Sarpo Mira cultivar, with each genotype undergoing three treatments corresponding to a weekly chemical fungicide spray regime, a decision support system-based spray regime and a control 'no spray' treatment. For the purposes of this study only the weekly chemical ('chemical') and no spray ('control') treatments were examined. This led to six treatments in total being considered; Desiree control, Desiree chemical, cisgenic Desiree control, cisgenic Desiree chemical, Sarpo Mira control and Sarpo Mira chemical. The cisgenic line was previously engineered to contain a single copy of the *Rpi-vnt1.1* gene (derived from *S. venturii*), which confers resistance to the late blight pathogen *Phytophthora infestans* [34, 35] and was provided to the AMIGA project via the DuRPh programme (http://www.DuRPh.nl) of Wageningen University. Each genotype × treatment plot measured 3 m × 3 m with plots separated on all sides by 3 m of grass. Each site contained 54 plots randomised in order across 6 replicating blocks with 9 plots (3 genotypes × 3 treatments) per block. From year-to-year plots were rotated through the 1 ha site to ensure that for each year plots were only positioned on land that was original grass pasture. This strategy was important to minimise the accumulation of soil-borne potato diseases in the soil as a result of repeat cropping but also from the nematode perspective it ensured that the 'starting point' for each plots was the same each year; by sowing them on original grass pasture. Plots received the same crop management protocols (with the exception of chemical fungicide treatments) indistinct of the genotype evaluated. Sites were prepared by deep ploughing and rotavating before standard commercial potato drills were formed through each block of nine plots.

Soil sampling and nematode extraction
A flowchart detailing the experiment design, genotypic characteristics of the three potato genotypes grown, soil sampling as well as sample preparation for molecular analysis is presented in Additional file 1: Figure S1. Soil samples were collected from the plant rhizosphere at the initiation of flowering, which was typically during the first 2 weeks of August of each year. For each of the 6 treatments, 7 plots were randomly selected (4 from site 1 and 3 from site 2) and within each of the seven plots (per treatment), one plant was selected and with soil still attached to the roots, carefully placed inside a bag for transfer to the laboratory. Upon arrival soil adhering to the roots was scraped into the same bag and the plant removed. The remaining soil in the bag was thoroughly mixed before 100 g was removed for placing in a labelled plastic bag which was sealed and stored at 4 °C. Nematodes were extracted by processing 100 g of the homogenized soil/plot (seven replicates/treatment) via

an Oostenbrink elutriator, followed by passage through a series of sieves (45, 90, 125 and 180 mesh size) and then a cotton wool filter. After a 48 h incubation period at room temperature, a volume of 50 ml was then recovered from the cotton wool filter, from which nematodes were collected into 10 ml following a 4 °C treatment for 24 h. Final volumes were subsequently stored at −80 °C. Across the 3 years (2013, 2014, 2015) of the study a total of 126 soil samples were processed in this manner.

DNA extraction

Each 10 ml sample was freeze dried overnight before DNA was extracted as per the Purelink Genomic DNA kit (Invitrogen/Cat No. 1820-01) with an adapted protocol for nematode DNA. Modifications included: 360 μl of Purelink Genomic buffer plus 40 μl of proteinase K was added to each tube which was then agitated at 55 °C overnight. The suspension was then centrifuged (13,000 rpm, 3 min) and the resulting supernatant processed as per kit's recommendations, with the exception that the DNA was eluted from the column using 100 μl sterile water. All eluted samples were stored at −40 °C.

Target sequence amplification and sequencing

The 5′ end of the 18 small subunit rDNA gene (~1000 bp) was amplified using a set of universal primers (SSU18A and SSU26R [36]). All PCR reactions were completed in a 50 μl volume containing 50 ng of DNA template, 5 μl of 10× PCR buffer, 1 μl of each primer (10 mM) and 200 μM dNTP, with cycling conditions of; 95 °C−5 min, 30 × (95 °C−30 s, 60 °C−60 s, 72 °C−5 min), 72 °C−10 min. Five individual PCR reactions were completed for each of the seven samples per treatment to ensure adequate generation of the target amplicon, after which the PCR reaction volumes for each set of seven samples (per treatment) was pooled to deliver a composite PCR sample for each treatment. Each composite sample was then cloned into *E. coli* (p-GEM, Promega) and 50 individual colonies (per treatment/year) were randomly selected and sent to an external provider for Sanger sequencing. Acquired sequences were analysed against the GenBank database using standard BLAST analysis with alignments and clustering completed with Clustal 1× and Mega. Owing to low DNA concentrations attained with some of the 2013 samples, a nested PCR approach was required and implemented with the secondary PCR (to that detailed previously) employing the SSU9R and S18 primers [36], which generate a nested fragment ~500 bp. All sequences were deposited in the NCBI GenBank under accession numbers: KY119383–KY119427, KY119428–KY119476, KY119477–KY119513, KY119514–KY119563, KY119564–KY119588, KY119589–KY119632, KY119633–KY119676, KY119677–KY119724,

KY119725–KY119770, KY119771–KY119811, KY119812–KY119853, KY119854–KY119901, KY119902–KY119944, KY119945–KY119986, KY119987–KY120026, KY120027–KY120072, KY120073–KY120120, KY120121–KY120164.

Nematode community analysis

To fully characterize the nematode community structure (NCS) in the respective rhizospheric samples, the NCS was analysed on the 18 composite samples in terms of indices, absolute values and qualitative taxonomy with both free living and plant parasitic groups evaluated. The indices analysis was divided into ecological measurement with the Maturity index (MI) (and its variant MIMO and $\sum$MIMO), in functional guild indices: enrichment index (EI), structure index (SI), chanel index (CH), bacterial feeding (BF) c−p = 1 and 2 and diversity indices: Shannon index (H), Shannon equitability (EH), Simpson index (D), Simpson index, probability of diversity (1-D) and the Simpson reciprocal index (1/D). The interaction structure (SI) and enrichment (EI) values were analysed through a graphical representation of the nematode faunal analysis and absolute values determining family, genus and species numbers. Sorensen coefficients for evaluating similarity between treatments were also calculated. A summary of all calculations realized is described in Additional file 2: Table S1. Finally, the taxonomy at family and genus level was evaluated to find possible bio-indicators of disturbance of the environment related to the different potato genotypes and chemical treatments applied.

Data and statistical analysis

To verify that the number of clones extracted from the composite samples (sampling effort) would provide sufficient information (how well the community has been sampled) on the total nematode richness (species and genus) in the rhizosphere, an individual-based rarefaction analysis was completed using the R package "stat" [37]. To estimate the effect on the nematode community of potato genotype, chemical treatment and the interaction between the two, a two-way analysis of variance (ANOVA) was performed with data blocked per year. When effects were significant, multiple comparisons between the means were made as per the LSD test, with differences at probability of $P \leq 0.01$ and $P \leq 0.05$ considered. All analyses were performed using GenStat software v.18. To show in more detail the number of family, genus and species, identity of observed species and genus and trophic groups richness by each interaction genotype and disease management per year (2013–2015) a two dimensional representation (heatmap) was performed for each case using the GenStat software v.18.

Results

Climatic conditions for the 2013, 2014 and 2015 field studies

In contrast to the 1st (2013) and 3rd (2015) year of the study, 2014 was generally characterized by higher soil temperatures and relative humidity (Table 1; Additional file 3: Figure S2). However, focussing on months for cultivation only (May to August), air and soil temperature along with relative humidity varied significantly across all 3 years (Additional file 3: Figure S2). In relation to rainfall, in 2014, May and August were the months in which the most precipitation was registered and for 2015 it was May and July. The 2013 field season experienced scarce rainfall, with significantly lower values (P < 0.05) of relative humidity, compared to 2014 and 2015, but with higher air and soil temperatures for June to August and July to August respectively (Table 1).

Rarefaction analysis

The rarefaction analysis, which was completed for both nematode species (Fig. 1) and genus (Fig. 2) for all treatments over the 3 years of the study, illustrated the levels of nematode richness identified through the adopted strategy. While the chemical treated cisgenic potato samples returned the highest genus and species richness through 2013 and 2014 this was not the case in the final year of the study, 2015. Across the six treatments studied, the cv. Desiree derived samples recorded the lowest degree of fluctuation between the two disease management strategies applied, in contrast to cv. Sarpo Mira and cisgenic Desiree.

Ecological succession indices

Mean maturity index (MI) values (Table 2) obtained for samples taken from untreated plots were higher (2.36 for Desiree; 2.05 for cisgenic Desiree; 2.13 for Sarpo Mira) than those recorded in the presence of the chemical fungicide treatment (1.93 for Desiree; 1.69 for cisgenic Desiree; 1.91 for Sarpo Mira). For the plant parasitic index (PPI) a similar trend (for control vs. weekly chemical treatment) was observed for cisgenic Desiree and Desiree only (2.83 vs. 1.10, Desiree; 2.63 vs. 1.83, cisgenic Desiree) and again with the PPI/MI (1.28 vs. 0.53, Desiree; 1.28 vs. 0.95, cisgenic Desiree) the modified maturity index (MIMO) (2.68 vs. 2.54, Desiree; 2.42 vs. 2.38, cisgenic Desiree) and the $\sum$MIMO (2.80 vs. 2.68, Desiree; 2.46 vs. 2.54 cisgenic Desiree). In contrast, a converse trend was noted with Sarpo Mira (Table 2).

Considering crop genotype as an individual factor, cisgenic Desiree derived samples recorded the lowest mean values for the MI (1.87), MIMO (2.40) and $\sum$MIMO (2.50) indices compared to its direct comparator cv. Desiree and the alternative conventionally bred variety Sarpo Mira, which obtained the higher mean values for the PPI (3.01), PPI/MI (1.54), MIMO (2.64) and $\sum$MIMO (2.79) indices. When ranked (lowest to highest) for the MIMO and $\sum$MIMO indices, genotypes ordered as cisgenic Desiree, Desiree, Sarpo Mira whereas for the PPI and PPI/MI indices genotypes ranked as Desiree, cisgenic Desiree, Sarpo Mira. In the case of the MI index, mean values delivered an ordered ranking of cisgenic Desiree, Sarpo Mira, Desiree. Examining the impact of disease management (independent of the potato genotype sown), the means values for MI (2.18/1.84), PPI (2.70/2.10) and PPI/MI (1.28/1.10) proportions were larger in the absence of the chemical fungicide treatment (Table 2). The opposite was noted for MIMO (2.53/2.57) and $\sum$MIMO (2.61/2.74).

Statistically, weak effects were noted for the effect of chemical treatment on the MI index (P < 0.13), as well as

Table 1 Comparative analysis of monthly rainfall (R), air temperature (AT), soil temperature (ST) and relative humidity (RH) measurements made from January to August for 2013, 2014 and 2015 at the field site in Oak Park, Carlow, Ireland

Month	Rainfall			Air temperature			Soil temperature			Relative humidity		
	F value	P value	Year[a]	F value	P value	Year[a]	F value	P value	Year[a]	F value	P value	Year[a]
January	3.09	<0.051	2014[b]	0	1		1.07	0.348		0	1	
February	16.9	<0.001	2014[b]	0.28	0.759		4.12	<0.02	2014	4.59	<0.013	2014[b]
March	0.09	0.912	2014	3.41	<0.038	2013[b], 2014	29.85	<0.001	2014[b], 2015	5.87	<0.004	2014[b]
April	0.85	0.429	2014	17.52	<0.001	2014[b], 2015	30.69	<0.001	2014[b], 2015	8.88	<0.001	2014[b]
May	2.21	0.115	2015	5.44	<0.006	2014[b], 2015	4.95	<0.009	2014[b]	6.86	<0.002	2014[b]
June	0.83	0.439	2014	65.25	<0.001	2013[b], 2015	4.5	<0.014	2014[b]	9.74	<0.001	2014[b], 2015
July	3.12	<0.049	2015[b]	2.42	0.095	2013	45.07	<0.001	2013[b], 2014	5.86	<0.004	2014[b]
August	0.30	0.742	2014	28.23	<0.001	2013[b]	22.03	<0.001	2013[b]	5.25	<0.007	2015[b]

[a] Years with the highest mean monthly value for the respective month and metric

[b] Respective year in which recorded month differed significantly (P < 0.05) from same month in other 2 years

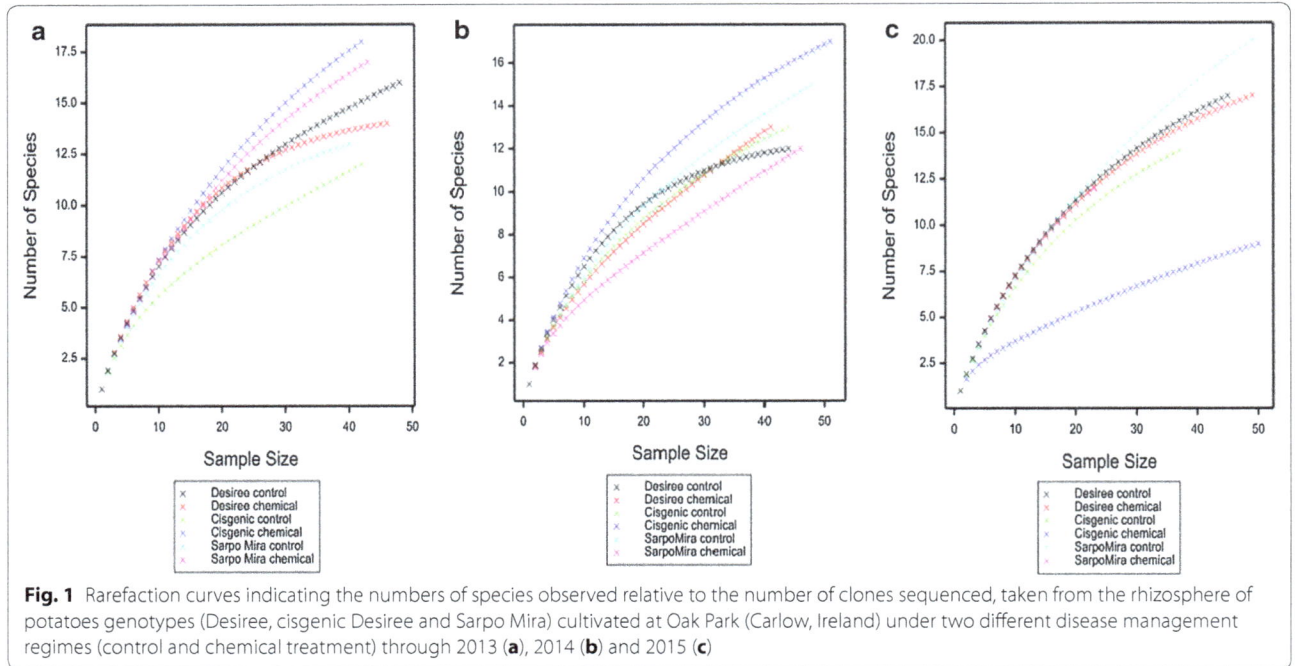

Fig. 1 Rarefaction curves indicating the numbers of species observed relative to the number of clones sequenced, taken from the rhizosphere of potatoes genotypes (Desiree, cisgenic Desiree and Sarpo Mira) cultivated at Oak Park (Carlow, Ireland) under two different disease management regimes (control and chemical treatment) through 2013 (**a**), 2014 (**b**) and 2015 (**c**)

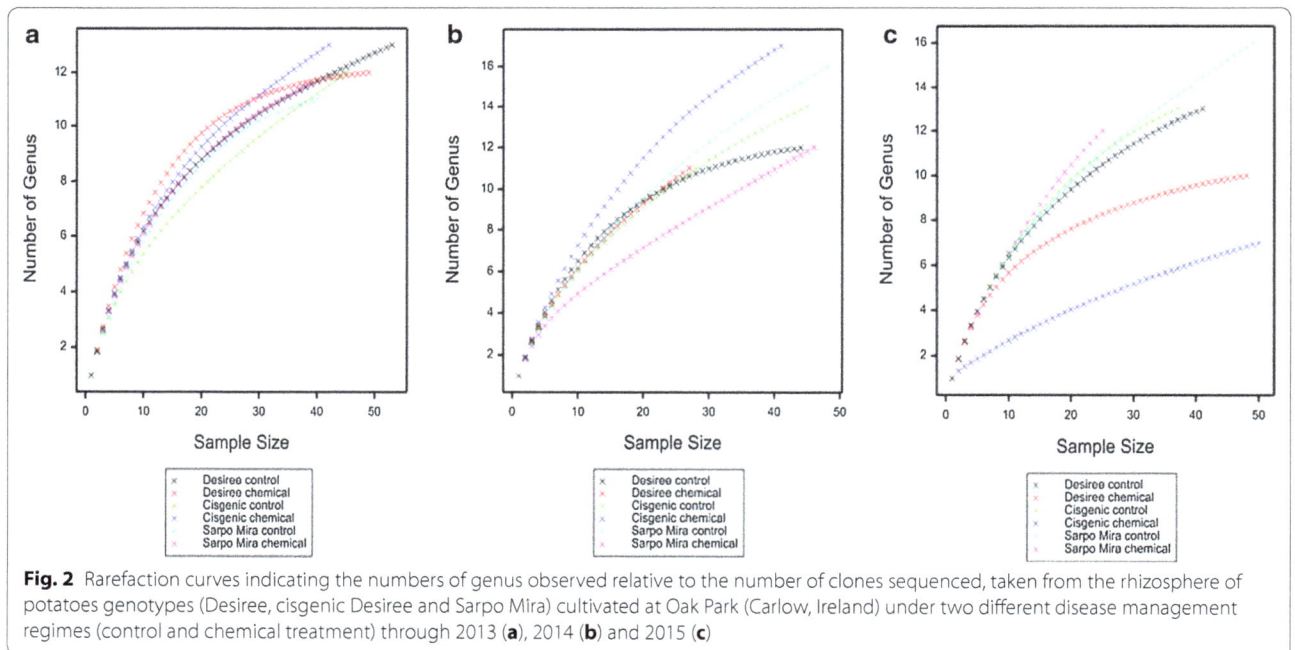

Fig. 2 Rarefaction curves indicating the numbers of genus observed relative to the number of clones sequenced, taken from the rhizosphere of potatoes genotypes (Desiree, cisgenic Desiree and Sarpo Mira) cultivated at Oak Park (Carlow, Ireland) under two different disease management regimes (control and chemical treatment) through 2013 (**a**), 2014 (**b**) and 2015 (**c**)

the interaction of genotype × chemical treatment on the PPI (P < 0.20) and the PPI/MI ratio (P ≤ 0.18). The year sampled had a weak effect on $\sum$MIMO (P < 0.08) while in the case of the MIMO index 2013 differed significantly from 2014 to 2015 (P < 0.01, Table 2).

Trophic groups

During the 3 years of the study, seven of the eight feeding groups proposed by Yeates [21] were identified; bacterial feeding (BF), plant feeding (PF), fungal feeding (FF), omnivorous (OM), predacious (PR), bacterial feeding

Table 2 Impact of potato genotype (Desiree, cisgenic Desiree, Sarpo Mira), disease management (control, chemical treatment) and year (2013, 2014, 2105) on nematode community ecological succession indices [maturity index (MI), plant parasite index (PPI), modified MI to include removing of the c–p = 1 family (MIMO) and removing of the c–p = 1 family but with inclusion of the PPI to generate $\sum$MIMO] from rhizospheric samples taken from Oak Park field site

Index	Potato genotype	Disease management	2015	2014	2013	Mean	Mean/ genotype	Mean/disease management
MI	Desiree	Control	3.05	1.76	2.27	2.36	2.14	2.18
	Desiree	Chemical	2.10	1.61	2.07	1.93		1.84
	Cisgenic Desiree	Control	2.03	2.15	1.97	2.05	1.87	
	Cisgenic Desiree	Chemical	1.16	2.13	1.79	1.69		
	Sarpo Mira	Control	2.63	1.76	2.00	2.13	2.02	
	Sarpo Mira	Chemical	1.86	1.56	2.30	1.91		
	Mean		2.14	1.83	2.07			
PPI	Desiree	Control	3.00	3.50	2.00	2.83	1.97	2.70
	Desiree	Chemical	0.00	0.00	3.31	1.10		2.10
	Cisgenic Desiree	Control	2.50	3.00	2.40	2.63	2.23	
	Cisgenic Desiree	Chemical	0.00	2.67	2.83	1.83		
	Sarpo Mira	Control	3.00	2.71	2.29	2.67	3.01	
	Sarpo Mira	Chemical	3.00	3.67	3.39	3.35		
	Mean		1.92	2.59	2.70			
PPI/MI	Desiree	Control	0.98	1.99	0.88	1.28	0.91	1.28
	Desiree	Chemical	0.00	0.00	1.60	0.53		1.10
	Cisgenic Desiree	Control	1.23	1.40	1.22	1.28	1.11	
	Cisgenic Desiree	Chemical	0.00	1.25	1.58	0.95		
	Sarpo Mira	Control	1.14	1.54	1.15	1.28	1.54	
	Sarpo Mira	Chemical	1.61	2.35	1.47	1.81		
	Mean		0.83	1.42	1.32			
MIMO	Desiree	Control	3.46	2.47	2.10	2.68	2.61	2.53
	Desiree	Chemical	3.00	2.39	2.23	2.54		2.57
	Cisgenic Desiree	Control	2.33	*2.94*	2.00	2.42	2.40	
	Cisgenic Desiree	Chemical	2.33	*2.80*	2.00	2.38		
	Sarpo Mira	Control	2.83	2.63	2.00	2.49	2.64	
	Sarpo Mira	Chemical	3.38	2.71	2.30	2.80		
	Mean		2.89	2.66	2.11**			
$\sum$MIMO	Desiree	Control	3.42	2.89	2.08	2.80	2.74	2.61
	Desiree	Chemical	3.00	2.39	2.64	2.68		2.74
	Cisgenic Desiree	Control	2.34	2.94	2.09	2.46	2.50	
	Cisgenic Desiree	Chemical	2.33	2.79	2.50	2.54		
	Sarpo Mira	Control	2.84	2.68	2.13	2.55	2.79	
	Sarpo Mira	Chemical	3.30	2.88	2.88	3.02		
	Mean		2.87	2.76	2.39			

* P < 0.05, ** P < 0.01, *** P < 0.001

or entomopathogenic (BF OR EN) and fungal feeding or entomopathogenic (FF OR EN). In the absence of a weekly chemical disease management treatment, up to 5, 6 and 6 trophic groups were present for each respective potato genotype (cisgenic Desiree, Desiree and Sarpo Mira) across the 3 years. In the presence of a weekly chemical disease management application, trophic group numbers were identified at up to 6, 5 and 6 for cisgenic Desiree, Desiree and Sarpo Mira respectively.

The variability across the 3 years of the field study is evident in Fig. 3. In 2013 (Fig. 3a) only five trophic groups (bacterial feeding (BF), plant feeding (PF), omnivorous (OM), fungal feeding (FF) and bacterial feeding or entomopathogens (BF or EN) were identified with BF and PF dominating more than 80% of the total recorded, with PF significantly dominating (P < 0.01) the chemical treated cultivar over the control samples for each cultivar. Six and seven trophic groups were recorded in 2014 and

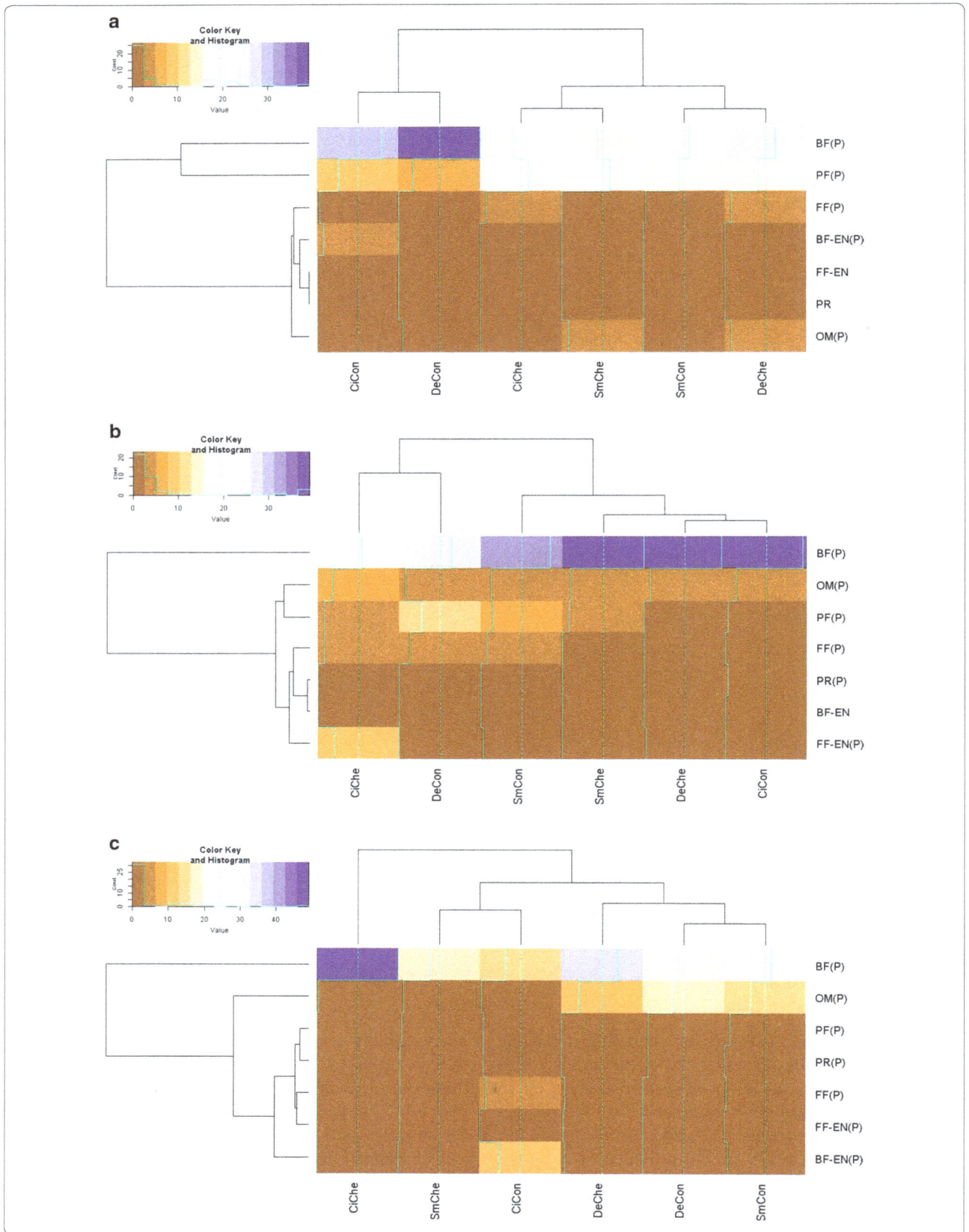

(See figure on previous page.)
Fig. 3 Heatmaps depicting impact of genotype (Desiree, cisgenic Desiree and Sarpo Mira), disease management (control, chemical) and year (2013, **a**; 2014, **b**; 2015, **c**) on the prevalence of nematodes from trophic groups representing bacterial feeding (BF), plant feeding (PF), fungal feeding (FF), omnivorous (OM), predacious (PR), fungal feeding or entomopathogens (FF or EN) and bacterial feeding or entomopathogens (BF or EN). Genotype × management interactions are labelled as: Desiree control [DeCon], Desiree chemical [DeChe], cisgenic Desiree control [CiCon], cisgenic Desiree chemical [CiChe], Sarpo Mira control [SmCon], Sarpo Mira chemical [SmChe]. (P) = present. For the colour key and histogram *X-axis* individual nematodes accounted for and *Y-axis* the times that the number (account) is repeated

2015 respectively, with a high population of BF followed by OM in both years.

With the exception of the PF group, which was statistically different between years ($P < 0.001$) and in regard to year × disease management ($P < 0.01$), no significant difference was recorded across the remaining groups for either cultivar/disease management/year studied. Across the study, no predator nematodes were identified in either the Sarpo Mira (control), Desiree (chemical) or cisgenic Desiree (chemical) rhizosphere samples. Across the 3 years examined, weak effects were observed for the impact of disease management on PR ($P < 0.13$) and year ($P < 0.07$) and the interaction of year × disease management on BF ($P < 0.13$) and year for OM ($P < 0.08$) but overall crop genotype had no significant impact on the occurrence of trophic groups observed ($P < 0.05$).

Functional guild indices

Examining the degree of colonizer–persister (cp) across the main trophic groups (BF, PF, FF, OM and PR); BF recorded 1–3, FF 2–3, PR recorded 1, 3–5, OM 4 and 5 and PF 2–4. Evaluating the diversity of nematode functional groups and their respective c–p classification, each index recorded a distinct response (Table 3). For EI, the highest mean was associated with chemical treatment (71) versus the absence of chemical fungicides (49), while the inverted trend occurred with BF_2 (29/51). In both cases the differential values were significant ($P < 0.05$). The influence of chemical applications led to the highest mean values recorded with the EI and BF_1 index (EI; 72, 78, 62 and BF_1; 93, 90, 67) compared to the respective control values (EI: 50, 56, 41 and BF_1; 64, 89, 66). In contrast, for BF_2 the highest values were recorded in the absence of chemical management (50, 44 and 59). At a crop genotype level, Sarpo Mira recorded the lowest mean EI value (52) but subsequently the highest BF_2 mean (48) and CH (17). The cisgenic Desiree genotype returned the lowest SI mean (41) but the highest BF_1 (89). Examining the influence of year in more detail, four of the five indices (EI, BF_1, BF_2 and SI) presented a significant difference ($P < 0.001$, $P < 0.004$, $P < 0.001$ and $P < 0.001$ respectively) across the 3 years of the study (Table 4). Examining the values in more detail, 2013 recorded the lowest mean values for the EI, BF_1 and S1

indices (20, 39 and 16 respectively) and highest with the BF_2 index (80).

Investigating potential associations between the mean values obtained for the EI, BF_2, BF_1 and SI indices and the recorded weather metrics (air temperature, soil temperature, relative humidity and rainfall) identified consistent polynomial associations for each of the indices studied (Fig. 4), with rainfall (Fig. 4a) and relative humidity (Fig. 4c) impacting similarly on mean index values and correspondingly for the variables of air (Fig. 4b) and soil (Fig. 4d) temperature. For the direct factors of rainfall and soil temperature, inverse associations (BF_2 vs. EI, BF_1 and SI) were observed for the indices relative to the factor studied (Fig. 4a, d). This trend was also observed for the indirect factors of relative humidity (Fig. 4c) and air temperature (Fig. 4b).

Examining treatment effects on the basal, structural and enrichment components of the soil food web identified a significant difference ($P < 0.001$) between EI and SI over time. The construction of nematode profiles for 2013 revealed that food webs for 5 of the 6 treatments (exception being cisgenic Desiree + chemical) positioned within quadrat D (Fig. 5), indicating a depleted and degraded food wed structure. For 2014, all six treatments were plotted to quadrat B, typical of an enrichment condition. In the case of 2015, the final year of the study, all treatments remained in quadrat B, with the exception of the Sarpo Mira control and the cisgenic Desiree + chemical, which positioned in quadrat C and A respectively (Fig. 5).

Nematode abundance and diversity indices

An alternative measure of disturbance considered was the impact of crop genotype and/or disease management treatments on nematode diversity, measured through the abundance of individual nematode family, genus and species and at a species level according to richness (H), evenness (EH) and (D, 1-D and 1/D) dominance indices on a yearly basis through the study. While samples collected from the cisgenic Desiree chemical treatment plots during 2014 and 2013 recorded higher numbers of nematode species and genera than the alternative treatments (Fig. 6), no statistical difference was detected. Taking into account the rare (less frequent-Shannon index)

Table 3 Effect of potato genotype (Desiree, cisgenic Desiree, Sarpo Mira), disease management (control, chemical treatment) and year (2013, 2014, 2105) on nematode trophic diversity indices [enrichment index (EI), bacterial feeding c–p = 2 (BF_2), bacterial feeding c–p = 1 (BF_1), chanel index (CI) and the structure index (SI)] studied based on rhizospheric samples taken from Oak Park field site

Index	Potato genotype	Disease management	2015	2014	2013	Mean	Mean/ genotype	Mean/disease management
EI	Desiree	Control	67	83	0	50	61	49*
	Desiree	Chemical	83	86	45	72		71
	Cisgenic Desiree	Control	61	92	14	56	67	
	Cisgenic Desiree	Chemical	97	81	56	78		
	Sarpo Mira	Control	33	87	4	41	52	
	Sarpo Mira	Chemical	95	92	0	62		
	Mean		73	87	20***			
BF_2	Desiree	Control	33	17	100	50	39	51*
	Desiree	Chemical	17	14	55	28		29
	Cisgenic Desiree	Control	39	8	86	44	33	
	Cisgenic Desiree	Chemical	3	19	44	22		
	Sarpo Mira	Control	67	13	96	59	48	
	Sarpo Mira	Chemical	5	8	100	38		
	Mean		27	13	80***			
CH	Desiree	Control	2	7	0	3	5	16
	Desiree	Chemical	3	1	16	7		6
	Cisgenic Desiree	Control	0	0	33	11	11	
	Cisgenic Desiree	Chemical	0	15	16	10		
	Sarpo Mira	Control	0	3	100	34	17	
	Sarpo Mira	Chemical	0	0	0	0		
	Mean		1	5	28			
BF_1	Desiree	Control	98	93	0	64	78	73
	Desiree	Chemical	97	99	84	93		83
	Cisgenic Desiree	Control	100	100	67	89	89	
	Cisgenic Desiree	Chemical	100	85	84	90		
	Sarpo Mira	Control	100	97	0	66	66	
	Sarpo Mira	Chemical	100	100	0	67		
	Mean		99	95	39***			
SI	Desiree	Control	89	54	17	53	54	47
	Desiree	Chemical	78	52	34	55		52
	Cisgenic Desiree	Control	53	77	0	43	41	
	Cisgenic Desiree	Chemical	44	70	0	38		
	Sarpo Mira	Control	74	64	0	46	55	
	Sarpo Mira	Chemical	82	67	41	63		
	Mean		70	64	16***			

* $P < 0.05$, ** $P < 0.01$, *** $P < 0.001$

and abundant (dominant-Simpson index) species per sample, the diversity indices returned similar patterns between treatments (Table 5). The mean H index values were >2 for all treatments, irrespective of year, disease management and potato genotype with no significance recorded between treatment; similarly, no significance was returned between treatments in regards to the evenness distribution (EH) of individuals per species present in samples, which was found to be closer to 1 than to 0

for each combination. The uniformity of the mean H and EH values across treatments is illustrated in Additional file 4: Figure S3. In contrast the probability that two nematodes randomly selected from within a sample belonged to the same species (D) was closer to 0 than 1. Lastly, the analyses recorded a statistically similar but high probability (0.84–0.89) of nematode diversity (1-D) across genotypes (per treatment per year) with the number of species (1/D) recorded between 7 and 9 per crop (Table 5).

Table 4 Analysis of variance for the effect of potato 'genotype' (Desiree, cisgenic Desiree, Sarpo Mira), disease management (control, chemical treatment) and year (2013, 2014, 2105) on the trophic diversity indices EI, BF_2, CH, BF_1 and SI (see Table 3 legend for explanation of abbreviations), in field site in Oak Park (Carlow, Ireland)

Treatment	EI	BF_2	CH	BF_1	SI
Genotype	1.06	0.93	0.45	1.2	1.66 (B) (P < 0.24)
Management	6.52 (P < 0.05)	5.94 (P < 0.05)	1	0.75	0.39
Genotype × management	0.002	0.0008	5.94	0.63	0.90
Year	23.10 (P < 0.001)	21.59 (P < 0.001)	2.49 (A) (P < 0.13)	10.17 (P < 0.004)	22.02 (P < 0.001)
t5%	–	–	3.827	–	2.87
t1%	–	–	4.855	–	3.827

A and B = t value compared against the critical value to 5 and 1% in the same column

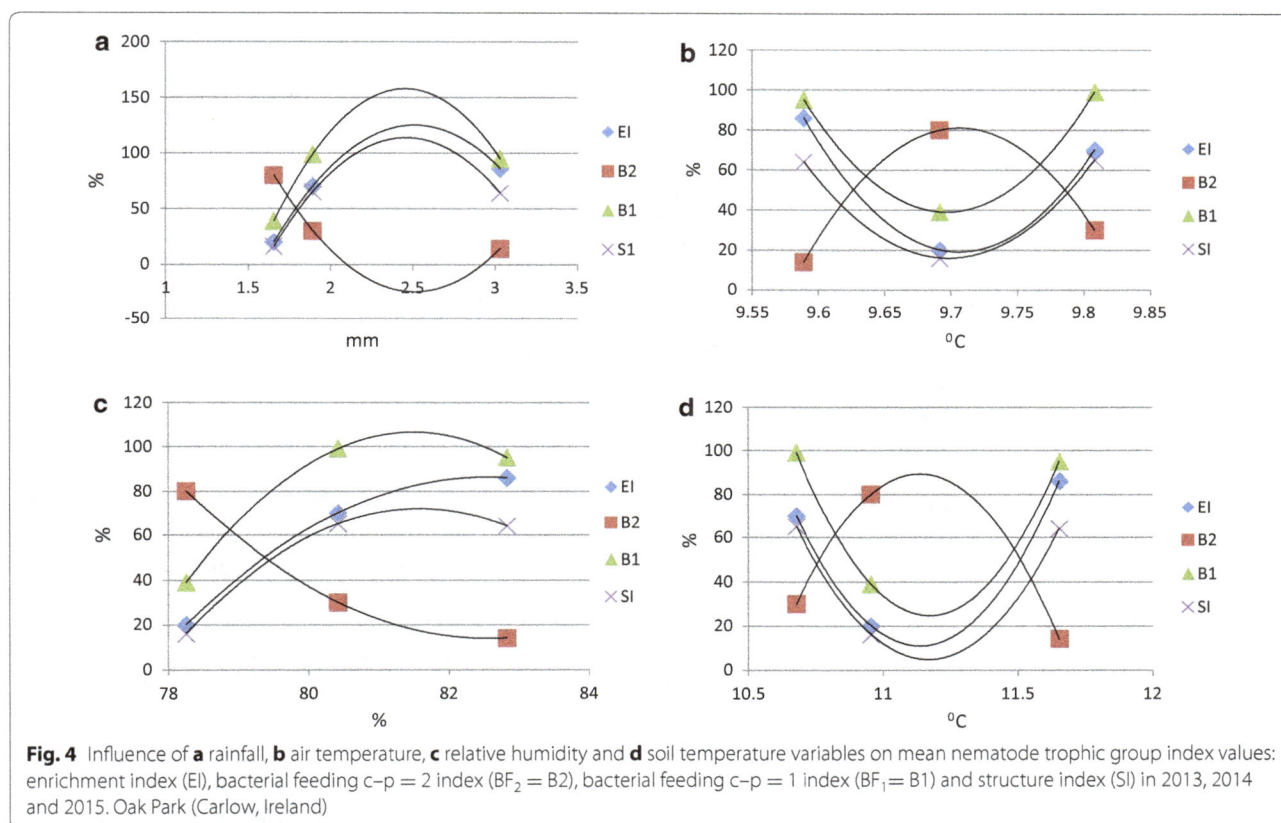

Fig. 4 Influence of **a** rainfall, **b** air temperature, **c** relative humidity and **d** soil temperature variables on mean nematode trophic group index values: enrichment index (EI), bacterial feeding c–p = 2 index (BF_2 = B2), bacterial feeding c–p = 1 index (BF_1= B1) and structure index (SI) in 2013, 2014 and 2015. Oak Park (Carlow, Ireland)

Nematode families and genus as a bio-indicator of environmental disturbance

Up to 29 distinct families were associated across all treatments evaluated over 2013, (Fig. 7a), 2014 (Fig. 7b) and 2015 (Fig. 7c). There was no significant difference (P < 0.05) between the family, and genus, nematode numbers of cisgenic Desiree [16 (family, control treatment) vs. 17 (family, chemical treatment) and 26 (genus, control treatment) vs. 27 (genus, chemical treatment)] and its comparator Desiree genotype [17 (family, control treatment) vs. 15 (family, chemical treatment) and 27 (genus, control) vs. 25 (genus, chemical)], irrespective of the absence/presence of disease management strategies. For Sarpo Mira, there was a decrease in numbers following chemical treatment [22 (family, control treatment) vs. 14 (family, chemical treatment) and 31 (genus, control treatment) vs. 26 (genus, chemical treatment)]. Sorensen coefficient values calculated for nematode families within each potato genotype (Table 6) indicated substantial overlap between treatments: 0.63 for Desiree control vs. chemical, 0.79 for cisgenic Desiree control vs. chemical and 0.67 for Sarpo Mira control vs. chemical

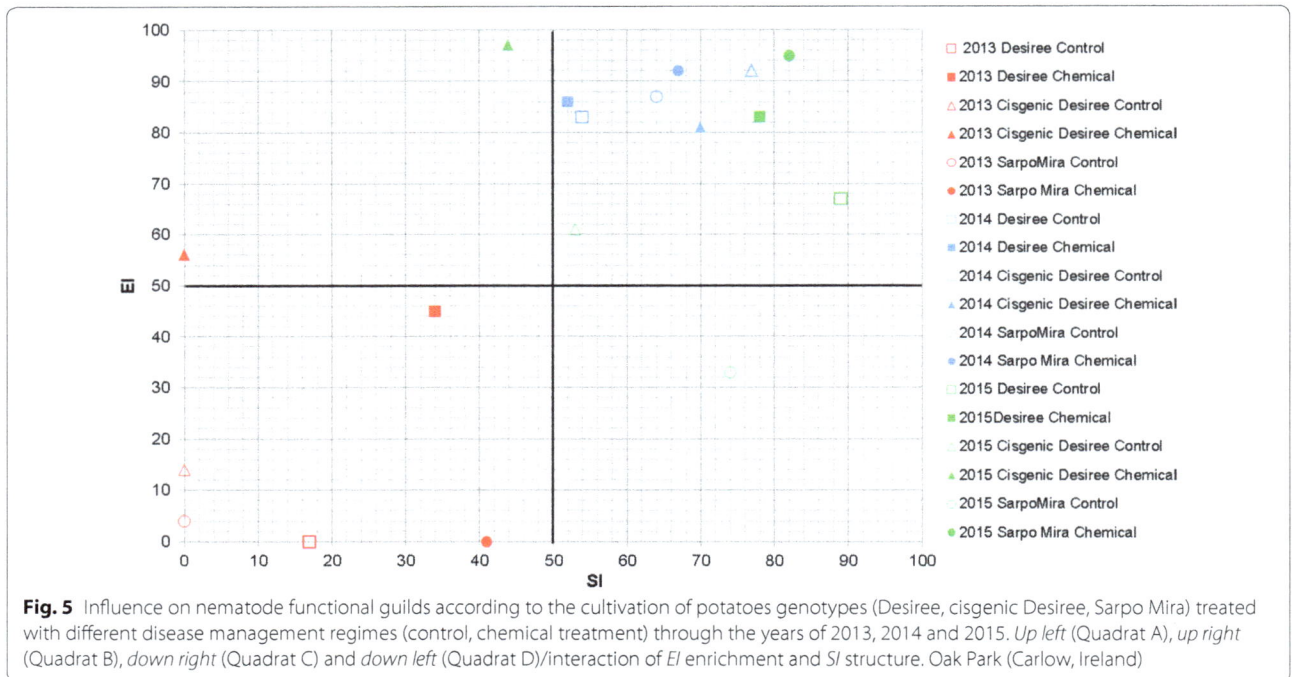

Fig. 5 Influence on nematode functional guilds according to the cultivation of potatoes genotypes (Desiree, cisgenic Desiree, Sarpo Mira) treated with different disease management regimes (control, chemical treatment) through the years of 2013, 2014 and 2015. *Up left* (Quadrat A), *up right* (Quadrat B), *down right* (Quadrat C) and *down left* (Quadrat D)/interaction of *EI* enrichment and *SI* structure. Oak Park (Carlow, Ireland)

during the 3 years. Factoring the influence of time, coefficient values were calculated for each respective permutation of genotype and disease management (Table 6). Examining equivalence at the family level, Sarpo Mira coefficient values were similar through the 3 years of the study (0.55–0.57) compared to the more variable Desiree (0.50–0.63) and cisgenic Desiree (0.43–0.73). Independent of the regime deployed, for 2013 a cisgenic Desiree vs. Desiree comparison returned a CC = 0.53, in contrast to 0.34 for cisgenic Desiree vs. SarpoMira. For 2014, values ranged from 0.37 to 0.47, while from the final year (2015), cisgenic Desiree and SarpoMira shared 50% of nematode families sampled.

The relative uniformity in regards to the distribution of nematode families across the treatments with respect to each year is illustrated in Fig. 7. In more detail, on a year-by-year basis 2013 was characterised by seventeen families (Fig. 7a) with the *Cephalobidae* abundant in all treatments evaluated with more *Cephalobidae* individuals noted in the control treatments independent of the crop evaluated (38, 32 and 20 for Desiree control,

cisgenic Desiree control and SarpoMira control respectively). Only a nominal number of the *Rhabditidae* family were recorded while seven families associated with plant feeding (*Tylenchulidae, Tylenchidae, Telotylenchidae, Trichodoridae, Merliniinae, Longidoridae, and Pratylenchidae*) were counted. For 2014 (Fig. 7b), 23 families were detected with an abundance of the *Rhabditidae* (8–21 members) family recorded along with the *Cephalobidae* family (6–14) at the same time and members of a third nematode bacterial feeding, the *Panagrolaimidae* (2–11) dominating especially in cisgenic Desiree control and SarpoMira chemical derived samples (Fig. 7b). As with 2013, seven plant feeding families were detected (*Heteroderidae, Hoplolaimidae, Merliniidae, Trichodoridae, Pratylenchidae, Tylenchidae, Telotylenchidae*). The *Heteroderidae* and *Hoplolaimidae* families were present in Desiree control samples and the *Hoplolaimidae* family was only found associated with the Sarpo Mira control sample. The presence of two FF or EN (*Neotylenchidae* and *Sphaerulariidae*) was also identified. For 2015 (Fig. 7c), 20 families were listed with a similar

(See figure on next page.)

Fig. 6 Heatmaps illustrating number of individual nematode **a** family, **b** genus and **c** species identified following extraction from the rhizosphere of potatoes genotypes (Desiree, cisgenic Desiree and Sarpo Mira) cultivated at Oak Park (Carlow, Ireland) under two different disease management regimes (control, chemical treatment) through 2013, 2014 and 2015. Genotype × management interactions are labelled as: Desiree control [DeCon], Desiree chemical [DeChe], cisgenic Desiree control [CiCon], cisgenic Desiree chemical [CiChe], Sarpo Mira control [SmCon], Sarpo Mira chemical [SmChe]. For the colour key and histogram *X-axis* number of individual nematodes accounted for and *Y-axis* the times that the number (account) is repeated

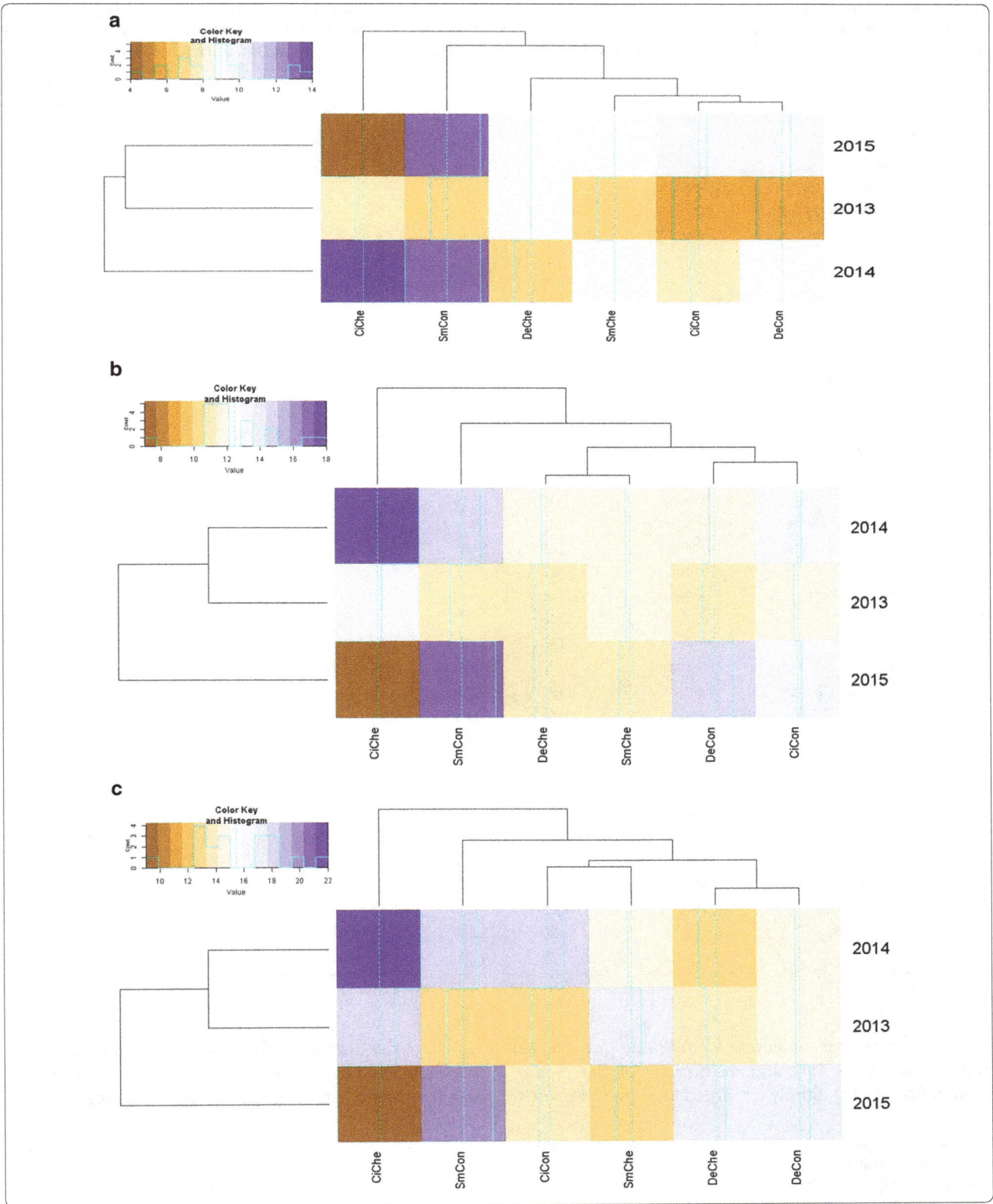

Table 5 **Nematode diversity, as per richness (H), evenness (EH) and dominance (D, 1-D, 1/D) indices, arising from samples taken from under different potatoes genotypes (Desiree, cisgenic Desiree, Sarpo Mira) treated with different disease management (control, chemical treatment) through the years of 2013, 2014 and 2015 at Oak Park (Carlow, Ireland)**

Index	Potato genotype	Disease management	2015	2014	2013	Mean	Mean/ genotype	Mean/disease management
H	Desiree	Control	2.54	2.51	2.46	2.50	2.43	2.43
	Desiree	Chemical	2.54	2.07	2.48	2.36		2.33
	Cisgenic Desiree	Control	2.27	2.62	2.20	2.36	2.30	
	Cisgenic Desiree	Chemical	1.34	2.8	2.58	2.24		
	Sarpo Mira	Control	2.55	2.46	2.26	2.42	2.40	
	Sarpo Mira	Chemical	2.44	2.23	2.47	2.38		
	Mean		2.28	2.45	2.41			
EH	Desiree	Control	0.67	0.66	0.64	0.66	0.64	0.64
	Desiree	Chemical	0.65	0.56	0.64	0.62		0.63
	Cisgenic Desiree	Control	0.63	0.69	0.57	0.63	0.61	
	Cisgenic Desiree	Chemical	0.34	0.75	0.69	0.59		
	Sarpo Mira	Control	0.65	0.63	0.61	0.63	0.65	
	Sarpo Mira	Chemical	0.76	0.58	0.66	0.67		
	Mean		0.62	0.65	0.64			
D	Desiree	Control	0.10	0.10	0.11	0.10	0.11	0.12
	Desiree	Chemical	0.10	0.18	0.09	0.12		0.15
	Cisgenic Desiree	Control	0.15	0.09	0.16	0.13	0.16	
	Cisgenic Desiree	Chemical	0.38	0.08	0.10	0.19		
	Sarpo Mira	Control	0.12	0.13	0.14	0.13	0.13	
	Sarpo Mira	Chemical	0.11	0.15	0.12	0.13		
	Mean		0.16	0.12	0.12			
1-D	Desiree	Control	0.90	0.90	0.89	0.90	0.89	0.88
	Desiree	Chemical	0.90	0.82	0.91	0.88		0.85
	Cisgenic Desiree	Control	0.85	0.91	0.84	0.87	0.84	
	Cisgenic Desiree	Chemical	0.62	0.92	0.90	0.81		
	Sarpo Mira	Control	0.88	0.87	0.86	0.87	0.87	
	Sarpo Mira	Chemical	0.89	0.85	0.88	0.87		
	Mean		0.84	0.88	0.88			
1/D	Desiree	Control	10.41	10.00	9.09	9.83	9.33	8.53
	Desiree	Chemical	9.80	5.56	11.11	8.82		8.38
	Cisgenic Desiree	Control	6.61	11.11	6.25	7.99	8.19	
	Cisgenic Desiree	Chemical	2.67	12.50	10.00	8.39		
	Sarpo Mira	Control	8.45	7.69	7.14	7.76	7.85	
	Sarpo Mira	Chemical	8.80	6.67	8.33	7.93		
	Mean		7.79	8.92	8.65			

ratio of members of the family *Rhabditidae* (average 12) and *Cephalobidae* (11) and occurrence of the *Panagrolaimidae* (1–11) family recorded across treatments. However, the distribution for *Rhabditidae* and *Cephalobidae* appeared dependent on the interaction of genotype and treatment (e.g. cisgenic Desiree chemical reached

(See figure on next page.)
Fig. 7 Heatmaps illustrating the distribution of nematode families relative to each potato genotype × management interaction (Desiree control [DeCon], Desiree chemical [DeChe], cisgenic Desiree control [CiCon], cisgenic Desiree chemical [CiChe], Sarpo Mira control [SmCon], Sarpo Mira chemical [SmChe]) for 2013 (**a**), 2014 (**b**) and 2015 (**c**). Oak Park (Carlow, Ireland). For the colour key and histogram *X-axis* number of individual nematodes accounted for and *Y-axis* the times that the number (account) is repeated

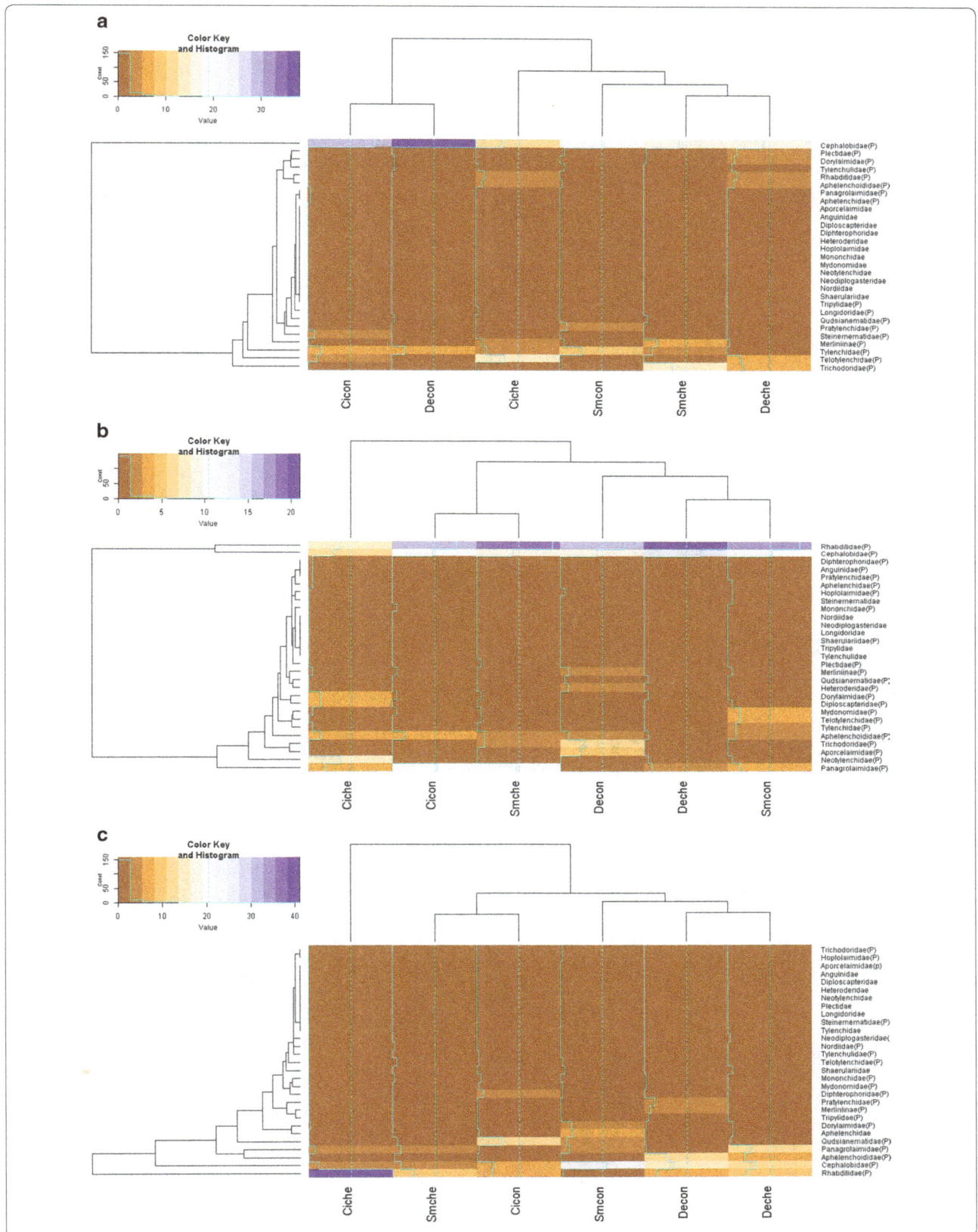

Table 6 Sorensen coefficients calculated on comparisons within[a] potato genotypes (control, chemical treatment) and between[b] potato genotypes for each individual year (2013, 2014 and 2015) and total during the 3 years for nematode families and genus identified from study completed at Oak Park (Carlow, Ireland)

	Comparison[a]	Year 2015	2014	2013	Total	Comparison[b]	Year 2015	2014	2013	Comparison	Year 2015	2014	2013
Family	Desiree control vs. chemical	0.63	0.50	0.53	0.63	Desiree vs. Cisgenic Desiree	0.48	0.37	0.53	Control vs. chemical	0.36	0.54	0.41
	Cisgenic Desiree control vs. chemical	0.43	0.73	0.43	0.79	Desiree vs. SarpoMira	0.49	0.47	0.62				
	Sarpo Mira control vs. chemical	0.55	0.57	0.55	0.67	Cisgenic Desiree vs. Sarpo Mira	0.50	0.36	0.34				
Genus	Desiree control vs. chemical	0.46	0.58	0.18	0.50	Desiree vs. Cisgenic Desiree	0.48	0.36	0.69	Control vs. chemical	0.32	0.48	0.31
	Cisgenic Desiree control vs. chemical	0.40	0.58	0.38	0.63	Desiree vs. SarpoMira	0.47	0.40	0.74				
	Sarpo Mira control vs. chemical	0.28	0.52	0.64	0.64	Cisgenic Desiree vs. Sarpo Mira	0.49	0.41	0.64				
							0.48	*0.39*	*0.69****				

Italic values indicate Sorensen coefficients for nematode genus identified in 2015, 2014 and 2013 between potato genotypes

* P < 0.05, ** P < 0.01, *** P < 0.001

a maximum of 41 members as represented by the blue coloured cells in Fig. 7c). The occurrence of BF or EN nematode family members (*Steinernematidae*) was noted in Desiree chemical and SarpoMira control and FF or EN (*Neotylenchidae*) in Desiree and SarpoMira chemical. Five plant feeding families were recorded (*Merliniidae, Telotylenchidae, Tylenchulidae, Hoplolaimidae and Trichodoridae*) in total.

Sixty-two individual genera were identified across the 3 years of the experiment, with 35, 34 and 31 individual genera identified per year, 2013, 2014 and 2015 respectively (Fig. 8a–c). In relation to the Sorensen coefficient within potato genotypes, 50% of genera identified were equivalent between the chemical and control treatment for Desiree and 63 and 64% for the same treatments with cisgenic Desiree and Sarpo Mira respectively (Table 6). As with the nematode family assessment, examining the coefficient values relative to each year of the study identified a broad range from 0.18 to 0.64 when comparing the impact of control vs. chemical treatment across the three potato genotypes studied. In addition, the overlap of genus between potato genotypes (irrespective of chemical treatment) ranged from 0.36 to 0.74, with a clear statistical difference (P < 0.001) between year (Table 6). Temporally, a large overlap (0.64–0.74) of genus was noted in 2013 between potato genotypes; which contrasted with 2014 (0.36–0.41) and 2015 (0.47–0.49). Differences on the presence/absence of specific genera were also evident. For example, *Clarkus* (*Dorylaimia*—2015) and *Pratylenchoides* (*Tylenchida*—2014) were isolated from the Desiree and cisgenic Desiree control plots, with *Clarkus* (*Dorylaimida*—2015) also isolated from SarpoMira—chemical treated plots (Fig. 8b, c).

Discussion
Conducted over 3 years the goal of this study was to develop an initial baseline on the level of nematode abundance and diversity related to specific potato cropping systems, thereby addressing a knowledge deficit that currently exists in the literature. In particular, it was hypothesised that the cultivation of GM cisgenic Desiree potatoes would not impact significantly on the abundance and/or diversity of non-target soil nematodes. Additional contributory factors that were investigated related to weather variability and the management protocols adopted in regards to the presence/absence of chemical control measures against *P. infestans*, the causative organism of potato late blight disease.

Succession ecological indices
Nematode ecological succession usually progresses in an orderly and predictable manner unless set back by an environmental disturbance such as cultivation, pollution or nutrient enrichment [22]. The maturity index as originally proposed [38] along with its modifications MIMO, $\sum$MIMO besides of PPI and ratio PPI/MI have been used previously for monitoring different kinds of disturbance [39, 40], including the cultivation of GM crops [41, 42]. In this 3-year study, and based on the protocol undertaken, the rhizospheric nematode community did not register any significant effect with the cultivation of the cisgenic Desiree line compared to its comparator, cv. Desiree in the presence or absence of fungicide management. Neither was there a significant difference between crop management or the crop cultivars Desiree and Sarpo Mira, which is significant in light of the disparate genetic background of both cultivars and the fact that cv. Sarpo Mira possesses five genetic sources of resistance to *P. infestans* [9].

Although no significant difference was noted in this study in regards to rhizospheric inhabiting nematodes, a similar outcome was reported in regards to the effect of transgenic insect resistant corn on nematode assemblages [41], which was not based on rhizospheric samples. While the results of this study relate to the rhizosphere, it is worth noting that the complexity of the interactions between roots, their exudates and associated soil microorganisms continues to be elucidated [43].

The imprecision of the MI as a quantitative tool has been discussed in previous studies [40, 44] since high MI values, equating to undisturbed conditions, are conditioned by rare K-selected persisters (less disturbed) but with high c–p (3–5) or predominant r-selected opportunistic colonizers with c–p = 1 (enriched) [40]. Therefore, high values in one scenario may mask an accurate estimation of what is actually occurring in regards to nematode diversity. On the other hand, the low MIMO index values obtained in 2013 (in comparison with 2014 and 2015) indicated that the 2013 nematode communities were experiencing an environmental stress, which was irrespective of potato genotype and chemical management applied. It must be acknowledged that the change in land management may have been an influencing factor, with the AMIGA site having previously been a low managed grass pasture for ~10 years, before being used for potato cultivation. However, it is important to note that the rotation strategy adopted in this study ensured that for each year plots were positioned on original grass pasture, thereby ensuring that each year had effectively the same 'starting point' in regard to the status of the ground on which the plots were sown. Nematode communities with c–p = 2 have been associated with a limitation of resources, adverse environment conditions or recent contamination [45]. Therefore, while we

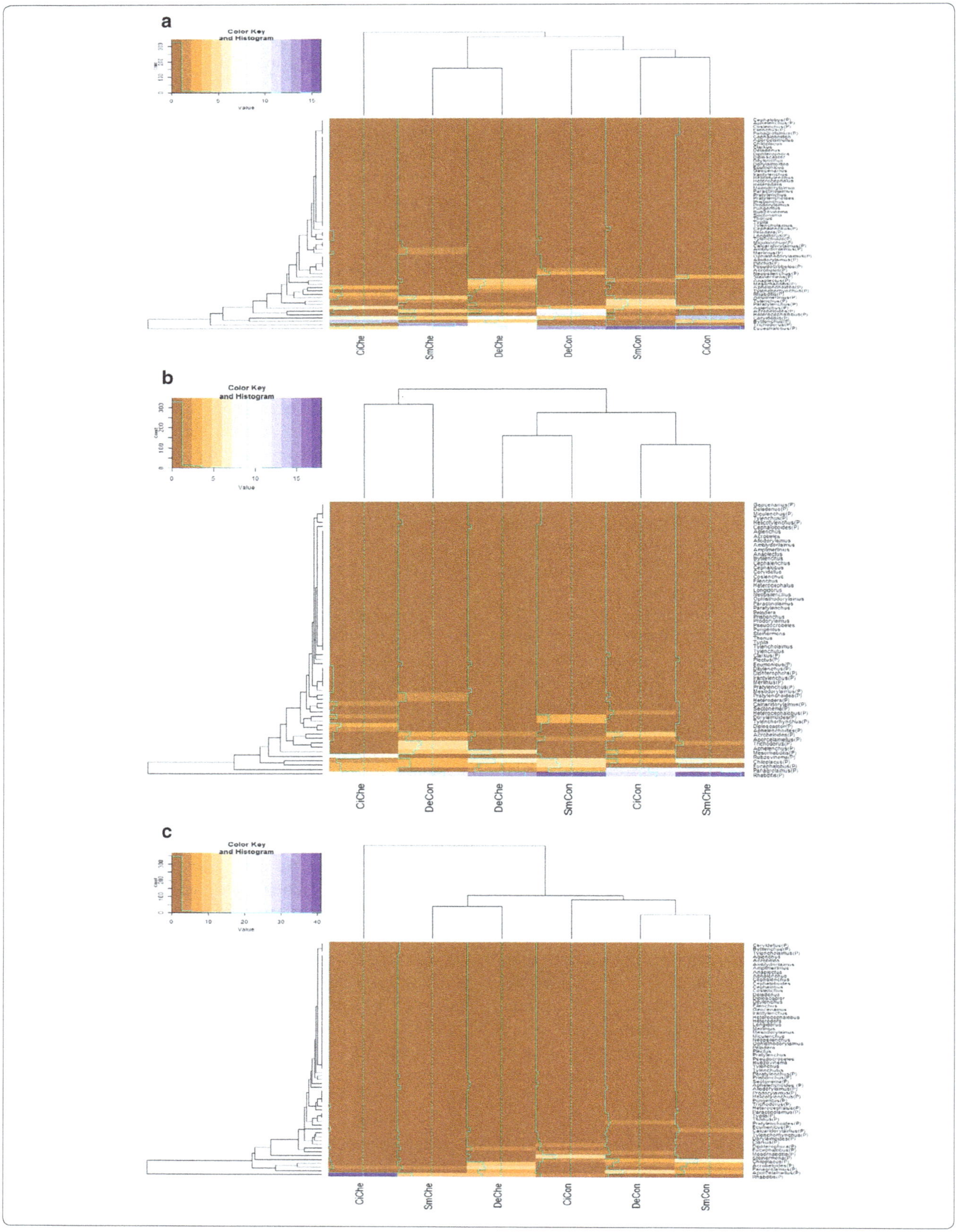
a
Color Key and Histogram
Value
CiChe
SmChe
DeChe
DeCon
SmCon
CiCon
b
Color Key and Histogram
Value
CiChe
DeCon
DeChe
SmCon
CiCon
SmChe
c
Color Key and Histogram
Value
CiChe
SmChe
DeChe
CiCon
DeCon
SmCon

hypothesise that the index values attained for 2013 were a product of the unfavourable weather conditions, which may have driven the increment in generalist opportunist nematodes (c–p = 2), it is not possible to determine what kind of nematode succession was present, since c–p = 2 are formed by both bacterial (remain of the primary succession) and fungal feeder (secondary succession or primary, depending on the nutrient status C:N ratio) [46]. A qualitative analysis of the maturity indices did indicate differences between the potato genetic background and their interaction with the disease management strategies (the no spray control vs. weekly chemical applications). For example, while for the MI, which encapsulates all free living nematodes, the three genotypes showed a similar tendency, when the members with a c–p = 1 are removed (MIMO) or included the PPI and the PPI/MI ratio, both Desiree and cisgenic Desiree reported comparable tendencies in contrast to Sarpo Mira, which has a different genetic background to that of cv. Desiree. This would indicate that both the Desiree genotype and the cisgenic Desiree genotype studied here interact with and regulate their respective rhizobiomes (likely via root exudates [43] in the same manner and the variability being recorded is inter- as opposed to intra-cultivar specific.

As concluded by Neher [22] the natural ecological succession can be setback by many factors, however, we point out that the level and type of response obtained will depend greatly on environmental conditions (rain and soil temperature as direct factors and environment temperature and relative humidity as indirect factors) in the moment that the experiment is carried out. Soil temperature and moisture have already been identified as primary abiotic factors impacting on nematode distribution and abundance [20]. Here the MI was found to be conditioned by weather conditions and weakly by disease management while it was the MIMO index which was more affected by climatic variation than potato genotype and crop management. Darby et al. [39] showed that the composition of nematode communities (when measured through MI) differ greatly between geographic locations with disparate weather conditions. In this study we detected over the 3 years an influence of weather conditions on community composition, although it is important to clarify that the study was completed at a single geographic location, hence reducing the variability associated with soils from distinct geographic places. Any

follow up study should include additional locations in order to comprehensively address this recorded trend.

Trophic groups

Trophic group absolute value, without distinguishing between c–p, is another method with which to investigate nematode trophic structures [27, 47, 48]. Here no statistical difference was identified in quantitative values between the cisgenic Desiree and either its genetic comparator, cv. Desiree or the alternative genotype Sarpo Mira, plus/minus chemical management practises. Qualitatively, differences were identified. The absence, presence, reduction or increment of trophic groups has been associated with the level of susceptibility or tolerance that some nematode groups experience [47]. Based on the analysis completed in this study, the presence/absence of PR appeared to have been more influenced by the application of chemical fungicides in the disease management regimes and the weather patterns than by the potato genotype. For example, the absence of PR and FF and the increase of BF and PF in all treatments (chemical and control) in 2013 in comparison to 2014 and 2015 could be associated more with limited resources (stress conditions) due to the scarcity of precipitation and the high air and soil temperatures, which occurred through 2013 and would have favoured those nematodes less sensitive to environmental disturbance. In contrast, the weather conditions of 2014 and 2015 were more supportive of an enrichment condition, which can be linked with the reduction of PF (and increase of BF and OM). This would arise from the activation of soil biological processes, hence increasing food resources for the nematode populations [27].

Here the application of fungicides through the chemical management practises, parallels a decrease in the PR group in both Desiree genotypes but it is not possible to associate it directly with a specific active ingredient as different fungicides were applied relative to plant growth stage and the incidence of late blight disease into the site. However, the work correlates with results reported by Smith et al. [49] who examined the impact of the Benomyl™ systemic fungicide on prairie tall grass and it also relates to the use of herbicide [27], which combined, reinforces the theory that the PR nematode group is highly sensitive to chemical disturbance. The only contradiction to this is the fact that the same response was not recorded with Sarpo Mira, suggesting that this cultivar can possibly counteract the negative impact on the PR group; possibly due to the extreme growth vigour of the cultivar, thereby presenting a larger biomass for the PR group to prey on [50]. In this study, cv. Sarpo Mira recorded a higher PPI and PPI/MI ratio under disease management strategies relative to Desiree and cisgenic Desiree. Of interest, Bonger and Ferris [31] determined that the occurrence and abundance of PPI is largely determined by the community structure, host status and critically the vigour of plants growing in the soil. This phenomenon is supported here where the presence of the PR group could be influenced by the reduction/elimination of the prey (PPI) taking into consideration plant vigour, the chemical applications or the interaction of both. In this study, the PPI index as previously discussed was influenced somewhat by potato genotype and disease management (but not weather conditions); the separation in trophic groups here suggests that PF is most likely influenced by all three factors. It is also worth considering that as the vigour of cv. Sarpo Mira induced a different nematode community structure, the inclusion of such a vigorous phenotype in an integrated pest management plan may generate a balanced community (PR and PPI) and hence induce consistent levels of suppression against distinct pathogenic nematodes. While there was no evidence of pathogenic nematodes in the field used in this study, bearing in mind a recent review [51] on the role of predacious nematodes in the biological control of plant parasitic nematodes, this phenomenon requires further study.

Indifferent to maturity indices, where basically the nematode community is separated into two groupings, colonizers and persisters, the separation of the nematode community using trophic groups provides a valuable insight into the complexities of the rhizosphere. As such, we hypothesise, based on this preliminary study that trophic groups such as PR can be influenced by disease management strategies by weather conditions and possible by the plant vigour and that FF or EN, BF or EN and OM are more influenced by weather conditions with BF affected by the interaction of weather and disease management.

Functional guild

The combined analysis of the response of the nematode community through its feeding type or trophic group [21] and its life history strategy [38], measured as its functional guild [46] is another way of evaluating the response of the nematode community to environmental disturbance factors. Comparing the disease management regimes independently of potato genotype showed a significant difference existed for the EI

and BF_2 indices in regard to presence/absence of fungicide applications. As shown previously with the trophic groups, fungicide applications also altered the structure of the soil food web. In this study, the weekly chemical fungicide treatments generated an enrichment condition given per an increment of BF_1 and reduction in CH. In contrast, the corresponding control treatment highlighted a more basal condition which included recovery from a moderate disturbance, through tillage and fertilizer operations as part of the standard management of the site.

Predation and competition among trophic levels provide "top–down" regulation of food web structure and function [45]. The significant differences noted here between years for the EI, BF_1, BF_2 and SI and the different treatment on the distinct quadrants would come to confirm not only the significance of year-to-year disturbances as per the Succession Ecological Indices and the Trophic Groups but also the type of disturbance. The information obtained through the assessments of the functional guilds support the conclusion of Cesarz et al. [52], stating that knowledge on functional guilds proves a better understanding about soil alterations.

Nematode taxonomy, abundance and diversity indices

The nematode community structure was examined in both a qualitative and quantitative manner. Based on the results from this study the association between the *Clarkus* genus (Family *Mononchidae*) and the entomopathogenic *Steinernema* (Family *Rhabditidae*), *Rubzovinema* (Family *Neotylenchidae*) and *Deladenus* (Family *Shaerulariidae*) genus may serve as a bioindicator of environmental disturbance through for example the application of chemical fungicide. While the *Trichodorus* (*Trichodoridae*) appear to be relatively tolerant to adverse weather conditions (dry weather) [53] and fungicide application, on the other hand plant parasitic nematodes occur in three widely separated orders: Triplonchida, Dorylaimida and Tylenchida. All triplonchid and dorylaimid plant parasitic nematodes are migratory ectoparasites of roots. Within the Tylenchida however, several different types of plant parasitism can be recognised [54]. In the first year (2013) of this study, migratory ectoparasites (1d) [21] of roots (*Longidoridae, Trichodoridae and Tylenchidae*) were identified dominating in samples derived from chemical treated plots. This contrasted with epidermal cell and root hair feeders (1e) [21] and algae lichen feeders (1f) [21] (*Tylenchidae*) as found in the non-chemical treated samples. As 2013 was characterized by unfavourable climatic conditions for nematode community structure as associated by an abundance of members of the *Cephalobidae* family, 2014 and 2015 were years where the *Rhabditidae* family dominated along with a distribution of Tylenchidas with distinct types of plant (1a–f) [21] and *Trichodorus* ectoparasites. The identification of the *Aporcelaimellus* genus during the years of higher rainfall (2014 and 2015) along with a general increase in the numbers of omnivorous nematodes, would support previous hypothesis by Porazinska et al. [55], whereby the correlation of soil moisture with the presence of omnivorous nematodes is more long term than temporary.

As distinct nematode family, genus or species respond in different ways to disturbed soil management practises [40, 56] or environmental perturbations [26, 46, 53], taxonomic analysis focussed at a genus level can be considered fundamental [40, 57], in regards to quantifying the impact of crop genotype cultivation on rhizospheric nematode diversity. From observations made in this study, such comparative taxonomic analysis to the level of genus were most sensitive in detecting temporal differences across the 3 years of the study, with the adverse conditions of 2013 appearing to induce a similar population of genus across the treatments.

In our study, sequencing a fragment of the 18 SSU rDNA gene sufficiently discriminated between nematode populations across the different disease management treatments and weather conditions. Indeed, the number of genera detected in this study exceed that recorded in previous GM-related studies that relied solely on morphological identification [42, 58]. Similarly, alternative DNA-based detection techniques (e.g. T-RFLP) have also proven versatile at capturing more information than classical morphological analysis [59]. Sample pooling to facilitate sequencing has been recently demonstrated in regards to the high throughput sequencing of soil nematode communities [60]. For the work presented here, the approach of sequencing clones from a unique composite sample did provide a detailed representation of nematode diversity as supported by the richness, high value of equitability (EH) where the distribution of species inside the samples was more than 60%. This point is supported by the other parameter affecting nematode diversity, Simpson diversity index (1-D), in which a probability up to 89% was obtained, indicating the high probability that two individuals randomly selected from the same sample belong to different species. Further support is provided by the completed rarefaction analysis, which indicated the high levels of nematode species and genus richness obtained relative to the sampling process adopted. In light of the recent advancements in next generation sequencing (NGS) technologies, and the ever-reducing

costs of applying these processes, future studies that use NGS will provide complementary insight into the observations made here in developing a robust database while also elucidating further the levels of nematode diversity within the potato rhizosphere.

Overall, this study has generated a baseline dataset accounting for nematode abundance and diversity for GM potato cultivation practices over 3 years. Capitalising on this resource, evaluations concluded that year of analysis exerted the largest impact on nematode diversity and that the cultivation of a cisgenic *P. infestans* resistant potato genotype had no significant effect on nematode diversity and community structure that was any greater than its comparator potato genotype cv. Desiree. Separately, the knowledge base generated here, provides an opportunity to develop specific bio-indicators to assist future environmental studies, specifically in regards to the cultivation of conventional/ genetically engineered potatoes and/or fungicide applications. Taking into consideration that the response of a bio-indicator is dependent upon the interaction of multiple factors (e.g. host genotype × phenotype, weather conditions, crop management practises, presence/ absence of crop pathogens), to build upon the outputs of this study, first steps should consider validating the output from this study across multiple locations for specific nematode families/genus as indicated here. From that, a paradigm should be established with multiple factors included, relative to the variables of the studies being used for data input and which are known to affect the environment, which is relevant to the bio-indicator. From here, the robustness of the emerging model can then be tested across an expanded trial system prior to its implementation as a diagnostic bio-indicator for environmental studies.

Conclusions

- Cultivation of the cisgenic Desiree line studied here had no significant effect on nematode community diversity and/or structure relative to that recorded for its comparator, cv. Desiree. Differences that were recorded were inter-genotype specific.
- Fungicide applications can influence nematode community structures and this can be exasperated by extreme weather conditions. However, it would appear that this effect can be countered by the vigour of the plant being treated.
- The Maturity indices are merely an indicator of a disturbed environment and require the inclusion of functional guild and taxonomic data to accurately quantify levels of disturbance in potato ecosystems.

Additional files

Additional file 1: Figure S1. Flowchart detailing the field design for the AMIGA study across 2013, 2014 and 2015, potato genotypes included and corresponding treatments (* IPM was additional strategy in the AMIGA project but which was not included as a treatment in this study). Postharvest molecular analysis is also detailed.

Additional file 2: Table S1. Summary of indices (Ecological succession, Functional guild and Diversity), graphic representation of food web structure and similarity coefficients employed to characterize the nematode community in the rhizosphere of three distinct potatoes genotypes (cv. Desire, cv. Sarpo Mira and cisgenic Desiree) submitted to two different disease management (control, chemical treatment), during field studies across 2013, 2014 and 2015 park, Ireland.

Additional file 3: Figure S2. (a) Mean daily rainfall values taken for Oak Park (Carlow, Ireland) during 2013, 2014 and 2015. (b) Mean daily temperature values taken for Oak Park (Carlow, Ireland) during 2013, 2014 and 2015. (c) Mean daily soil temperature values (at 30 cm depth) taken for Oak Park (Carlow, Ireland) during 2013, 2014 and 2015. (d) Mean daily relative humidity values for Oak Park (Carlow, Ireland) during 2013, 2014 and 2015.

Additional file 4: Figure S3. Uniformity of values across the Shannon Diversity (H) and Shannon Equitability (EH) indices for samples taken from under different potatoes genotypes (Desiree, cisgenic Desiree, Sarpo Mira) treated with different disease management regimes (control, chemical treatment) through the years of 2013, 2014 and 2015 at Oak Park (Carlow, Ireland).

Authors' contributions

Conceived and designed the experiments: VOC and EM. Performed the experiments VOC and SP. Analyzed the data: VOC and EM. Wrote the paper: VOC and EM. All authors read and approved the final manuscript.

Author details

[1] Dept. Crop Science, Teagasc, Oak Park, Carlow, Ireland. [2] Present Address: Plant Biology and Crop Science, Rothamsted Research Station, West Common, Harpenden, Hertfordshire AL5 2JQ, UK.

Acknowledgements

We thank the DuRPh programme for providing the cisgenic Desiree line (A15-031) to the work programme. We also thank the student summer staff of the Dept. Crop Science who assisted in crop management tasks during each growing season and the farm staff at Teagasc Oak Park for performing soil preparation tasks prior to sowing.

Competing interests

The authors declare that they have no competing interests.

Funding

This is publication No. 18 produced within the framework of the project Assessing and Monitoring the Impacts of Genetically Modified Plants on Agro-ecosystems (AMIGA), funded by the European Commission in the Framework programme 7. THEME [KBBE.2011.3.5-01].

References

1. Food and Agricultural Organisation of the United Nations statistical database. http://faostat.fao.org/faostat/collections?version=ext&hasbulk=0&subset=agriculture. Accessed 10 Feb 2016.
2. Haverkort AJ, Boonekamp PM, Hutten R, Jacobsen E, Lotz LAP, Kessel GJT, Visser R, van Der Vossen E. Societal costs of late blight in potato and prospects of durable resistance through cisgenic modification. Potato Res. 2008;51:47–57.
3. Fry WE, Birch PRJ, Judelson HS, Grünwald NJ, Danies G, Everts KL, Gevens AJ, Gugino BK, Johnson DA, Johnson SB, et al. Five reasons to consider *Phytophthora infestans* a reemerging pathogen. Phytopathology. 2015;105(7):966–81.
4. Commission E. EC regulation No 1107/2009. Concerning the placing of plant protection products on the market and repealing Council Directives 79/117/EEC and 91/414/EEC. 2009.
5. Cooke DEL, Cano LM, Raffaele S, Bain RA, Cooke LR, Etherington GJ, Deahl KL, Farrer RA, Gilroy EM, Goss EM, et al. Genome analyses of an aggressive and invasive lineage of the Irish potato famine pathogen. PLoS Pathog. 2012;8(10):e1002940.
6. Gebhardt C, Valkonen JPT. Organisation of genes controlling disease resistance in potato. Annu Rev Phytopathol. 2001;39:79–102.
7. Vleeshouwers VGAA, Raffaele S, Vossen JH, Champouret N, Oliva R, Segretin ME, Rietman H, Cano LM, Lokossou A, Kessel G, et al. Understanding and exploiting late blight resistance in the age of effectors. Annu Rev Phytopathol. 2011;49(1):507–31.
8. Mullins E. Engineering for disease resistance: persistent obstacles clouding tangible opportunities. Pest Manag Sci. 2015;71(5):645–51.
9. Rietman H, Bijsterbosch G, Cano LM, Lee H-R, Vossen JH, Jacobsen E, Visser RGF, Kamoun S, Vleeshouwers VGAA. Qualitative and quantitative late blight resistance in the potato cultivar Sarpo Mira is determined by the perception of five distinct RXLR effectors. Mol Plant Microbe Interact. 2012;25(7):910–9.
10. Fry WE. *Phytophthora infestans*: the plant (and R gene) destroyer. Mol Plant Pathol. 2008;9:385–402.
11. Haas BJ, Kamoun S, Zody MC, Jiang RHY, Handsaker RE, Cano LM, Grabherr M, Kodira CD, Raffaele S, Torto-Alalibo T, et al. Genome sequence and analysis of the Irish potato famine pathogen *Phytophthora infestans*. Nature. 2009;461(7262):393–8.
12. Rodewald J, Trognitz B. Solanum resistance genes against *Phytophthora infestans* and their corresponding avirulence genes. Mol Plant Pathol. 2013;14(7):740–57.
13. Jacobsen E, Schouten H. Cisgenesis, a new tool for traditional plant breeding, should be exempted from the regulation on genetically modified organisms in a step by step approach. Potato Res. 2008;51(1):75–88.
14. Haesaert G, Vossen JH, Custers R, De Loose M, Haverkort A, Heremans B, Hutten R, Kessel G, Landschoot S, Van Droogenbroeck B, et al. Transformation of the potato variety Desiree with single or multiple resistance genes increases resistance to late blight under field conditions. Crop Prot. 2015;77:163–75.
15. Haverkort AJ, Boonekamp PM, Hutten R, Jacobsen E, Lotz LAP, Kessel GJT, Vossen JH, Visser RGF. Durable late blight resistance in potato through dynamic varieties obtained by cisgenesis: scientific and societal advances in the DuRPh project. Potato Res. 2016;59(1):35–66.
16. Jones JDG, Witek K, Verweij W, Jupe F, Cooke D, Dorling S, Tomlinson L, Smoker M, Perkins S, Foster S. Elevating crop disease resistance with cloned genes. Philos Trans R Soc B Biol Sci. 2014;369:20130087.
17. EC. Directive 2001/18/EC of the European Parliament and of the Council of 12th march 2001 on the deliberate release into the environment of genetically modified organisms and repealing Council Directive 90/220/EC. Off J Eur Communities. 2001;L106:1–39.
18. EFSA. Guidance on the environmental risk assessment of genetically modified plants. EFSA J. 2010;8(11: 1879):111.
19. Arpaia S, Messéan A, Birch NA, Hokannen H, Härtel S, van Loon J, Lovei G, Park J, Spreafico H, Squire GR, et al. Assessing and monitoring impacts of genetically modified plants on agro-ecosystems: the approach of AMIGA project. Entomologia. 2014;2(1). doi:10.4081/entomologia.2014.154
20. Bakonyi G, Nagy P, Kovacs-Lang E, Kovacs E, Barabas S, Repasi V, Seres A. Soil nematode community structure as affected by temperature and moisture in a temperate semiarid shrubland soil nematode community structure as affected by temperature. Appl Soil Ecol. 2007;37:31–40.
21. Yeates GW, Bongers T, De Goede RGM, Freckman DW, Georgieva SS. Feeding habitats in soil families and genera—an outline for soil ecologist. J Nematol. 1993;25:315–31.
22. Neher DA. Soil community composition and ecosystem process. Agrofor Syst. 1999;45:185–99.
23. Briar SS, Barker C, Tenuta M, Entz MH. Soil nematode responses to crop management and conversion to native grasses. J Nematol. 2012;44(3):245–54.

24. Pan K, Gong P, Wang J, Wang Y, Liu C, Li W, Zhang L. Application of nitrate and ammonium fertilizers alter soil nematode food webs in a continuous cropping system in South-western Sichuan China. Eurasian J Soil Sci. 2015;4:287–300.

25. Vandegehuchte ML, Sylvain ZA, Reichmann LG, de Tomasel CM, Nielsen UN, Wall DH, Sala OE. Responses of a desert nematode community to changes in water availability. Ecosphere. 2015;6:1–15.

26. Vervoort MTW, Vonk JA, Mooijman PJW, Van den Elsen SJJ, Van Megen HHB, Veenhuizen P, Landeweert R, Bakker J, Mulder C, Helder J. SSU ribosomal DNA-based monitoring of nematode assemblages reveals distinct seasonal fluctuations within evolutionary heterogeneous feeding guilds. PLoS ONE. 2012;7(10):e47555.

27. Zhao J, Neher DA, Fu S, Li Z, Wang K. Non-target effect of herbicides on nematode soil assemblages. Pest Manag Sci. 2013;69:679–84.

28. Höss S, Nguyen HT, Menzel R, Pagel-Wieder S, Miethling-Graf R, Tebbe CC, Jehle JA, Traunspurger W. Assessing the risk posed to free-living soil nematodes by a genetically modified maize expressing the insecticidal Cry3Bb1 protein. Sci Total Environ. 2011;409(13):2674–84.

29. Griffiths BS, Caul S, Thompson J, Birch ANE, Scrimgeour C, Cortet J, Foggo A, Hackett CA, Krogh PH. Soil microbial and faunal community responses to Bt maize and insecticide in two soils. J Environ Qual. 2006;35(3):734–41.

30. Yeates GW. Nematodes as soil indicators: functional and biodiversity aspect. Biol Fertil Soils. 2003;37:199–210.

31. Bongers H, Ferris H. Nematode community structure as a bioindicator in environmental monitoring. Trends Ecol Evol. 1999;14(6):224–8.

32. Shannon CE, Warren W. A mathematical model of communication. Urbana: University of Illinois Press; 1949.

33. Neher DA, Campbell CL. Sampling for regional monitoring of nematode communities in agricultural soils. J Nematol. 1996;28(2):196–208.

34. Foster SJ, Park T-H, Pel M, Brigneti G, Śliwka J, Jagger L, van der Vossen E, Jones JDG. Rpi-vnt1.1, a Tm-22 homolog from *Solanum venturii*, confers resistance to potato late blight. Mol Plant Microbe Interact. 2009;22(5):589–600.

35. Pel MA, Foster SJ, Park T-H, Rietman H, van Arkel G, Jones JDG, Van Eck HJ, Jacobsen E, Visser RGF, Van der Vossen EAG. Mapping and cloning of late blight resistance genes from *Solanum venturii* using an interspecific candidate gene approach. Mol Plant Microbe Interact. 2009;22(5):601–15.

36. Blaxter ML, De Ley P, Garey JR, Liu LX, Scheldeman P, Vierstraete A, Vanfleteren JR, Mackey LY, Dorris M, Frisse LM, et al. A molecular evolutionary framework for the phylum Nematoda. Nature. 1998;392(6671):71–5.

37. R. A language and environment for statistical computing. http://www.r-project.org/. Accessed 17 May 2016.

38. Bongers T. The maturity index: an ecological measure of environmental disturbance based on nematode species composition. Oecologia. 1990;83:14–9.

39. Darby B, Neher DA, Belnap J. Soil nematode community are ecologically more mature beneath late-than early-successional stage biological soil crust. Appl Soil Ecol. 2007;35:203–12.

40. Porazinska DL, Duncan LW, McSorley R, Graham JH. Nematode communities as indicators of status and processes of a soil ecosystem influenced by agricultural management practices. Appl Soil Ecol. 1999;13(1):69–86.

41. Li X, Liu B. A 2-year field study shows little evidence that the long term planting of transgenic insect-resistance cotton affect the community structure of soil nematodes. PLoS ONE. 2013;8:1–15.

42. Yang B, Chen H, Liu X, Ge F, Chen Q. Bt cotton planting does not affect the community characteristics of rhizosphere soil nematodes. Appl Soil Ecol. 2014;73:156–64.

43. Mommer L, Kirkegaard J, van Ruijven J. Root–root interactions: towards a rhizosphere framework. Trends Plant Sci. 2016;21(3):209–17.

44. Neher DA, Wu J, Barbercheck ME, Anas O. Ecosystem type affects interpretation of soil nematode community measures. Appl Soil Ecol. 2005;30(1):47–64.

45. Ferris H, Borgers T, de Goude RM. A framework for soil food web diagnostics: extension of the nematode faunal analysis concept. Appl Soil Ecol. 2001;18:13–29.

46. Bongers T, Bongers M. Functional diversity of nematodes. Appl Soil Ecol. 1998;10:239–51.

47. Bernard E. Soil nematode diversity. Biol Fertil Soils. 1992;1:99–103.

48. Li Q, Yong J, Wen-Ju L. Effect of heavy metal on soil nematode community in the vicinity of metallurgical factory. J Environ Sci. 2006;18:323.

49. Smith MD, Hartnett DC, Rice CW. Effects of long-term fungicide applications on microbial properties in tallgrass prairie soil. Soil Biol Biochem. 2000;32:935–46.

50. Kostenko O, Duyts H, Grootemaat S, De Deyn GB, Bezemer TM. Plant diversity and identity on predatory nematodes and their prey. Ecol Evol. 2015;5(4):836–47.

51. Khan MN, Kim BC. A review on the role of predatory soil nematodes in the biological control of plant parasitic nematodes. Appl Soil Ecol. 2007;35(2):370–9.

52. Cesarz S, Reich PB, Scheu S, Ruess L, Schaefer M, Eisenhauer N. Nematode functional guilds, no trophic groups, reflect shifts in soil food webs and process in response to interacting global change. Pedobiologia. 2015;58:23–32.

53. Zhao J, Shao Y, Wang X, Neher DA, Xu G, Li Z. Sentinel soil invertebrate taxa as bioindicators for forest management practices. Ecol Indic. 2013;24:236–9.

54. Tytgat T, De Meutter J, Gheysen G, Coomans A. Sedentary endoparasitic nematodes as a model for other plant parasitic nematodes. Nematology. 2000;2(1):113–21.

55. Porazinska DL, McSorley R, Duncan LW, Graham JH, Wheaton TA, Parsons LR. Nematode community composition under various irrigation schemes in a citrus soil ecosystem. J Nematol. 1998;30(2):170–8.

56. Yeates GW, Bongers T. Nematode diversity in agroecosystems. Agric Ecosyst Environ. 1999;74:113–35.

57. Zhao J, Neher D. Soil nematode genera that predict specific types of disturbance. Appl Soil Ecol. 2013;64:135–41.

58. Liu Y, Li J, Steward CN Jr, Luo Z, Xiao N. The effects of the presence of Bt-transgenic oilseed rape in wild mustard populations on the rhizosphere nematode and microbial communities. Sci Total Environ. 2015;530–531:263–70.

59. Palomares-Ruiz JE, Castillo P, Montes-Borrego M, Muller H, Landa BB. Nematode community populations in the rhizosphere of cultivated olive differs according to the plant genotype. Soil Biol Biochem. 2012;45:168–71.

60. Sapkota R, Nicolaisen M. High-throughput sequencing of nematode communities from total soil DNA extractions. BMC Ecol. 2015;15(1):1–8.

5

Testing the potential significance of different scion/rootstock genotype combinations on the ecology of old cultivated olive trees in the southeast Mediterranean area

Oz Barazani[1]*[ID], Yoni Waitz[1], Yizhar Tugendhaft[2,3], Michael Dorman[4], Arnon Dag[2], Mohammed Hamidat[5], Thameen Hijawi[5], Zohar Kerem[3], Erik Westberg[6] and Joachim W. Kadereit[6]

Abstract

Background: A previous multi-locus lineage (MLL) analysis of SSR-microsatellite data of old olive trees in the southeast Mediterranean area had shown the predominance of the Souri cultivar (MLL1) among grafted trees. The MLL analysis had also identified an MLL (MLL7) that was more common among rootstocks than other MLLs. We here present a comparison of the MLL combinations MLL1 (scion)/MLL7 (rootstock) and MLL1/MLL1 in order to investigate the possible influence of rootstock on scion phenotype.

Results: A linear regression analysis demonstrated that the abundance of MLL1/MLL7 trees decreases and of MLL1/MLL1 trees increases along a gradient of increasing aridity. Hypothesizing that grafting on MLL7 provides an advantage under certain conditions, Akaike information criterion (AIC) model selection procedure was used to assess the influence of different environmental conditions on phenotypic characteristics of the fruits and oil of the two MLL combinations. The most parsimonious models indicated differential influences of environmental conditions on parameters of olive oil quality in trees belonging to the MLL1/MLL7 and MLL1/MLL1 combinations, but a similar influence on fruit characteristics and oil content. These results suggest that in certain environments grafting of the local Souri cultivar on MLL7 rootstocks and the MLL1/MLL1 combination result in improved oil quality. The decreasing number of MLL1/MLL7 trees along an aridity gradient suggests that use of this genotype combination in arid sites was not favoured because of sensitivity of MLL7 to drought.

Conclusions: Our results thus suggest that MLL1/MLL7 and MLL1/MLL1 combinations were selected by growers in traditional rain-fed cultivation under Mediterranean climate conditions in the southeast Mediterranean area.

Keywords: Akaike information criterion (AIC) selection model, Environmental conditions, Multi-locus lineage analysis, Olive oil quality, Selection

Background

The history of fruit tree domestication is strongly linked to grafting, which provided an easy technique for clonal reproduction of trees with desirable properties that are difficult to propagate vegetatively. It is generally accepted that the domestication of several fruit trees such as apple, plum and others was not possible without the development of the grafting technique [1]. In addition to being useful or necessary for propagation, rootstocks can influence the size of scions and increase their vigor. Physiological investigation of the interactions between scions and rootstocks in grapevine showed that grafting can reduce the toxic effects of salinity by the ability of the rootstock to limit the uptake of Na^+ and Cl^- ions by the scion [2]. In several crop species, e.g., in peach [3] and grapevine [4], rootstocks have been shown to reduce leaf

*Correspondence: barazani@agri.gov.il
[1] Institute of Plant Sciences, Israel Plant Gene Bank, Agricultural Research Organization, 75359 Rishon LeZion, Israel
Full list of author information is available at the end of the article

chlorosis caused by iron deficiency. Rootstocks also have been selected to increase tree tolerance to abiotic stresses such as drought and soil pH [1], and increase the resistance against soil pathogens [5].

Genetic comparison of suckers and the canopy of old olive trees in the Iberian peninsula [6] and the southeast Mediterranean area [7] provide strong evidence for grafting as a common practice in olive cultivation in the past [8, 9]. Diez et al. suggested that by grafting of scions on wild growing trees, natural populations of *Olea europaea* subsp. *europaea* var. *sylvestris* were transformed into olive groves [6], and the use of individual 'wild' olive trees as vigorous rootstocks in traditional olive cultivation has also been suggested [9]. However, evidence on the potential contribution of the rootstock to olive tree fitness and phenotypic properties is very limited and is based on recent experimental systems using combinations of known cultivars [10–13]. In these experiments, particular combinations of rootstocks and scions were shown to decrease the harmful effects of excessive boron concentrations in the soil [11], and to increase resistance to *Verticillium* wilt [10, 12].

Previously we reported that most old olive trees in the southeast Mediterranean area are grafted. In addition, a multi-locus lineage (MLL) analysis had shown that most of the scions (ca. 90%) belong to a single MLL (MLL1), presumably representing the Souri cultivar, and that most of the rootstocks probably originated from plant individuals resulting from sexual reproduction [7]. However, we also identified an MLL (MLL7) that was more common than other MLLs in rootstocks of grafted old olive trees and was present in 23% of the trees analysed [7]. This led to the hypothesis that olive cultivation in the region may have involved selection not only of a specific scion but also of a specific rootstock. Traditional olive groves in the southeast Mediterranean area are distributed along a geographic gradient of diverse climatic, topographic and edaphic conditions [14, 15]. As rootstocks might have been selected for improvement of the root system in stressful conditions and/or for their influence on phenotypic properties of the scion, we aimed to investigate the contribution of the most common rootstock (MLL7) to the fitness and phenotype of olive trees in different environments. We here use a model selection procedure based on the Akaike information criterion (AIC) to investigate the potential advantage of the MLL1/MLL7 combination by quantifying the impact of a number of environmental variables on several agriculturally important phenotypic traits.

Methods

In our previous study [7] we reported on a total of 249 old olive trees in which both suckers and scions were collected from the same trees and genotyped. Identification of scion and rootstock MLLs was performed with leaf samples taken from tree canopies (i.e. scions) and from suckers that developed from the trunk base [7]. Thus, a comparison between scion and sucker of the same tree enabled us to differentiate between three genetic groups (GG): 1) GG1 included trees in which the common Souri cultivar (MLL1) was grafted on MLL7 (49 trees); 2) GG2, in which both suckers and scions were assigned to MLL1 (62 trees); and 3) GG3 included those trees in which the common Souri cultivar was grafted on single-occurrence rootstock MLLs that probably originated from sexual reproduction (117 trees). Trees of the second group (GG2) were either the result of vegetative propagation of MLL1 or of grafting of MLL1 scions on MLL1 truncheons. The analysis included a total of 228 old olive trees from 31 groves with various environmental conditions in the southeast Mediterranean area.

Phenotypic characterization

Fruits were collected during the harvest in a single season in 2008 and were used for morphological evaluation, oil extraction and evaluation of content and quality of the oil. Morphological evaluation included the weight and dimension of 10 fruits and stones of each tree. The Abencor system (MC2 Ingenieria Y Sistemas, Spain) was used to extract oil from 1 kg of fruits of each of the investigated trees [16], and the relative content of oil and paste water was determined after Soxhlet chemical extraction as previously described [17]. Fatty acid (FFA) profiles were determined following [18], and the ratio between monounsaturated fatty acids (MUFA) and polyunsaturated fatty acids (PUFA) was determined. Following Ben-Gal et al. [18], the peroxide value (milliequivalents of active oxygen per kilogram oil; mequiv kg^{-1}), acidity (% free oleic acid) and total content of polyphenols (mg kg^{-1} oil) in the oil were also determined.

Environmental parameters

Average annual rainfall and elevation data were gathered from the Geographic Information System center database (Hebrew University of Jerusalem) using the lati-/longitudinal coordinates of groves. Daily temperatures for 2008 were collected using the MODIS remote sensing of surface and canopy temperatures (http://modis-land.gsfc.nasa.gov/); T_{max} and T_{min} were calculated according to Blum et al. [19] and used to determine the number of growing degree days (GDD) during the period from first flowering to fruit harvest (mid-April to mid-November), following the equation $\frac{T_{max}+T_{min}}{2} - T_{base}$ (9.1 °C) [20]). Calcium carbonate content in the soil was used to assess edaphic conditions, as calcareous soil is one of the limiting factors of agricultural practice in the region [15]. Soil samples, five in each grove, were collected from three soil depths (0–30, 30–60 and 60–90 cm), and soil analysis was conducted by the Gilat Extension Services Laboratory and Research Center (Israel); the results are presented as the average value of the three layers analysed (% of $CaCO_3$).

Statistical analysis

Linear regression was used to examine the abundance of trees belonging to the two scion/sucker combinations MLL1/MLL7 (GG1) vs. MLL1/MLL1 (GG2) along gradients of environmental conditions. Linear models were also used to examine which environmental factors explain phenotypic variation. To evaluate the environmental effects, we used a model selection procedure based on the Akaike information criterion (AIC), corrected for small sample sizes [21]. Models where the given trait (e.g. oil content, peroxide value, etc.) was explained using all possible combinations of environmental factors (e.g. GDD, $CaCO_3$, etc.), genetic groups (GG1 vs. GG2) and the interactions of each environmental factor with genetic groups were evaluated based on AIC. Old olive trees belonging to GG3 were not included in the AIC analysis as they do not represent a genetically homogenous group. The AIC analysis thus included a total of 111 trees. Factors present in the most parsimonious model (i.e. with the lowest AIC) and their direction of influence were then summarized. The inclusion of interaction terms enabled us to understand whether the genetic groups (GG1 in comparison to GG2) differ in their response to the different environmental factors (i.e. presence of a genotype-by-environment interaction). All statistical analyses were done using R [22].

Results

The abundance of olive trees with different genotype combinations in groves along environmental gradients

Traditional rain-fed olive groves in the southeast Mediterranean area are scattered through geographical districts that vary in climatic conditions and soil texture and chemistry. For this study, olive groves were selected to represent an aridity gradient from north to south, ranging from relatively mesic Mediterranean climate sites (≥450 mm rainfall) to semi-arid (350–450 mm) and arid conditions with less than <350 mm rainfall per year (ASH) (Table 1, Fig. 1). We included groves of relatively high elevation in the Samaria and Judean Mts. (440 to ca. 720 m a.s.l.) to lower elevation in the inner plain (110–290 m a.s.l.), the Carmel and the coastal plain (MAK, 84 m a.s.l. and ASH 23 m a.s.l., respectively). Variation in the average annual maximum and minimum temperatures (expressed as growing-degree day, GDD) was also found among sites. The range of $CaCO_3$ content in the soil (2.0–57.7%) represents the variability of edaphic

conditions in the region (Table 1). Linear regression did not show any significant correlation between any of the environmental parameters (data not shown).

Mapping the relative number of trees of all three MLL combinations in the investigated groves showed that trees of the MLL1/MLL7 combination (GG1) are more abundant in the northern (Galilee) and central parts (Samaria and Judean Mts.) of the cultivation area of olives in the southeast Mediterranean area (Fig. 1). The linear regression showed a significant association between average rainfall and the proportion of trees belonging to GG1 ($F_{1,29} = 7.068$, $P = 0.0126$) and those of GG2 ($F_{1,29} = 6.929$, $P = 0.0135$), but this proportion was not associated with any of the other environmental parameters considered, i.e. GDD, elevation (m a.s.l) and the relative content of $CaCO_3$ in the soil (Table 2). As expected when estimating the effect of single environmental factor in ecological studies, the significant association between average rainfall and the proportion of GG1 and GG2 trees ($R^2 = 0.19$) suggests that other unknown environmental factor(s) contribute to the abundance of these two MLL combinations along the aridity gradient. Nevertheless, the abundance of trees of GG1 was higher at more mesic locations, while the proportion of trees belonging to GG2 increased with increasing aridity (Fig. 2).

Testing the potential contribution of genotype combination to tree phenotype

ANOVA post hoc tests did not show significant differences between trees belonging to the GG1 and GG2 combinations in any of the eight phenotypic traits measured (Additional file 1: Figure S1, Additional file 2: Table S1). When taking into account environmental effects and genetic group in the AIC model selection procedure, the content of paste water in the fruit, the peroxide value and stone length were positively associated with grafting, while the ratio of MUFA/PUFA and stone width were negatively influenced by grafting of MLL1 on MLL7 (Table 3). Note that the use of positive or negative 'influence' or 'effect' refers to their statistical term in the model, and thus does not necessarily reflect beneficial (positive) or detrimental (negative) effects. Our results further showed that the environmental parameters influenced most of the fruit traits of GG1 and GG2 trees in a similar way (Table 3 and Additional file 3). Exceptions to this were the positive effects of elevation and $CaCO_3$ content on paste water and peroxide value in trees of GG1, respectively, and the negative effects of these environmental parameters on these traits in trees of GG2. Oil content and fruit and stone dimension and weight were positively influenced by average rainfall in both GG1 and GG2 trees. Similar negative effects of average rainfall were found in acidity and paste water content, while

Table 1 Environmental data of the investigated groves arranged along an aridity gradient; GDD data for ASH is missing

	Geographic region	CaCO$_3$ (%)	Soil type	GDD	Average annual rainfall (mm)	Elevation (m a.s.l.)
YRM	Galilee	25.67	Terra Rosa	3365	324.90	341.67
YRR	Galilee	25.67	Terra Roas	3365	324.90	341.67
RAI	Galilee	2.00	Terra Rosa	3655	391.60	379.92
RAR	Galilee	2.67	Terra Rosa	3736	391.60	347.13
ZAL	Galilee	4.33	Terra Rosa	4002	391.60	92.69
DIH	Galilee	34.00	Terra Rosa	3769	391.60	385.75
KAM	Galilee	18.67	Terra Rosa	3631	391.60	329.75
KZE	Galilee	8.00	Heavy soil	4014	391.60	111.38
MAK	Carmel	14.00	Terra Rosa	3425	388.90	84.27
EJB	Samaria	57.67	Terra Rosa	2861	317.50	248.49
ETS	Samaria	28.33	Rendzina	3745	454.50	242.04
EWIK	Samaria	37.00	Terra Rosa	3738	454.50	339.05
ETD	Samaria	45.33	Terra Rosa	3786	454.50	337.72
EQA	Samaria	55.33	Terra Rosa	3573	408.10	286.00
ENZ	Samaria	48.67	Terra Rosa	3446	356.10	373.01
ENB	Samaria	37.67	Heavy soil	2481	356.10	522.56
ESB	Samaria	38.67	Terra Rosa	3372	408.10	289.72
ESK	Samaria	41.00	Terra Rosa	3124	408.10	337.72
ERK	Samaria	38.00	Rendzina	2709	466.10	460.75
HAD	Inner plain	35.33	Rendzina	3821	276.00	111.17
ERB	Judean Mts	54.33	Rendzina	3334	466.10	439.33
EJBA	Judean Mts.	33.67	Terra Rosa	2650	382.90	575.53
JER	Judean Mts.	46.00	Terra Rosa	4077	353.40	723.71
EIK	Judean Mts.	15.33	Terra Rosa	3297	382.90	630.54
YAL	Judean Mts.	38.67	Terra Rosa	3350	287.10	615.27
EBB	Judean Mts.	33.00	Terra Rosa	2133	359.70	726.96
AZK	Inner plain	39.33	Rendzina	3650	320.40	272.65
BNR	Inner plain	24.33	Rendzina	3714	320.40	292.64
EHS	Judean Mts.	25.33	Terra Rosa	2898	301.40	547.66
ASH	Coastal plain	34.33	Sand	–	193.90	23.65
AMZ	Inner plain	42.00	Rendzina	3608	344.50	328.08

the average rainfall did not have any effect on the other oil quality characters in the two groups of trees (i.e. total polyphenol concentration, peroxide values and MUFA/PUFA). Elevation had a stronger effect on MUFA/PUFA in trees of the GG1 combination than in those belonging to GG2 and on peroxide values in GG2 trees than in trees of GG1. Similarly, CaCO$_3$ had a stronger influence on stone weight in GG1 trees and on stone length in trees belonging to GG2. GDD had a positive effect on peroxide values in both groups of trees, but its effect was more pronounced in trees of GG2. In addition, the negative effect of GDD on stone weight was stronger in trees belonging to GG1 than in those belonging to GG2. The adjusted R^2 values of the AIC model showed the lowest value for total concentration of polyphenols (0.03) and the highest for peroxide value (0.40) and stone weight (0.55) (Table 3 and Additional file 3).

Discussion

Old olive trees in the southeast Mediterranean area that belong to the common Souri cultivar have rootstocks of different genotype (MLL). Our previous results indicated that 23% of the rootstocks belong to one multi-locus lineage (i.e. MLL7) [7]. Mapping of the trees in which the local MLL1 variety is grafted on the common MLL7 (GG1), and those in which scion and suckers both belong to MLL1 (GG2) showed that the first group of trees (GG1) are more abundant in the northern and central parts of the olive cultivation area in the region (Fig. 1). In addition, the linear regression analysis indicated a significant association between the relative number of GG1 trees at each grove and the average annual precipitation (Fig. 2). Thus, old olive trees growing in groves with different environmental conditions (Table 1), but with the same MLL in their fruit bearing part (MLL1) and with

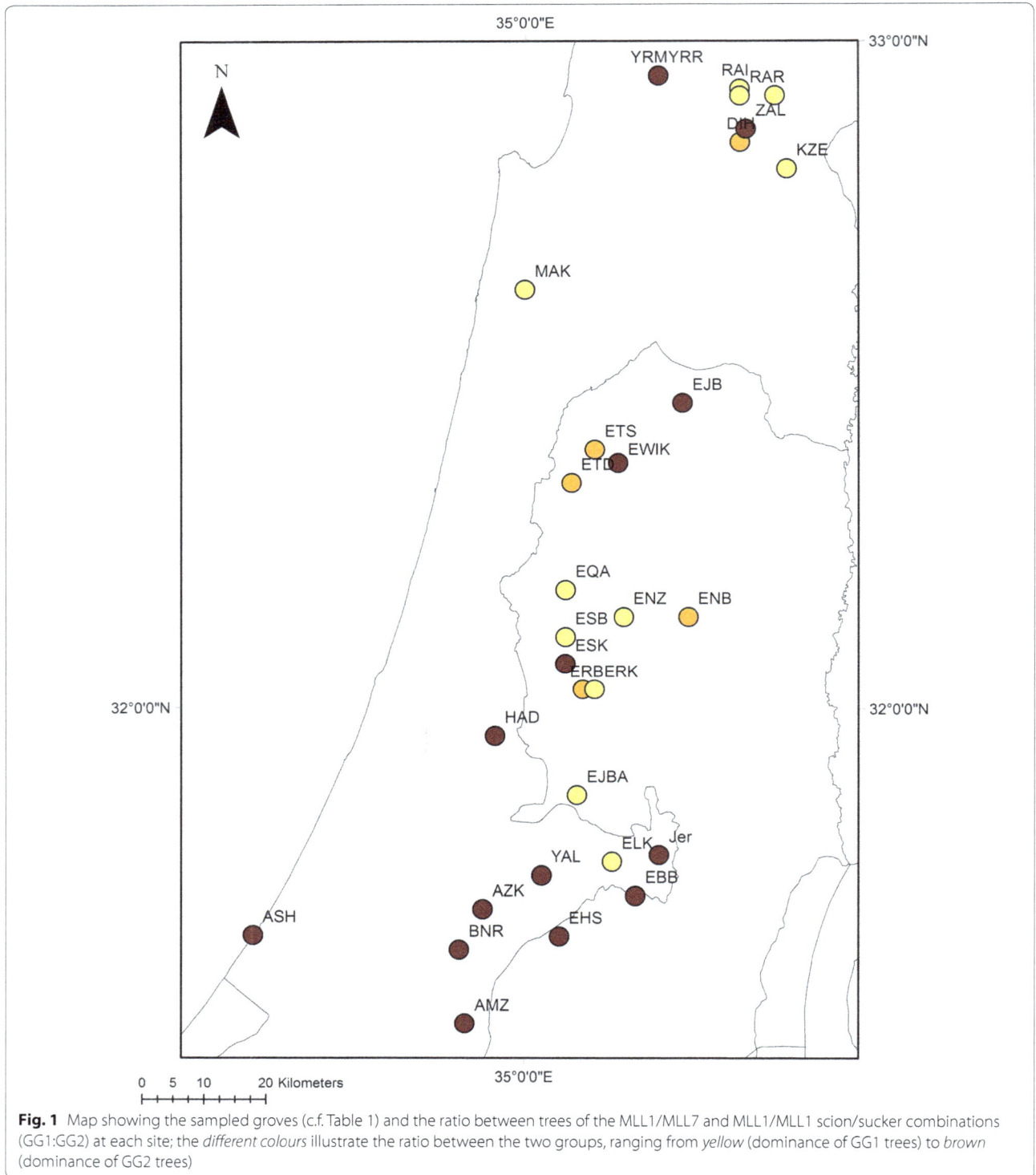

Fig. 1 Map showing the sampled groves (c.f. Table 1) and the ratio between trees of the MLL1/MLL7 and MLL1/MLL1 scion/sucker combinations (GG1:GG2) at each site; the *different colours* illustrate the ratio between the two groups, ranging from *yellow* (dominance of GG1 trees) to *brown* (dominance of GG2 trees)

rootstocks belonging to either MLL1 or MLL7, offer an opportunity to examine the possible contribution of the rootstock to traits of the crop, which more commonly is studied in common garden experiments.

Our sampling included trees growing in a range of soils that vary in their $CaCO_3$ content (2.0–57.7%) (Table 1). It has been suggested that grafting of olives is recommended for growing trees in problematic soils, such as

Table 2 Linear regression between environmental parameters and the relative number of old olive trees belonging to GG1 and GG2

	GG1			GG2		
	R^2	$F_{1,29}$	P	R^2	$F_{1,29}$	P
Average rainfall	0.1960	7.0680	0.0126	0.1928	6.9290	0.0135
Elevation	4.26×10^{-5}	0.0012	0.9722	0.1148	3.7610	0.0623
GDD	0.01120	0.3285	0.5710	0.0061	0.1716	0.6818
Calcium	0.07734	2.4310	0.1298	0.0211	0.6257	0.4353

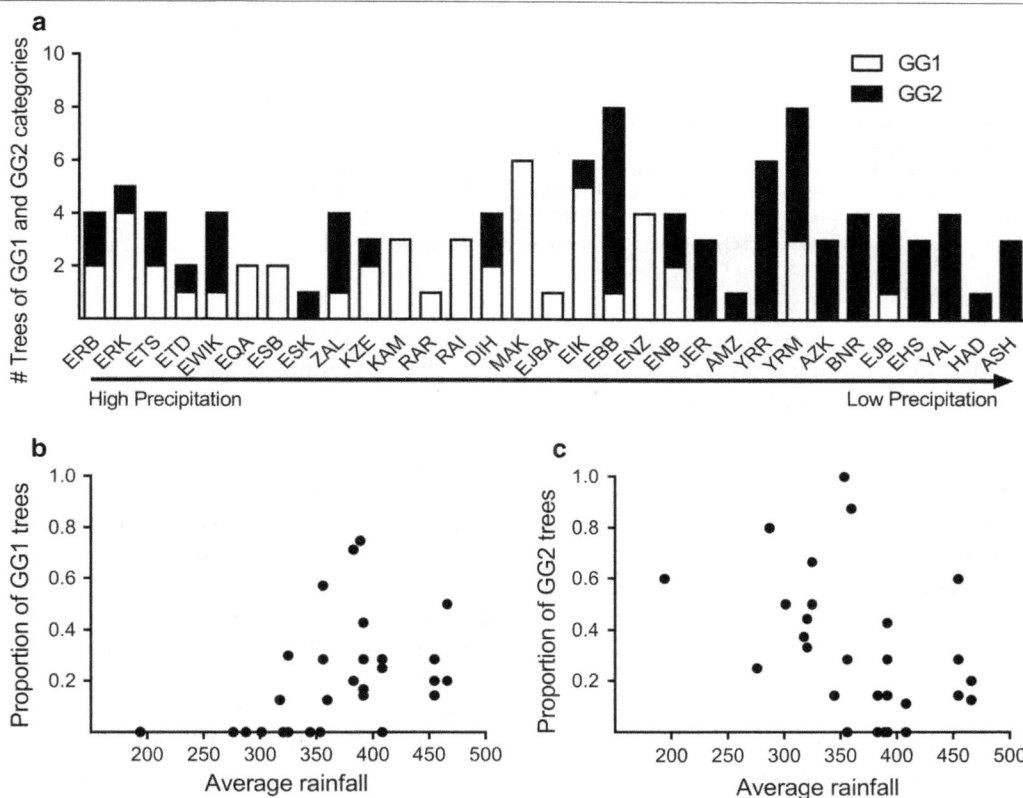

Fig. 2 Relative proportions of trees belonging to GG1 and GG2 along an aridity gradient showing the predominance of trees of GG1 in groves in geographic locations with relatively higher average annual rainfall and their absence from the driest groves (**a**). Relation between average annual rainfall and the proportion of trees of GG1 (**b**) and GG2 (**c**); the relative number of trees was calculated from the total number of trees at each site (i.e. including all three categories: GG1, GG2 and GG3)

soils with high $CaCO_3$ contents or saline soils [13]. We thus hypothesized that grafting on the specific MLL7 can provide an advantage in the calcareous soils of the region, where soil pH is high and availability of water and micronutrients is limited [15]. Supporting this hypothesis, the AIC model selection procedure showed differential positive and negative responses of peroxide values to the effect of $CaCO_3$ in GG1 and GG2 trees, respectively

(Table 3). Also, a strong positive influence of elevation on the oleic acid content in the oil, and thus on MUFA/PUFA, of GG1 trees in comparison with GG2 trees was found (Table 3). As peroxides are produced from the oxidation of fatty acids, high levels of peroxide are undesirable, causing rancid taste and reducing the oil shelf-life [23]. Equally, oleic acid (MUFA) is a major determinant of mouth-feel, aroma and shelf life of olive oil [24]. As

Table 3 Predicted direction of influence for the effects of environmental variables on the phenotypic traits of the investigated trees belonging to GG1 and GG2; model summaries are provided in Additional file 3

	GG1[a]	Average rainfall		Elevation		CaCO$_3$		GDD		Adjusted R^2
		GG1	GG2	GG1	GG2	GG1	GG2	GG1	GG2	
Oil content		+	+			+	+	–	–	0.21
Paste water	+	–	–	+	–	–	–			0.45
Acidity value		–	–	+	+	–	–	+	+	0.22
Peroxide value	+			+	++	+	–	+	++	0.40
Polyphenols conc.				–	–			–	–	0.03
MUFA/PUFA	–			++	+			+	+	0.08
Stone weight	–	+	+			++	+	––		0.55
Stone length	+	+	+			+	++	–		0.17
Stone width		+	+	+	+	+	+	–		0.22
Fruit weight		+	+	+	+					0.06
Fruit width		+	+					–		0.08

Positive and negative association is given by the coefficients of the most parsimonious model (Additional file 3) for the given trait as either positive (+ green cells), negative (– red cells) or zero (blank grey cells; when the variable was not present in the final model). Interaction between genotype combination and environmental effects is indicated by the differences between the two columns for a given environmental variable. For example, the effect of CaCO$_3$ on peroxide value is positive in GG1 but negative in GG2 due to the presence of an interaction term in the final model

[a] Results of general linear model explaining the effect of genotype combination on phenotypic traits

both taste and shelf life are likely to have been quality criteria important for early farmers, our results, showing differential responses of the two groups of scion/sucker combination in peroxide values and oleic acid contents, suggest a potential contribution of the MLL7 rootstock to olive oil quality under different edaphic conditions.

Results of the AIC-based model selection indicated that all environmental parameters in general had a similar influence on fruit and stone size and weight, and hence on oil content in the fruits, in the two genotype combinations (Table 3). In addition, average rainfall influenced the quality traits (as well as all other phenotypic characters) in GG1 and GG2 trees in a similar way (zero, positive or negative). However, elevation as well as GDD had a stronger positive influence on peroxide values in GG2 than in GG1 trees, and elevation had stronger positive effect on the MUFA/PUFA ratio in GG1 trees as compared to GG2 trees (Table 3). Thus these results provided further evidence, in addition to the effect of elevation and CaCO$_3$ described above, that some environmental parameters have a differential influence on oil quality in GG1 vs. GG2 trees. The decreasing abundance of trees belonging to GG1 (i.e. common MLL1 grafted on MLL7) along an aridity gradient (Fig. 2) may imply that grafting on MLL7 may increase sensitivity of olive trees to drought. Support of this hypothesis may be provided by a recent study which showed that young trees that were produced from the common MLL1 showed higher drought tolerance than trees of the Barnea cultivar (no

comparison with MLL1/MLL7 was made), suitable for intensive agricultural conditions based on controlled irrigation [25].

Conclusions

Overall, our results seem to imply that grafting of the common Souri cultivar (MLL1) on the MLL7 rootstock was governed by two opposing forces. On the one hand, growers in the past may have chosen to graft the common cultivar (MLL1) on MLL7 in order to enhance crop performance under certain environments. As MLL7 was used for grafting much more commonly than other MLLs [7], it appears that not grafting alone but grafting on MLL7 has this positive effect, which suggests deliberate rootstock selection. On the other hand, this advantage of grafting on MLL7 is countered by the plausible sensitivity of MLL7 to drought, so that this rootstock was less used in increasingly arid environments. If the scion/rootstock combination MLL1/MLL1 should represent non-grafted trees, it could be concluded that the grafting technique itself was less used in arid environments. Recent studies that assessed oil quality of the Souri cultivar under different irrigation regimes demonstrated the better performance of the cultivar under rain-fed condition [26, 27]. As these oil quality traits reflect the nutritional value of the olive oil, its taste and oxidative status [23, 24], it seems that the scion/rootstock combination MLL1/MLL7 was ideally adapted to conditions in a southeast Mediterranean climate.

Additional files

Additional file 1: Figure S1. Box plot comparisons of phenotypic traits in old olive trees of the MLL1/MLL7 (GG1) and MLL1/MLL1 (GG2) scion/ sucker combinations. Traits included the oil content in the fruits, waste water content obtained in the oil extraction process, four oil quality characteristics and three morphological properties of the fruits and stones.

Additional file 2: Table S1. Quantitative phenotypic data for each tree belonging to GG1 and GG2; md represent missing data.

Additional file 3. Results of the model-selection procedure.

Abbreviations
AIC: Akaike information criterion; FFA: free fatty acid; GDD: growing degree days; GG: genetic group; MLL: multi-locus lineage; MUFA: monounsaturated fatty acids; PUFA: polyunsaturated fatty acids; SSR: simple sequence repeats.

Authors' contributions
OB, AD, ZK, TH and JWK conceived this study. YT and MH collected the samples and performed the phenotypic characterization. EW conducted the genetic analysis. YW and MD conducted the statistical analysis. OB and JWK wrote the manuscript with contributions from all co-authors. All co-authors approved submission to BMC Ecology. All authors read and approved the final manuscript.

Author details
¹ Institute of Plant Sciences, Israel Plant Gene Bank, Agricultural Research Organization, 75359 Rishon LeZion, Israel. ² Department of Fruit Tree Sciences, Institute of Plant Sciences, Agricultural Research Organization, Gilat Research Center, 85280 M.P. Negev 2, Israel. ³ Institute of Biochemistry, Food Science and Nutrition, Faculty of Agricultural, Food and Environmental Quality Sciences, The Hebrew University of Jerusalem, 76100 Rehovot, Israel. ⁴ Department of Geography and Environmental Development, Ben-Gurion University of the Negev, 84105 Beer-Sheva, Israel. ⁵ Arab Agronomist Association, Al Nahda St., Ramallah and Al-Bireh Governorate, 4504 Al-Bireh, Palestine. ⁶ Institut für Spezielle Botanik und Botanischer Garten, Johannes Gutenberg-Universität Mainz, 55099 Mainz, Germany.

Acknowledgements
We thank all farmers who allowed us to sample trees in their orchards. We also thank Mr. I. Zipori for his valuable contribution to this study.

Competing interests
The authors declare that they have no competing interests.

Funding
This study was partially supported by the German Research Foundation's (DFG) trilateral program (Grant No. KA 635/14).

References
1. Mudge K, Janick J, Scofield S, Goldschmidt EE. A history of grafting. In: Janick J, editor. Horticultural reviews, vol. 35. Hoboken: Wiley; 2009. p. 437–93.
2. Dag A, Ben-Gal A, Goldberger S, Yermiyahu U, Zipori I, Or E, David I, Netzer Y, Kerem Z. Sodium and chloride distribution in grapevines as a function of rootstock and irrigation water salinity. Am J Enol Viticult. 2015;66:80–4.
3. Sotomayor C, Ruiz R, Castro J. Growth, yield and iron deficiency tolerance level of six peach rootstocks grown on calcareous soil. Cienc Investig Agrar. 2014;41:403–9.
4. Covarrubias JI, Retamales C, Donnini S, Rombola AD, Pastenes C. Contrasting physiological responses to iron deficiency in Cabernet Sauvignon grapevines grafted on two rootstocks. Sci Horticulturae. 2016;199:1–8.
5. King SR, Davis AR, Liu WG, Levi A. Grafting for disease resistance. HortScience. 2008;43:1673–6.
6. Diez CM, Trujillo I, Barrio E, Belaj A, Barranco D, Rallo L. Centennial olive trees as a reservoir of genetic diversity. Ann Bot. 2011;108:797–807.
7. Barazani O, Westberg E, Hanin N, Dag A, Kerem Z, Tugendhaft Y, Hmidat M, Hijawi T, Kadereit JW. A comparative analysis of genetic variation in rootstocks and scions of old olive trees—a window into the history of olive cultivation practices and past genetic variation. BMC Plant Biol. 2014;14:146.
8. Foxhall L. Olive cultivation in ancient Greece: seeking the ancient economy. Oxford: Oxford University Press; 2007.
9. Zohary D, Hopf M. Domestication of plants in the old world. 2nd ed. Oxford: Oxford Science Publications, Claredon Press; 1994.
10. Bubici G, Cirulli M. Control of *Verticillium* wilt of olive by resistant rootstocks. Plant Soil. 2012;352:363–76.
11. Chatzissavvidis C, Therios I, Antonopoulou C, Dimassi K. Effects of high boron concentration and scion-rootstock combination on growth and nutritional status of olive plants. J Plant Nutr. 2008;31:638–58.
12. Soriano A, Martin M, Piedra A. Grafting olive cv. Cornicabra on rootstocks tolerant to *Verticillium dahliae* reduces their susceptibility. Crop Prot. 2003;22:369–74.
13. Therios I. Olives, vol. 18. Wallingford: CABI; 2009.
14. Goldreich Y. The climate of Israel: observation, research and application. New York: Springer; 2003.
15. Singer A. The Soils of Israel. Berlin: Springer; 2007.
16. Ben-David E, Kerem Z, Zipori I, Weissbein S, Basheer L, Bustan A, Dag A. Optimization of the Abencor system to extract olive oil from irrigated orchards. Eur J Lipid Sci Tech. 2010;112:1158–65.
17. Dag A, Kerem Z, Yogev N, Zipori I, Lavee S, Ben-David E. Influence of time of harvest and maturity index on olive oil yield and quality. Sci Horticulturae. 2011;127:358–66.
18. Ben-Gal A, Dag A, Basheer L, Yermiyahu U, Zipori I, Kerem Z. The influence of bearing cycles on olive oil quality response to irrigation. J Agr Food Chem. 2011;59:11667–75.
19. Blum M, Lensky IM, Nestel D. Estimation of olive grove canopy temperature from MODIS thermal imagery is more accurate than interpolation from meteorological stations. Agr For Meteorol. 2013;176:90–3.
20. Snyder RL. Hand calculating degree days. Agr For Meteorol. 1985;35:353–8.
21. Johnson JB, Omland KS. Model selection in ecology and evolution. Trends Ecol Evol. 2004;19:101–8.
22. R Core Team: R. A language and environment for statistical computing. R Foundation for Statistical Computing. In: Vienna. https://www.R-project.org/; 2016.
23. Mariotti M. Virgin olive oil: definition and standards. In: Peri C, editor. The extra-virgin olive oil handbook. Chichester: Wiley; 2014. p. 11–9.
24. Mariotti M, Peri C. The composition and nutritional properties of extra-virgin olive oil. In: Peri C, editor. The extra-virgin olive oil handbook. Chichester: Wiley; 2014. p. 21–34.
25. Tugendhaft Y, Eppel A, Kerem Z, Barazani O, Ben-Gal A, Kadereit JW, Dag A. Drought tolerance of three olive cultivars alternatively selected for rain fed or intensive cultivation. Sci Horticulturae. 2016;199:158–62.
26. Ben-Gal A, Dag A, Yermiyahu U, Tsipori I, Presnov E, Faingold I, Kerem A. Evaluation of irrigation in a converted, rain fed olive orchard: the transition year. Acta Hortic. 2008;792:99–106.
27. Dag A, Ben-Gal A, Yermiahu U, Basheer L, Yogev N, Kerem Z. The effect of irrigation level and harvest mechanization on virgin olive oil quality in a traditional rain-fed 'Souri' olive orchard converted to irrigation. J Sci Food Agric. 2008;88:1524–8.
28. Barazani O, Keren-Keiserman A, Westberg E, Hanin N, Dag A, Ben-Ari G, Fragman-Sapir O, Tugendhaft Y, Kerem Z, Kadereit JW. Genetic variation of naturally growing olive trees in Israel: from abandoned groves to feral and wild? BMC Plant Biol. 2016;16:261.

6

Microbial diversity in the floral nectar of *Linaria vulgaris* along an urbanization gradient

Jacek Bartlewicz[1]* [iD], Bart Lievens[2], Olivier Honnay[1] and Hans Jacquemyn[1]

Abstract

Background: Microbes are common inhabitants of floral nectar and are capable of influencing plant-pollinator interactions. All studies so far investigated microbial communities in floral nectar in plant populations that were located in natural environments, but nothing is known about these communities in nectar of plants inhabiting urban environments. However, at least some microbes are vectored into floral nectar by pollinators, and because urbanization can have a profound impact on pollinator communities and plant-pollinator interactions, it can be expected that it affects nectar microbes as well. To test this hypothesis, we related microbial diversity in floral nectar to the degree of urbanization in the late-flowering plant *Linaria vulgaris*. Floral nectar was collected from twenty populations along an urbanization gradient and culturable bacteria and yeasts were isolated and identified by partially sequencing the genes coding for small and large ribosome subunits, respectively.

Results: A total of seven yeast and 13 bacterial operational taxonomic units (OTUs) were found at 3 and 1 % sequence dissimilarity cut-offs, respectively. In agreement with previous studies, *Metschnikowia reukaufii* and *M. gruessi* were the main yeast constituents of nectar yeast communities, whereas *Acinetobacter nectaris* and *Rosenbergiella epipactidis* were the most frequently found bacterial species. Microbial incidence was high and did not change along the investigated urbanization gradient. However, microbial communities showed a nested subset structure, indicating that species-poor communities were a subset of species-rich communities.

Conclusions: The level of urbanization was putatively identified as an important driver of nestedness, suggesting that environmental changes related to urbanization may impact microbial communities in floral nectar of plants growing in urban environments.

Keywords: Nectar yeasts, Urbanization, *Metschnikowia*, *Acinetobacter*, Nestedness, Nectar microbial communities, *Linaria vulgaris*

Background

Numerous studies have shown that microbes are common inhabitants of floral nectar [1–8]. Nectar inhabiting microbes (NIMs) have been shown to modify important physicochemical properties of floral nectar, such as sugar and amino acid composition [9, 10], to alter nectar odor [11, 12], and even to increase the temperature of the flower itself [13]. Changes in physicochemical properties of floral nectar can, in turn, alter the attractiveness of a given flower to pollinators, resulting in increased visitation rates and plant reproductive success [14, 15].

Most research so far has shown that microbial diversity in floral nectar is low and often dominated by a limited number of culturable species [4, 7, 8, 16, 17]. However, the few studies that have investigated variation in nectar microbial communities among plant populations have shown that the distribution of NIMs is not uniform, but can vary substantially between populations. For example, bacterial communities in the floral nectar of the summer asphodel (*Asphodelus aestivus*) changed significantly

*Correspondence: bartlewiczjacek@gmail.com
[1] Biology Department, Plant Conservation and Population Biology, KU Leuven, Kasteelpark Arenberg 31, 3001 Heverlee, Belgium
Full list of author information is available at the end of the article

along an aridity gradient [18]. Similarly, microbial communities in the floral nectar of several populations of the spring-flowering forest herb *Pulmonaria officinalis* showed large within-population variation and low among-population similarity, indicating that NIM community assembly may to some extent be context-dependent [5]. These results further suggest that both variation in the local species pool of microbes and in local environmental conditions can shape NIM communities.

All studies investigating the diversity and abundance of NIMs so far have focused on plant populations occurring in natural environments, and virtually nothing is known about the composition of nectar microbial communities in urban environments. Transformation of natural landscapes by urbanization can, however, be expected to have a profound impact on nectar microbial communities. Given that most NIMs are vectored from one flower to the next by insect pollinators, and that nectar yeasts in particular are thought to rely exclusively on them to colonize new flowers [19, 20], impoverishment of pollinator communities can negatively affect colonization rates and subsequent dispersal of NIMs and therefore affect NIM community composition. The typically increasing cover of impervious surfaces in urban environments has, for example, been shown to be correlated with decreased nesting density in *Bombus vosnesenskii* and with decreased abundance of other wild bees [21, 22]. It has also been shown that different bee species respond differently to urbanization, which results in varied pollinator guilds along an urban gradient [23]. Furthermore, plant populations may typically decrease in size and become more spatially isolated as a result of urbanization. The increased spatial isolation and reduced population size of co-flowering plant species in urban environments can impede the exchange of microorganisms between populations, which, in turn, results in a nested species distribution pattern, where only the most common species are present in the most isolated or the smallest habitat patches (see [24] for review of the nestedness concept).

To test the general hypothesis that microbial communities in nectar change along an urbanization gradient, we investigated microbial communities in the floral nectar of the late-flowering herb *Linaria vulgaris* (yellow toadflax), a species that occurs in both urban and rural habitats. After collecting nectar from twenty populations across an urbanization gradient and surveying its microbiota using culture dependent methods and Sanger sequencing, we specifically addressed the following questions:

1. How does urbanization affect microbial incidence in floral nectar of *L. vulgaris*?

2. Does urbanization lead to impoverished NIM communities, with the most heavily urbanized sites having the most impoverished communities?

3. Are impoverished NIM communities subsets of more species-rich communities?

Results and discussion
Yeast diversity and occurence

Following cultivation and isolation a total of 140 yeast isolates was obtained. When the sequences corresponding to these isolates were clustered according to a 3 % dissimilarity cutoff, seven fungal OTUs were identified (Table 1). Following BLAST analysis, these OTUs corresponded to five different validly named species from four families: *Metschnikowiaceae*, *Dothioraceae*, *Sporidiobolales* and *Tremellales*, indicating that *Ascomycota* were more frequently represented than *Basidiomycota*. Most isolates were related to *Metschnikowia gruessi* (101 isolates) and to a smaller degree to *M. reukaufii* (36 isolates), while other species were represented by single isolate (Table 1). Three out of the seven OTUs ($OTU_{0.03}Y2$, $OTU_{0.03}Y3$, and $OTU_{0.03}Y4$) found were identified as *M. reukaufii*. The Chao2 estimator predicted a slightly higher number of OTUs, 12.7, while the ICE estimator predicted 18.61 OTUs (Fig. 1).

Yeasts were found in all studied *L. vulgaris* populations. On average, there were two yeast OTUs present per population (sd: 0.79), with *M. gruessi* present in 95 % of the populations, and *M. reukaufii* present in 70 % of the populations. There was no correlation between yeast incidence and ISI_{500} (Spearman-$\rho = 0.06$, $p = 0.78$) or plant population size (Spearman-$\rho = 0.02$, $p = 0.98$). The size of each population and impervious surface index pertaining to it were not inter-correlated (Spearman-$\rho = -0.32$, $p = 0.15$).

Thus, our results showed that yeast communities in urban environments were mainly dominated by the yeasts *M. reukaufii* and *M. gruessi*. However, unlike in many natural environments [17, 25–27] *M. gruessi* was more abundant than *M. reukaufii*. Assessment of the phenotypic landscape of both yeast species has recently shown that both species displayed a significantly different physiological profile, most likely facilitating co-occurrence of both species [28]. Moreover, comparison of utilization profiles in single vs mixed cultures indicated that *M. reukaufii* generally grows better in sucrose solutions, and that *M. gruessi* grows better in mixed cultures in glucose and fructose solutions [28]. On the other hand, out of the yeast strains isolated from 19 different nectars, the strains isolated from the nectar of *Plantaginaceae*, the family to which *L. vulgaris* belongs, exhibited the most dissimilar phenotypes for *M. gruessi* and *M. reukaufii*. This might mean that the nectar of Plantaginaceae

Table 1 Yeast operational taxonomic units (OTUs) identified in the nectar of 20 *L. vulgaris* populations sampled along an urbanization gradient

OTU 3 %	118	Phylum	Family	Closest match with GenBank entries	Accession number	Score	E value	Length	Identity (%)
OTU$_{0.03}$Y1	22	Ascomycota	Metschnikowiaceae	*Metschnikowia gruessi*	JX067745	665.914	0	360	100
OTU$_{0.03}$Y2	10	Ascomycota	Metschnikowiaceae	*Metschnikowia reukaufii*	KM281795	682.534	0	369	100
OTU$_{0.03}$Y3	4	Ascomycota	Metschnikowiaceae	*Metschnikowia reukaufii*	FJ455114	675.147	0	365	100
OTU$_{0.03}$Y4	1	Ascomycota	Metschnikowiaceae	*Metschnikowia reukaufii*	JN642530	584.662	$3.18e^{-163}$	332	98.5
OTU$_{0.03}$Y5	1	Ascomycota	Dothioraceae	*Aureobasidium pullulans*	KP710217	774.886	0	419	100
OTU$_{0.03}$Y6	1	Basidiomycota	Sporidiobolales	*Sporobolomyces roseus*	AM160644	250.48	$1.32e^{-62}$	135	100
OTU$_{0.03}$Y7	1	Basidiomycota	Tremellales	*Cryptococcus aureus*	JN004200	1059.25	0	583	99.5

The sequences were grouped into OTUs based on 97 % identity at the large ribosomal subunit gene. The BLAST search was conducted in June 2015, excluding uncultured/environmental samples. Only the closest matches are reported

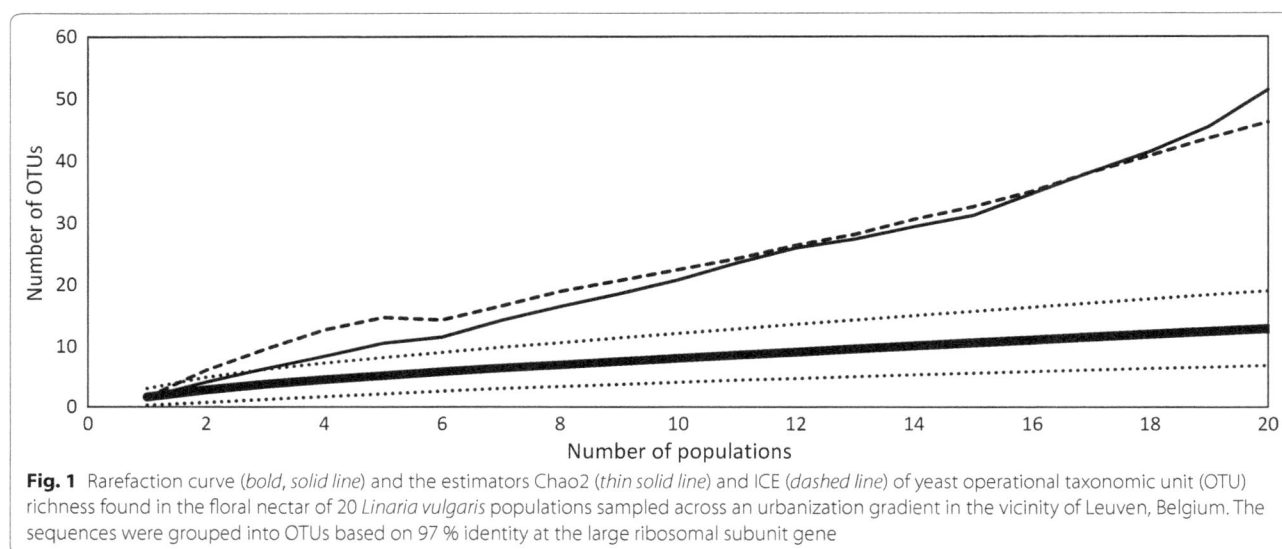

Fig. 1 Rarefaction curve (*bold, solid line*) and the estimators Chao2 (*thin solid line*) and ICE (*dashed line*) of yeast operational taxonomic unit (OTU) richness found in the floral nectar of 20 *Linaria vulgaris* populations sampled across an urbanization gradient in the vicinity of Leuven, Belgium. The sequences were grouped into OTUs based on 97 % identity at the large ribosomal subunit gene

exhibits strong differential constraints on yeast species growth, with which *M. gruessi* copes better in the specific case of *L. vulgaris*.

Bacterial diversity and occurence

In total, 52 bacterial isolates were obtained after cultivation. Thirteen bacterial OTUs were identified when clustered according to 1 % dissimilarity cutoff (Table 2). Based on BLAST results, these OTUs corresponded to 13 validly named species. The ICE and Chao2 estimators predicted a much higher number of 46.37 and 51.48 OTUs, respectively (Fig. 2). Similarly, the rarefaction curves showed that additional sampling would yield a higher number of OTUs, and this effect was more pronounced than in the case of yeast, probably due to higher number of bacterial singletons. The rarefaction curves for bacteria showed a similar level of saturation, however, to

what has been found in earlier culture-dependent studies (i.e. 4, 5, 20). *Acinetobacter nectaris* comprised over 50 % (29 isolates) of bacterial isolates, with *Rosenbergiella epipactidis* coming second (10 isolates). The remaining OTUs were represented by 1–3 isolates, coming mostly from the *Enterobacteriaceae*, but also from *Pseudomonadaceae*, *Microbacteriaceae*, *Bacillaceae* and *Sphingomonadaceae* (Table 2).

Bacteria were present in all but four of the studied populations, with an average of 1.65 OTUs per population (sd: 1.26). *A. nectaris* was the most frequently found species and occurred in 60 % of the populations, while *R. epipactidis* was present in 35 % of the populations. Bacterial incidence was also not correlated to ISI_{500} (Spearman-$\rho = -0.16$, $p = 0.49$), but it was significantly correlated to plant population size (Spearman-$\rho = 0.45$, $p = 0.04$) (Table 3).

Table 2 Bacterial operational taxonomic units (OTUs) identified in the nectar of 20 *L. vulgaris* populations sampled along an urbanization gradient

OTU 1 %	Number of isolates	Phylum	Family	Closest match with GenBank entries	Accession number	Score	E value	Length	Identity (%)
$OTU_{0.01}B1$	1	Actinobacteria	Microbacteriaceae	*Microbacterium testaceum*	KP642087	1796.06	0	972	100
$OTU_{0.01}B2$	1	Actinobacteria	Microbacteriaceae	*Rathayibacter festucae*	NR_042574	1768.36	0	957	100
$OTU_{0.01}B3$	1	Firmicutes	Bacillaceae	*Lysinibacillus odysseyi*	NR_113881	1921.64	0	1043	100
$OTU_{0.01}B4$	1	Proteobacteria	Sphingomonadaceae	*Sphingomonas faeni*	KM891564	1773.9	0	960	100
$OTU_{0.01}B5$	1	Proteobacteria	Enterobacteriaceae	*Pantoea ananatis*	KC139412	1919.79	0	1048	99.7
$OTU_{0.01}B6$	1	Proteobacteria	Enterobacteriaceae	*Pantoea vagans*	KP099965	1757.28	0	957	99.8
$OTU_{0.01}B7$	1	Proteobacteria	Enterobacteriaceae	*Pectobacterium carotovorum*	GU129979	1628.02	0	885	99.9
$OTU_{0.01}B8$	1	Proteobacteria	Enterobacteriaceae	*Ewingella americana*	KM891553	1842.23	0	997	100
$OTU_{0.01}B9$	1	Proteobacteria	Pseudomonadaceae	*Pseudomonas viridiflava*	NR_117825	1899.48	0	1028	100
$OTU_{0.01}B10$	3	Proteobacteria	Pseudomonadaceae	*Pseudomonas moraviensis*	KP165022	1892.09	0	1024	100
$OTU_{0.01}B11$	10	Proteobacteria	Enterobacteriaceae	*Rosenbergiella epipactidis*	NR_126303	1816.38	0	989	99.8
$OTU_{0.01}B12$	1	Proteobacteria	Moraxellaceae	*Acinetobacter boissieri*	NR_118409	1855.16	0	1004	100
$OTU_{0.01}B13$	29	Proteobacteria	Moraxellaceae	*Acinetobacter nectaris*	JQ771134	1879.16	0	1023	99.8

The sequences were grouped into OTUs based on 99 % identity at the small ribosomal subunit gene. The BLAST search was conducted in June 2015, excluding uncultured/environmental samples. Only the closest matches are reported

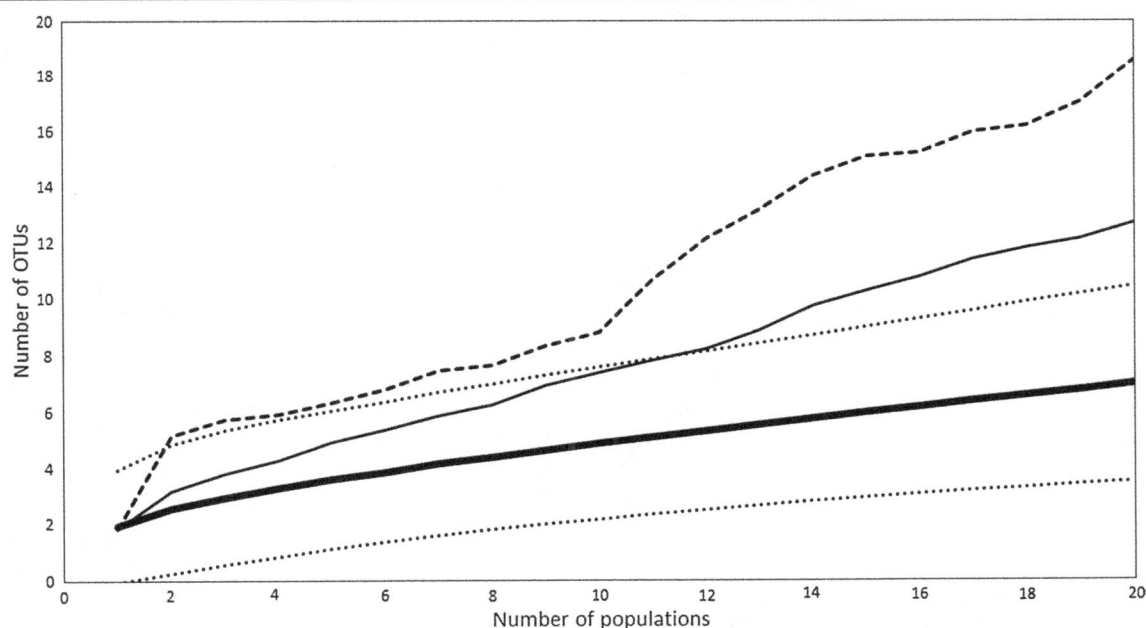

Fig. 2 Rarefaction curve (*bold, solid line*) and the estimators Chao2 (*thin solid line*) and ICE (*dashed line*) of bacterial operational taxonomic unit (OTU) richness found in the floral nectar of 20 *Linaria vulgaris* populations sampled across an urbanization gradient in the vicinity of Leuven, Belgium. The sequences were grouped into OTUs based on 99 % identity at the small ribosomal subunit gene

Overall, the incidence of bacteria in *L. vulgaris* (which were present in 48 % of the nectar samples) was higher than the previously reported average of 19.9 % for Mediterranean species, the 6.5 % for *Linaria* *viscosa* in Spain [29], and closer to the 53.5 % reported from South African plant species [4]. Interestingly, we found a high incidence of the recently described bacterial species *A. nectaris* [29]. This bacterium was

Table 3 Properties of the 20 sampled *Linaria vulgaris* populations

Popula-tion	Longitude	Latitude	ISI_{500} (%)	Population size	Nested-ness rank	Yeast $OTUs_{0.03}$ richness	Total yeast incidence (%)	Bacterial $OTUs_{0.01}$ richness	Total bacterial incidence (%)	Total microbial richness
A	4.709365	50.864275	36.7	400	3	3	100	3	100	6
B	4.704492	50.863418	33.84	450	20	2	100	0	0	2
C	4.700154	50.86323	42.35	100	14	2	100	1	40	3
D	4.676421	50.885377	44.85	15	18	2	20	0	0	2
E	4.715524	50.867437	27.44	60	13	2	60	0	0	2
F	4.713976	50.85837	18.9	520	11	2	80	2	60	4
G	4.72284	50.850707	44.72	420	19	2	100	0	0	2
H	4.7281	50.843644	39.17	300	15	2	40	1	80	3
I	4.723781	50.852871	32.26	70	12	1	60	2	80	3
J	4.685015	50.888567	31.16	150	4	1	40	4	80	5
K	4.725668	50.858904	19.26	200	17	1	100	1	40	2
L	4.727276	50.861677	22.98	83	6	3	80	2	40	5
M	4.725735	50.860678	19.78	500	2	3	80	2	100	5
N	4.801452	50.822292	7.06	500	7	2	60	3	40	5
O	4.803312	50.823027	12.99	500	16	1	40	1	20	2
P	4.784239	50.826077	7.81	170	9	2	60	1	40	3
Q	4.79446	50.825404	5.98	400	10	1	40	2	60	3
R	4.790337	50.825352	6.65	1250	1	3	80	3	80	6
S	4.794894	50.823779	6.99	160	5	1	100	4	80	5
T	4.805983	50.852606	5.36	50	8	3	80	1	20	4

The approximate coordinates of each population are given in decimal degrees. The degree or urbanization of the surroundings of each population was assessed using ISI_{500}, the impervious surface index within a radius of 500 m. For each of the sampled populations, the bacterial and yeast species richness (number of OTUs) and incidence (frequency of occurrence) are presented

found originally in the nectar of Mediterranean plant communities [4]. However, our results indicate that it is not solely restricted to the Mediterranean area, but that it can also be found frequently in the floral nectar of plants growing in North-Western Europe. *A. nectaris* was previously isolated in Belgium in some orchid species, but by no means was it a nectar dominating bacterial species [6]. In another study on nectar microbial communities of *P. officinalis* in Belgium, it was not found either [5]. In Israel, Fridman et al. [30] found bacteria belonging to the *Acinetobacter* genus in at least half of the samples from each of the three Mediterranean plant species they studied. *A. nectaris* was later detected in nectar of *Asphodelus aestivus* in the same country [18]. This suggests that despite its broad distributional range, it might be plant species specific, either due to the nectar properties it requires, or due to its introduction route, which is unknown so far. It is noteworthy, however, that of the two recently described nectar specialist *Acinetobacter* taxa found in this study, *A. nectaris* was represented by 29 isolates, but *A. boissieri*, only by one isolate, which parallels the situation in the *Metschnikowia* genus, when one or the other of

the two nectar specialist yeasts is usually dominant in any given plant species.

The second most abundant bacterium inhabiting the nectar of *L. vulgaris* was *R. epipactidis*, also a species previously isolated from nectar [31, 32]. Perhaps its lower abundance could be explained by its less effective dissemination method: it is speculated that thrips serve as its vectors [18]. Interestingly, although bacterial-yeast co-occurrence was common at the population level, it was less so at the plant level. This could be due to priority effects and that nectar colonized by yeasts could be less accessible to bacteria. However to confirm this, further co-occurrence studies on the nectary level are needed, as well as competition experiments [33].

Spatial distribution of yeasts and bacteria

Yeast OTU richness, measured as the number of OTUs per plant population, was not related to urbanization level (Spearman-$\rho = 0$, $p = 1.00$) nor to population size (Spearman-$\rho = 0.005$, $p = 0.98$) (Fig. 3). The number of bacterial OTUs in each *L. vulgaris* population was also not related to plant population size (Spearman-$\rho = 0.30$, $p = 0.19$), but was marginally negatively correlated to the

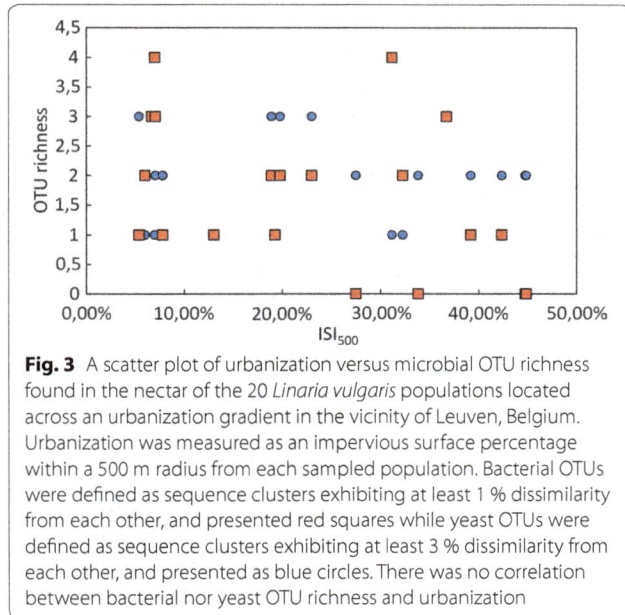

Fig. 3 A scatter plot of urbanization versus microbial OTU richness found in the nectar of the 20 *Linaria vulgaris* populations located across an urbanization gradient in the vicinity of Leuven, Belgium. Urbanization was measured as an impervious surface percentage within a 500 m radius from each sampled population. Bacterial OTUs were defined as sequence clusters exhibiting at least 1 % dissimilarity from each other, and presented red squares while yeast OTUs were defined as sequence clusters exhibiting at least 3 % dissimilarity from each other, and presented as blue circles. There was no correlation between bacterial nor yeast OTU richness and urbanization

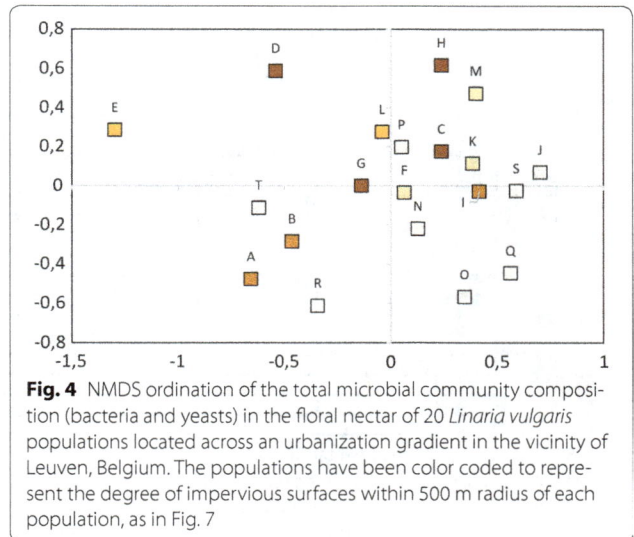

Fig. 4 NMDS ordination of the total microbial community composition (bacteria and yeasts) in the floral nectar of 20 *Linaria vulgaris* populations located across an urbanization gradient in the vicinity of Leuven, Belgium. The populations have been color coded to represent the degree of impervious surfaces within 500 m radius of each population, as in Fig. 7

ISI_{500} pertaining to this population (Spearman $\rho = -0.43$, $p = 0.056$) (Fig. 3). When bacterial and yeast OTU numbers were combined, the correlation with ISI_{500} was not significant (Spearman-$\rho = -0.36$, $p = 0.11$).

Co-occurrence of bacteria and yeasts was common at the population level, with all populations harboring yeasts, and only four lacking bacteria. A smaller degree of co-occurrence was observed at the plant level, where yeast and bacteria were found together in 33 % of the samples. A Mantel test showed no correlation between geographic distance and community dissimilarity for neither bacterial and yeast community composition separately (r = -0.08, $p = 0.83$ and r = 0.03, $p = 0.34$, respectively), nor when they were combined (r = -0.08, $p = 0.82$), NMDS also revealed no apparent clustering (Fig. 4). NMDS axis 1 was not correlated to ISI_{500} (Spearman-$\rho = 0.2$, $p = 0.39$), but NMDS axis 2 was (Spearman-$\rho = 0.47$, $p = 0.03$). Finally, the NIM communities showed significant nestedness ($N = 0.83$, $p < 0.01$; NODF = 42.19, $p < 0.01$), implying that species-poor communities were a subset of species-rich communities (Fig. 5). The nestedness rank was significantly correlated to ISI_{500} (Spearman-$\rho = 0.49$, $p = 0.02$), but not to plant population size (Spearman-$\rho = -0.18$, $p = 0.42$) (Fig. 6) . The nestedness rank correlation with ISI_{500} has to be treated with caution, since the rarefaction curves were not completely saturated, especially for bacteria (Figs. 1, 2). Additional OTUs would likely appear if the number of samples increased.

In contrast to our hypothesis, we did not find any clear changes in microbial incidence per se in the nectar of *L.*

vulgaris across the studied urbanization gradient. OTU richness of urban and non-urban nectar communities was similar for yeasts, whereas urban bacterial nectar communities were marginally poorer than non-urban ones. Similarity in community composition did also not decrease with geographic distance, and no apparent clustering was detected, suggesting that the species pools from which NIMs colonize their nectary microhabitats are homogeneously throughout the area. These results further suggest that pollinators were still present in urban environments in sufficient numbers to effectively disperse bacteria and yeast. Moreover, the most urbanized plant populations we studied were still located in sites with only 45 % of impervious surface coverage within 500 m. It is therefore not impossible that the effect of such moderate urbanization on pollinator guilds is not drastic, and that microbial frequency would decrease sharply in more extremely urbanized areas.

On the other hand, the studied communities showed significant nestedness, indicating that OTU-poor communities were a subset of OTU-rich communities, and that generalist species occurred everywhere and specialist species were restricted to particular populations. The significant albeit weak correlation between urbanization and nestedness rank, and the lack of such correlation with plant population size may imply that the urban communities were a subset of the rural communities. Because many of the *L. vulgaris* populations located in sites with very low impervious surface coverage within 500 m were geographically relatively close to each other (Fig. 7), and because rarefaction analyses revealed potentially incomplete sampling, this correlation could be a result of some unknown spatial effect unrelated to urbanization.

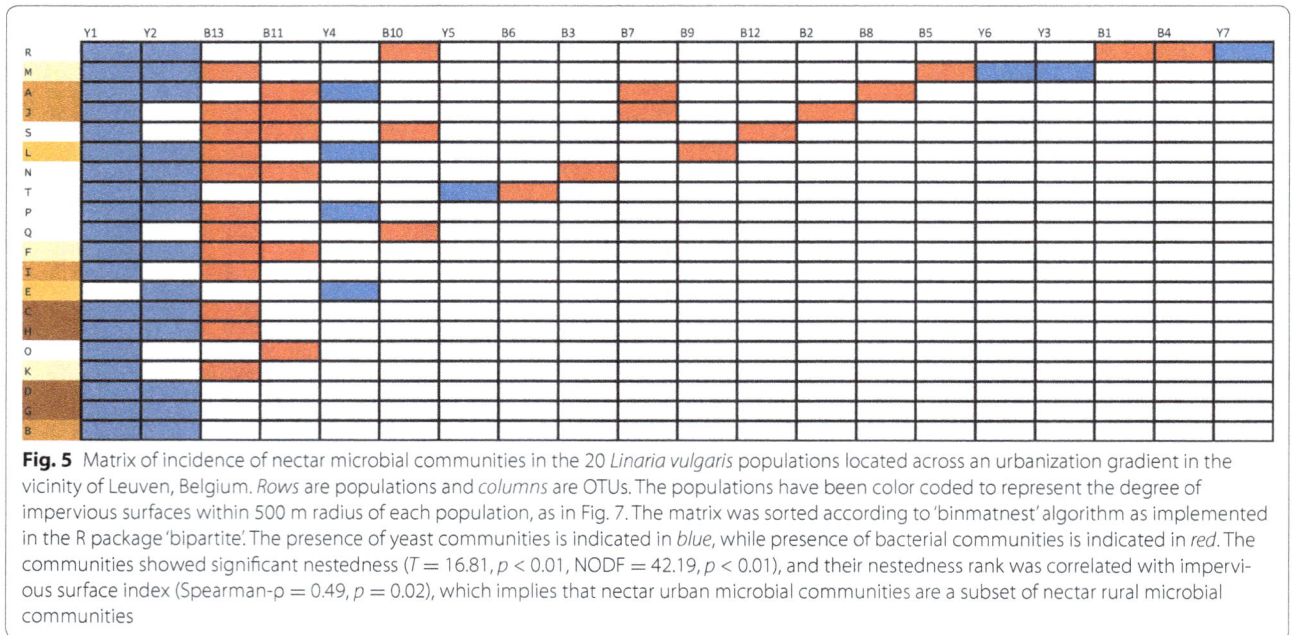

Fig. 5 Matrix of incidence of nectar microbial communities in the 20 *Linaria vulgaris* populations located across an urbanization gradient in the vicinity of Leuven, Belgium. *Rows* are populations and *columns* are OTUs. The populations have been color coded to represent the degree of impervious surfaces within 500 m radius of each population, as in Fig. 7. The matrix was sorted according to 'binmatnest' algorithm as implemented in the R package 'bipartite'. The presence of yeast communities is indicated in *blue*, while presence of bacterial communities is indicated in *red*. The communities showed significant nestedness ($T = 16.81$, $p < 0.01$, NODF $= 42.19$, $p < 0.01$), and their nestedness rank was correlated with impervious surface index (Spearman-$\rho = 0.49$, $p = 0.02$), which implies that nectar urban microbial communities are a subset of nectar rural microbial communities

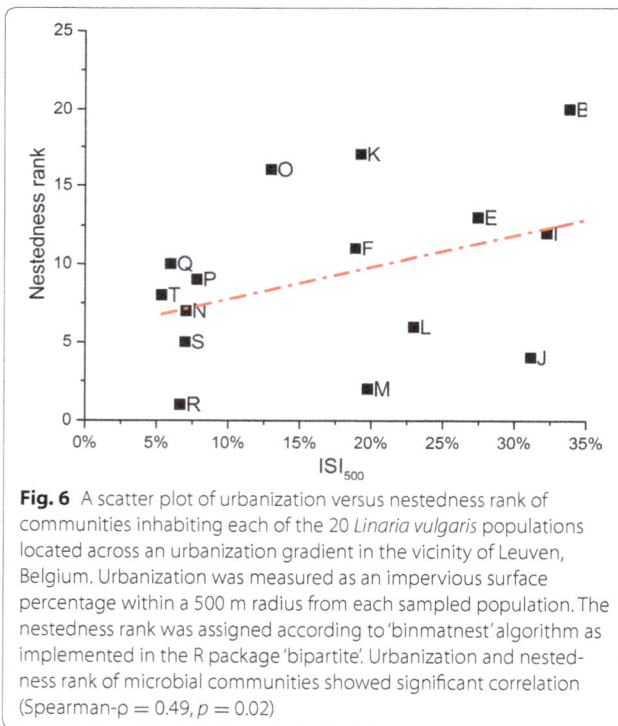

Fig. 6 A scatter plot of urbanization versus nestedness rank of communities inhabiting each of the 20 *Linaria vulgaris* populations located across an urbanization gradient in the vicinity of Leuven, Belgium. Urbanization was measured as an impervious surface percentage within a 500 m radius from each sampled population. The nestedness rank was assigned according to 'binmatnest' algorithm as implemented in the R package 'bipartite'. Urbanization and nestedness rank of microbial communities showed significant correlation (Spearman-$\rho = 0.49$, $p = 0.02$)

However, such correlation could also suggest that populations in urban environments may be less frequently visited by pollinators due the higher isolation of *Linaria* populations in urban environments or the lack of other co-flowering plant species that attract pollinators,

which in turn limits the exchange of less prevalent species. Alternatively, these results may suggest that nectar conditions in urban environments are less suitable for microbial growth than in natural environments and that only generalist species are capable of growing in the floral nectar of plants growing in urban environments. This result is consistent with the hypothesis that nectar yeasts should be capable of exploiting a wide range of nectar microhabitats, as they rank as two first entries in the nestedness matrix (Pozo et al. [17]). Interestingly, most bacteria, including nectar specialists, rank further in the nestedness matrix, which suggests that they are less adapted to different nectar types than yeasts.

Conclusion

Overall, our results show that nectar yeasts are able to grow well in anthropogenically transformed environments. They grow equally well in autumn-flowering species *L. vulgaris* as in species that bloom earlier in the season on which previous studies were focused: *Helleborus foetidus*, *Pulmonaria officinalis*, *Delphinium nuttalianum* or *Asphodelus aestivus*. Moreover, our results indicate that not only nectar yeast communities, but also nectar bacterial communities are species poor and dominated by two specialist genera that seem to have a broad geographic range. Although urbanization did not seem to affect overall microbial richness nor incidence, microbial communities inhabiting the nectar of urban populations of *L. vulgaris* were nested in the communities inhabiting the nectar of its rural populations, although this result

Fig. 7 Map showing the locations of the 20 *Linaria vulgaris* populations studied in the vicinity of Leuven, Belgium. These populations have been color coded to represent the degree of impervious surfaces within 500 m radius of each population. From these populations, five shoots were chosen at random, and a single inflorescence was taken from each plant. Nectar was collected from four flowers belonging to each inflorescence

has to be treated with caution due to sampling limitations. Nevertheless, this might mean that urbanization may increase the isolation of plant populations harboring nectar microhabitats, resulting in reduced prevalence of rare species. Further research in this field will need to take into account two aspects that were outside of the scope of this study. Firstly, visitation rates, community composition and abundance of pollinators will have to be established more precisely in the chosen study system to give reliable estimations of their influence on nectar communities. Secondly, it is known that but a fraction of microbes can be successfully cultivated [34]. This is also reflected by the fact that rarefaction curves in studies relying on culture dependent methods are frequently not saturated, which decreases the level of confidence one can put into conclusions drawn from OTU richness and diversity. Perhaps more pronounced effects of urbanization could be detected with a more sensitive technique to measure their community composition, for instance through next-generation sequencing.

Methods

Study species and study area

Linaria vulgaris is a self-incompatible, perennial herb that is characterized by an extensive root system. It produces zygomorphic, bright yellow flowers with an orange nectar guide and a long spur between summer and late autumn [35]. The species is mainly pollinated by long-tongued bumblebees, such as *B. hortorum* and *B. pascuorum* [36, 37]. Short-tongued species, such as *B. terrestris*, behave like nectar robbers and access nectar illegitimately by biting a hole in the spur of a *Linaria* flower [37]. The main sugar found in *L. vulgaris* nectar is sucrose, but glucose, fructose and even trace amounts of raffinose have also been found [35].

Twenty *L. vulgaris* populations that grew in locations exhibiting different levels of urbanization were examined (Table 3). No sampling permissions were required to access the populations and sample them. The study area consisted of the city of Leuven (situated about 25 km east of Brussels, Belgium) and its immediate suburbs that gradually shift into areas with a more rural character. To assess the level of urbanization of the locations where each *L. vulgaris* population grew, for every of these locations we calculated an impervious surface index within a radius of 500 m (ISI_{500}), as described previously [38] to serve as a proxy for urbanization (Fig. 7). In some cases, the 500 m radii overlapped. Some suburban areas also showed considerable amount of impervious surfaces within the 500 m radius, and no suitable populations were found in a 3 km stretch between populations P and H. Populations T, D and J were located outside of a linear transect encompassing all of the other populations. Given these deviations, our design was an approximation of sampling along an idealized urban–rural gradient. For each population, we also assessed its size by counting the number of inflorescences. Although the number of inflorescences does not necessarily reflect the total number of genets in the population, it does reflect the number of flowers that is perceived by pollinators.

Sampling and cultivation of microorganisms

From each population, five plants were randomly chosen, and a single inflorescence per plant was brought to the laboratory for further processing on the same day. Given that NIMs are often patchily distributed among flowers within an inflorescence [39], nectar from four flowers from one inflorescence was pooled. Within 8 h of collection, nectar was harvested. To this end, the spur was first cut with a sterile scalpel in order to avoid pollen contamination, and then 0.5 µl nectar was extracted using a micropipette. A total of 2 µl, pooled from four flowers, was diluted in 98 µl of water and kept at 4 °C. Within 24 h, 2 µl of the diluted nectar was plated on tryptone soy agar (TSA) supplemented with 0.01 % chloramphenicol or 0.01 % actidione in order to suppress the growth of bacteria and yeasts, respectively. The plates were incubated at 25 °C for 7 days. Subsequently, one colony of each morphotype per plate was isolated and purified on TSA. It has been shown in previous studies on nectar microbiota that identical morphotypes generally correspond to identical species [5, 6]. After purification, each isolate was again transferred to TSA after which it was stored at −80 °C in 96 well plates containing 40 % glycerol.

DNA extraction

Genomic DNA was extracted according to the procedure described by Pozo et al. [17]. Subsequently, partial ribosomal RNA genes were amplified using the primer pairs 27F/1492R and NL1/NL4 for bacteria and yeasts, respectively [4, 40]. When amplification failed using 27F/1492R, the forward primer 63F [41] was used instead of 27F. PCR amplification was performed using a Bio-Rad T100 thermal cycler in a reaction volume of 20 µl containing 0.5 µM of each primer, 0.25 mM of each dNTP, 1.25 units Takara Taq DNA polymerase, 1 × Takara Taq PCR buffer (Clontech Laboratories, Palo Alto, CA, USA), and 5 ng genomic DNA (as measured by a Nanodrop spectrophotometer). The reaction mixture was initially denatured at 94 °C for 2 min, followed by 35 cycles of 45 s at 94 °C, 45 s at 55 °C (yeasts) or 59 °C (bacteria), and 45 s at 72 °C, with a final extension at 72 °C for 10 min.

Subsequently, obtained amplicons were sequenced by Macrogen Inc. using the same reverse primers as those used for amplification.

Data analysis

Obtained sequences were individually trimmed for quality, using a minimum Phred score of 20, and, in cases of ambiguous base calls, manually edited based on the obtained electropherograms. Subsequently, each sequence was assigned a taxonomic identity based on BLAST [42] results using the GenBank nucleotide (nt) database [43], excluding environmental/uncultured samples. Next, the bacterial and yeast sequences were separately aligned using the MUSCLE algorithm implemented in Geneious 7R. These alignments have been used to create genetic distance matrices, which served as a basis to assign the sequences to OTUs using Mothur v 1.32.1 at 1 % and 3 % dissimilarity cutoffs for bacteria and yeasts, respectively. For each OTU, a representative sequence as determined by Mothur has been deposited in GenBank (accession numbers: KT347518–KT347530 for bacteria and KU900119–KU900125 for yeast).

Using EstimateS v 9.1.0, rarefaction analyses were conducted to assess our sampling effort. EstimateS was also used to calculate the ICE and Chao2 estimators of species richness [44]. For each *L. vulgaris* population, OTU richness and incidence were calculated. Incidence in a given population was expressed as percentage of samples that contained microorganisms. OTU richness was defined as the total number of OTUs per given population. Both of these variables were related to the urbanization measure ISI_{500} and plant population size using a Spearman rank correlation.

To visualize differences in microbial community composition among populations, we applied the non-metric multidimensional scaling (NMDS) ordination technique using the vegan package [45] in R software. As distance measure, we used the Bray-Curtis coefficient. EstimateS software was used to create a dissimilarity in community composition matrix, using the Jaccard estimator. A Mantel test using the R software was then conducted to relate the dissimilarities in community composition with geographical distances between populations. The significance of the Mantel test was assessed by performing 9999 permutations.

Lastly, we tested the hypothesis that NIM communities were significantly nested, i.e. that species-poor communities were a subset of the more rich ones and that rare OTUs were only present in the most OTU rich communities [46–48]. Two different measures to estimate the degree of nestedness were applied. We first calculated $N = (100-T)/100$, where T is the matrix temperature, a measure of matrix disorder that varies between 0°

(perfectly nested) and 100° (perfectly non-nested). Values of N close to one thus indicate a high degree of nestedness. However, because T may be dependent on the size and shape of the species presence matrix, we also calculated a second nestedness measure, based on overlap and decreasing fill (NODF), correcting for these flaws [48]. Conversely to T, high values of NODF signify a nested matrix structure.

Two different null models implemented in ANINHADO were used to test the significance of nestedness [47]. In the first null model, each cell in the interaction matrix has the same probability of being occupied. This null model is very general and does not take into account the fact that the number of connections per species may vary substantially. A more conservative null model would therefore be a model in which the probability of drawing an interaction is proportional to the degree of specialization [49]. In this null model, the probability of each cell being occupied is the average of the probabilities of occupancy of its row and column [48]. Finally, to identify the potential drivers behind the nestedness of our dataset, a nestedness rank was assigned to each sampling site according to the 'binmatnest' algorithm in the R package *bipartite*. Afterwards, plant population size and ISI_{500} was related to nestedness rank using the Spearman rank correlation.

Abbreviations
BLAST: basic local alignment search tool; ISI_{500}: impervious surface index within 500 m radius; OTU: operational taxonomic unit; NIMs: nectar inhabiting microbes; NMDS: non metric multidimensional scaling; NODF: nestedness metric based on overlap and decreasing fill; PCR: polymerase chain reaction; TSA: tryptone soya agar.

Authors' contributions
Conceived and designed the experiments: JB, BL, OH, HJ. Conducted the experiments: JB. Performed the analysis: JB. Wrote the paper: JB, BL, OH, HJ. All authors read and approved the final manuscript.

Author details
[1] Biology Department, Plant Conservation and Population Biology, KU Leuven, Kasteelpark Arenberg 31, 3001 Heverlee, Belgium. [2] Laboratory for Process Microbial Ecology and Bioinspirational Management (PME and BIM), Department of Microbial and Molecular Systems (M2S), KU Leuven, Campus De Nayer, Fortsesteenweg 30A, 2860 Sint-Katelijne Waver, Belgium.

Acknowledgements
The authors would like to thank Dr. Maria Pozo, the Editor and two Anonymous reviewers whose comments greatly improved this manuscript. The authors would also like to thank Belgian science policy office for providing funding necessary to carry out this investigation within the SPatial and Environmental determinants of Eco-Evolutionary dynamics (SPEEDY) project.

Competing interests
The authors declare that they have no competing interests.

References
1. De Vega C, Herrera CM, Johnson SD. Yeasts in floral nectar of some South African plants: quantification and associations with pollinator type and sugar concentration. S Afr J Bot. 2009;75:798–806.
2. Herrera CM, De Vega C, Canto A, Pozo M. Yeasts in floral nectar: a quantitative survey. Ann Botany. 2009;103:1415–23.
3. Golonka AM, Vilgalys R. Nectar inhabiting yeasts in Virginian populations of *Silene latifolia* (*Caryophyllaceae*) and coflowering species. Am Mid Nat. 2013;169:235–58.
4. Alvarez-Pérez S, Herrera CM, De Vega C. Zooming-in on floral nectar: a first exploration of nectar-associated bacteria in wild plant communities. FEMS Microbiol Ecol. 2012;80:591–602.
5. Jacquemyn H, Lenaerts M, Brys R, Willems K, Honnay O, Lievens B. Among-population variation in microbial community structure in the floral nectar of the bee-pollinated forest herb *Pulmonaria officinalis* L. PLoS ONE. 2013;8:e56917.
6. Jacquemyn H, Lenaerts M, Tyteca D, Lievens B. Microbial diversity in the floral nectar of seven *Epipactis* (Orchidaceae) species. Microbiol Open. 2013;2:644–58.
7. Lievens B, Hallsworth JE, Pozo MI, Belgacem ZB, Stevenson A, Willems KA, Jacquemyn H. Microbiology of sugar-rich environments: diversity, ecology and system constraints. Environ Microbiol. 2015;17:278–98.
8. Pozo MI, Lievens B, Jacquemyn H. Impact of microorganisms on nectar chemistry, pollinator attraction and plant fitness. In: Peck RL, editor. Nectar: production, chemical composition and benefits to animals and plants. 1st ed. New York: Nova Publishers; 2015. p. 1–45.
9. Peay KG, Belisle M, Fukami T. Phylogenetic relatedness predicts priority effects in nectar yeast communities. Proc Biol Sci. 2012;279:749–58.
10. Herrera CM, Garcia IM, Perez R. Invisible floral larcenies: microbial communities degrade floral nectar of bumble bee-pollinated plants. Ecology. 2008;89:2369–76.
11. Goodrich KR, Zjhra ML, Ley CA, Raguso RA. When flowers smell fermented: the chemistry and ontogeny of yeasty floral scent in pawpaw (*Asimina triloba* : Annonaceae). Int J Plant Sci. 2006;167:33–46.
12. Pozo MI, De Vega C, Canto A, Herrera CM. Presence of yeasts in floral nectar is consistent with the hypothesis of microbial-mediated signaling in plant-pollinator interactions. Plant Signal Behav. 2009;4:1102–4.
13. Herrera CM, Pozo MI. Nectar yeasts warm the flowers of a winter-blooming plant. Proc R Soc Lond. 2010;277:1827–34.
14. Schaeffer RN, Irwin RE. Yeasts in nectar enhance male fitness in a montane perennial herb. Ecology. 2014;95:1792–8.
15. Vannette RL, Gauthier M-PL, Fukami T. Nectar bacteria, but not yeast, weaken a plant–pollinator mutualism. Proc Biol Sci. 2013;280:20122601. doi:10.1098/rspb.2012.2601.
16. Herrera CM, Canto A, Pozo M, Bazaga P. Inhospitable sweetnes: nectar filtering of pollinator borne inocula leads to impoverished, phylogenetically clustered yeast communities. Proc R Soc Lond. 2010;277:747–54.
17. Pozo MI, Herrera CM, Bazaga P. Species richness of yeast communities in floral nectar of southern Spanish plants. Microb Ecol. 2011;61:82–91.
18. Samuni-Blank M, Izhaki I, Laviad S, Bar-Massada A, Gerchman Y, Halpern M. The role of abiotic environmental conditions and herbivory in shaping bacterial community composition in floral nectar. PLoS ONE. 2014;9:e99107.
19. Lachance M-A, Starmer WT, Rosa CA, Bowles JM, Barke J, Stuart F, Janzen DH. Biogeography of the yeasts of ephemeral flowers and their insects. FEMS Yeast Res. 2001;1:1–8. doi:10.1111/j.1567-1364.2001. tb00007.x.
20. Aizenberg-Gershtein Y, Izhaki I, Halpern M. Do honeybees shape the bacterial community composition in floral nectar? PLoS ONE. 2013;8:e67556. doi:10.1371/journal.pone.0067556.
21. Jha S, Kremen C. Resource diversity and landscape-level homogeneity drive native bee foraging. Proc Natl Acad Sci USA. 2013;110:555–8.
22. Fortel L, Henry M, Guilbaud L, Guirao AL, Kuhlmann M, Mouret H, Rollin O, Vaissiere BE. Decreasing abundance, increasing diversity and changing structure of the wild bee community (Hymenoptera: Anthophila) along an urbanization gradient. PLoS ONE. 2014;9:e104679.
23. Banaszak-Cibicka W, Żmihorski M. Wild bees along an urban gradient: winners and losers. J Insect Conserv. 2012;16:331–43. doi:10.1007/ s10841-011-9419-2.

24. Ulrich W, Almeida-Neto M, Gotelli NJ. A consumer's guide to nestedness analysis. Oikos. 2009;118:3–17.

25. Giménez-Jurado G. *Metschnikowia gruessi* sp. nov., the teleomorph of *Nectaromyces reukaufii* but not a *Candida reukaufii*. Syst Appl Microbiol. 1992;15:432–8.

26. Brysch-Herzberg M. Ecology of yeasts in plant–bumblebee mutualism in Central Europe. FEMS Microbiol Ecol. 2004;50:87–100.

27. Álvarez-Pérez S, Herrera C. Composition, richness and nonrandom assembly of culturable bacterial-microfungal communities in floral nectar of Mediterranean plants. FEMS Microbiol Ecol. 2013;83:685–99.

28. Pozo MI, Herrera CM, Van den Einde W, Verstrepen K, Lievens B, Jacquemyn H. The impact of nectar chemical features on phenotypic variation in two related nectar yeasts. FEMS Microbiol Ecol. 2015;91:fiv055.

29. Álvarez-Pérez S, Lievens B, Jacquemyn H, Herrera C. *Acinetobacter nectaris* sp. nov. and *Acinetobacter boissieri* sp. nov., two novel bacterial species isolated from floral nectar of wild Mediterranean insect-pollinated plants. Int J Syst Evol Microbiol. 2013;63:1532–9.

30. Fridman S, Izhaki I, Gercham Y, Halpern M. Bacterial communities in floral nectar. Environ Microbiol Rep. 2012;4:97–104.

31. Halpern M, Fridman S, Atamna-Ismaeel N, Izhaki I. *Rosenbergiella nectarea* gen. nov., sp. nov., in the family *Enterobacteriaceae*, isolated from floral nectar. Int J Syst Evol Microbiol. 2013;63:4259–65.

32. Lenaerts M, Alvarez-Perez S, De Vega C, Van Assche A, Johnson SD, Willems KA, Herrera CM, Jacquemyn H, Lievens B. *Rosenbergiella australoborealis* sp. nov., *Rosenbergiella collisarenosi* sp. nov. and *Rosenbergiella epipactidis* sp. nov., three novel bacterial species isolated from floral nectar. Syst Appl Microbiol. 2014;37:402–11.

33. Vannette RL, Fukami T. Historical contingency in species interactions: towards niche-based predictions. Ecol Lett. 2014;17:115–24.

34. Stewart EJ. Growing unculturable bacteria. J Bacteriol. 2012;194:4151–60.

35. Nepi M, Pacini E, Nencini C, Collavoli E, Franchi GG. Variability of nectar production and composition in *Linaria vulgaris* (L.) Mill. (*Scrophulariaceae*). Plant Syst Evol. 2003;238:109–18.

36. Corbet SA, Bee J, Dasmahapatra K, Gale S, Gorringe E, La Ferla B, Moorhouse T, Trevail A, Van Bergen Y, Vorontsova M. Native or exotic? Double or single? Evaluating plants for pollinator-friendly gardens. Ann Botany. 2001;87:219–32.

37. Stout JC, Allen JA, Goulson D. Nectar robbing, forager efficiency and seed set: bumblebees foraging on the self incompatible plant *Linaria vulgaris* (*Scrophulariaceae*). Acta Oecolo. 2000;21:277–83.

38. Bartlewicz J, Vandepitte K, Jacquemyn H, Honnay O. Population genetic diversity of the clonal self-incompatible herbaceous plant species *Linaria vulgaris* along an urbanization gradient. Biol J Linnean Soc. 2015. doi:10.1111/bij.12602.

39. Pozo MI, Herrera CM, Alonso C. Spatial and temporal distribution patterns of nectar-inhabiting yeasts: how different floral microenvironments arise in winter-blooming *Helleborus foetidus*. Fungal Ecol. 2014;11:173–80.

40. O'Donnell K. Fusarium and its near relatives. *In:* REYNOLDS, D. & TAYLOR, J. (eds.) *The Fungal Holomorph: Mitotic, Meiotic and Pleomorphic Speciation in Fungal Systematics*. Wallingford: CAB International: 1993. p. 000-000.

41. Marchesi JR, Sato T, Weightman AJ, Martin TA, Fry C, Hiom SJ, Wade WG. Design and Evaluation of Useful Bacterium-Specific PCR Primers That Amplify Genes Coding for Bacterial 16S rRNA. Appl Environ Microb. 1998;64:795–9.

42. Altschul SF, Gish W, Miller W, Myers EW, Lipman DJ. Basic local alignment search tool. J Mol Biol. 1990;215:403–10.

43. Benson DA, Cavanaugh M, Clark K, Karsch-Mizrachi I, Lipman DJ, Ostell J. Sayers EW. Nucleic Acids Res D. 2013. doi:10.1093/nar/gks1195.

44. Colwell RK EstimateS: statistical estimation of species richness and shared species from samples. 2013; Version 9. – User's Guide and application at < http://purl.oclc.org/estimates > .

45. Oksanen J, Blanchet FG, Kindt R, Legendre P, Minchin PR, O'Hara RB, Simpson GL, Solymos P, Stevens MHH, Wagner H. Vegan: Community Ecology Package. 2013; R package version 2.0-10. http://CRAN.R-project.org/package=vegan.

46. Atmar W, Patterson BD. The measure of order and disorder in the distribution of species in fragmented habitat. Oecologia. 1993;96:373–82.

47. Guimaraes PR, Guimaraes P. Improving the analyses of nestedness for large sets of matrices. Environ Modell Softw. 2006;21:1512–3.

48. Almeida-Neto M, Guimaraes P, Guimaraes PR, Loyola RD, Ulrich W. A consistent metric for nestedness analysis in ecological systems: reconciling concept and measurement. Oikos. 2008;117:1227–39.

49. Bascompte J, Jordano P, Melián CJ, Olesen JM. The nested assembly of plant-animal mutualistic networks. Proc of the Nat Aca of Sci USA. 2003;100:9383–7.

7

Shrubby cinquefoil (*Dasiphora fruticosa* (L.) Rydb.) mapping in Northwestern Estonia based upon site similarities

Kalle Remm*[ID] and Liina Remm

Abstract

Background: Different methods have been used to map species and habitat distributions. In this paper, similarity-based reasoning—a methodological approach that has received less attention—was applied to estimate the distribution and coverage of *Dasiphora fruticosa* for the region in the Baltic states where grows the most abundant population of this species.

Methods: Field observations, after thinning to at least 50 m interval, included 1480 coverage estimations in the species presence locations and 8317 absence locations. Species coverage for the 750 km^2 of directly unobserved area was calculated using machine learning in the similarity-based prediction system Constud. Separate predictive sets of site features (e.g. land cover, soil type) and exemplar weights were calibrated for spatial partitions of the study area (probable presence region, unclear region, proved absence region). A modified version of the Gower's distance metric, as used in Constud, is described.

Results: The resulting maps depicted the predicted coverage, the certainty of decision when predicting presence or absence, and the mean similarity to the exemplar locations used while predicting. Coverage prediction errors were smaller in the unclear partition—where the species was mostly absent—than in the probable presence partition, where coverage ranged from 0 to 90%.

Conclusions: We call for methodological comparisons using the same data set.

Keywords: *Dasiphora fruticosa*, Similarity-based reasoning, Species distribution mapping, Gower's distance metric

Background

Detailed distribution data are needed in order to monitor changes in species' distributions, for conservation, territorial planning, and species and habitat management, but it is impractical and expensive to conduct detailed field observations over large areas. For a detailed distribution map that covers hundreds of square kilometres, knowledge regarding the limiting and the favourable factors that affect the target species is required. The likelihood of a species being present or absent at unobserved locations can then be predicted using a statistical model [1–4], or alternatively, according to similarity of exemplar sites [5].

Exemplars are the cases selected out from a training data set by machine learning or by expert decision. Maps of a species estimated distribution are also important for further monitoring efforts, since predictions help to identify areas in need of urgent future sampling [6, 7].

Similarity-based—also known as case-based—reasoning is an alternative to statistical regression models and classification methods [8]. The use of similarity-based reasoning is widespread in the fields of image recognition [9], medicine [10], web and text mining [11], engineering [12], meteorology [13], site classification [14], and other subjects where large databases of previous cases exist and case studies dominate over highly formalized rules. Case-based systems reuse previous experiences at a low level of generalisation, do not produce models based on generalized statistical relationships and can be continuously

---
*Correspondence: kalle.remm@ut.ee
Institute of Ecology and Earth Sciences, University of Tartu, Vanemuise 46, 51014 Tartu, Estonia

updated with new knowledge, as new cases may be added to the case-base. There is no need to change a model if additional training data becomes available. Similarity-based reasoning is classified as machine learning if an iterative fitting of exemplars, feature weights, and other parameters, precedes the inference.

Similarity-based distribution mapping assumes a species (or other phenomenon) to occur in locations similar to those where the species has already been recorded. The principal difference of case-based reasoning from niche-based distribution models lies in not assuming and applying a niche as a theoretical abstraction. Case-based systems infer directly from the most similar exemplars, not using any theoretical model, except rules how to calculate similarity. Similarity-based methods are rarely used for species distribution and habitat suitability mapping, though some examples exist. T. H. Booth compared the climatic similarity of locations to identify sites suitable for introduction a tree species outside its natural range [15]. Carpenter et al. introduced DOMAIN—a similarity-based algorithm for modelling potential distributions of plant and animal species [16]. Clark compared characteristics of black bear sites with the variate mean values of all visited sites using the Mahalanobis distance statistic to map habitat suitability for the bear [17]. Skov introduced a software application for creating site similarity-based plant species distribution maps [18]. De Siqueira et al. mapped summary of environmental distances as similarity measures in a 16-dimensional environmental space to the known occurrence point of a rare plant species [19]. Remm and Remm created maps depicting the similarity of each location to the observed presence and absence sites of ten orchid species using software system Constud [20]. The methodological advancement compared to the previous similarity-based habitat mappings was machine learning of optimal weights for site features and for observed locations.

Shrubby cinquefoil (*Dasiphora fruticosa* (L). syn. *Potentilla fruticosa*, Rosaceae) is a perennial flowering shrub, mainly known for being decorative cultivar. This species has widespread natural populations in mountainous regions of Asia and across North America (except in the south); its distribution in Europe is more fragmentary, being found only in Pyrénées, Maritime Alps, Rhodope Mountains, Crimea, Ireland, Great Britain, Öland and Gotland (Sweden), north-western Estonia, and one location in Latvia [21, 22]. Shrubby cinquefoil cultivars differ genetically from natural Northern European populations [21]. Shrubby cinquefoil probably had a continuous distribution over the present boreal and nemoral zones after the Late Glacial Maximum (c. 13,000–10,000 years ago), but was eliminated from large areas owing to soil leaching, peat accumulation, and forestation [23, 24].

More recently, ploughing probably destroyed many populations.

The populations of *D. fruticosa* have mostly patchy spatial pattern owing to vegetative spread by sprawling stems and rare establishment of seedlings under natural conditions [21, 25]. As a field mapping object, shrubby cinquefoil is easy to detect visually, especially from the end of June to September when it is in flower. In Estonia, the bushes are usually about half a metre high, although on rare occasions it can grow to more than a metre in open places, and can even reach more than 1.5 m when leaning on juniper (*Juniperus communis*) bushes.

Shrubby cinquefoil is a protected plant in Estonia. The only sustainable population is between Tallinn, Keila, and Paldiski [21], where it is mainly found on alvar grasslands spread over Middle and Upper Ordovician limestone, and forms the largest natural population in the Baltic states. During the previous decades, urban sprawl around Estonia's capital Tallinn has encroached upon unique *D. fruticosa* alvars. On residential or industrial building plots, only a few natural *D. fruticosa* bushes have been retained according to our field experience. Although the species population within this area is currently viable, its health and continued existence needs attention and monitoring.

The Estonian Nature Information System, held at the Ministry of the Environment, contained 18 sites where the species was known to be present before this project was started, 15 of which fall within the region covered by this study. The species presence sites registered by the Ministry are publicly available from http://xgis.maaamet.ee/xGIS/XGis?app_id=UU62A. However, it is unclear to what extent these records represent the current distribution in spatial detail, since: (1) absence sites are not recorded; and (2) the spatial intensity of observations used to delineate species occurrence polygons in the national database is unknown.

The aim of this study was to create a detailed similarity-based map of the distribution of *D. fruticosa* in the study area, by combining the site features selected during a previous study [26]. In addition to the map, a novel technological approach, which included study area partitioning, will be proposed for spatially detailed distribution mapping.

Data

Study area

The study area covering 819 km^2 was located at the known *D. fruticosa* natural occurrence area in North-western Estonia (Fig. 1). According to a database of landscape categories, limestone, lacustrine and marine plains, and mires, dominate the study area [27]. The ground elevation above sea level of the study area was up to 58 m, with most (64%) lying between 20 and 40 m.

Fig. 1 Location of the study area (*red rectangle*). Redrawn from the public wms service http://kaart.maaamet.ee/wms/alus?

According to the Estonian National Topographic Database held by the Estonian Land Board, forest covers 45%, cultivated land 22%, natural grassland 13%, private yards 5%, unmanaged open land (in this region mainly alvar grassland) 4%, scrubland 2%, and inland waters 1% of the study area. Most residents within the study area live in towns. Approximately 28% (242 km^2) of the area has no permanent residents; more than half of the territory is sparsely populated with 1 to 100 inhabitants in 497 one square kilometre grid cells.

The study area contains the Vääna Landscape Reserve (4.1 km^2), created to protect inter alia the alvars where grows the largest natural population of *D. fruticosa* in the Baltic states (Figs. 2, 3). Similar alvar sites can be found elsewhere in western and northern Estonia, but *D. fruticosa* does not grow there.

Fieldwork

Field observations were made during the summers of 2008–2014 conducting walking tours across the terrain, preferably in regions representing land cover and soil categories, which, according to so far collected field data, could be suitable for the species, but where the density of our observations was lower and the presence or absence of *D. fruticosa* unclear.

The coordinates of the movement tracks, *D. fruticosa* occurrence locations, and deliberately selected typical absent locations were recorded using a Garmin Vista HC+ GPS recorder. *D. fruticosa* coverage percentage within 10 m was visually estimated at each occurrence location. The total length of the observation trails was at least 700 km (Fig. 3). Most (78%) of the study area contains observed sites within a radius of 1 km and 96.4% within a radius of 2 km; although, at the more detailed

Fig. 2 Field of flowering *D. fruticosa* at Vääna Landscape Reserve

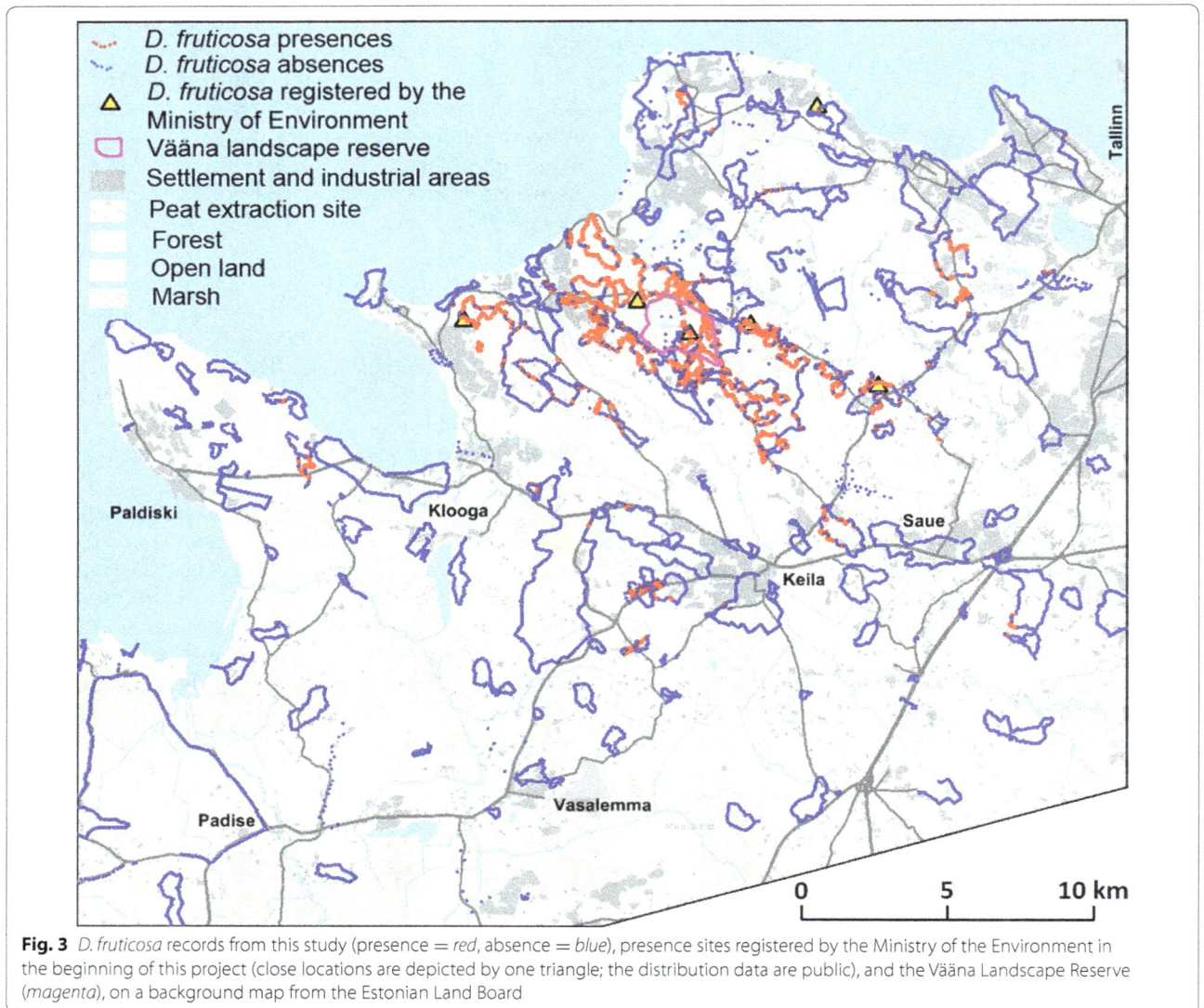

Fig. 3 *D. fruticosa* records from this study (presence = *red*, absence = *blue*), presence sites registered by the Ministry of the Environment in the beginning of this project (close locations are depicted by one triangle; the distribution data are public), and the Vääna Landscape Reserve (*magenta*), on a background map from the Estonian Land Board

100×100 m output grid level, only 8.3% (68.3 km^2) is covered by the direct observations. For the rest of the study area (750.7 km^2), the predicted occurrence and coverage of the species were calculated according to site similarity (see "Methods" section).

The GPS track recorder automatically stores coordinates along the route at variable interval according to the change in movement direction. The raw data contained 58,842 automatically recorded track points and 5687 actively recorded observation locations (2854 presence sites and 2833 absence sites). The raw records were thinned to ensure a distance of at least 50 m between each accepted site, in order to level out the density of the observations and avoid spatially close records. The track points at more than 50 m from the closest recorded presence location were considered, in addition to the

deliberately selected absence exemplar sites, as absence records. Thinning resulted in 8317 absence and 1480 presence locations at a spatial interval of at least 50 m. The species cover estimations in the thinned locations are freely available as an archived dataset [28]. More details regarding the field observations and data thinning are given in [26], which is based on the same data set.

Data layers and site features

The possible number of numerical features for any geographical location approaches infinity, if to consider different reference time, spatial and thematic generalization levels and neighbourhood extent options for deriving numerical features from data sources. The more indicative site features for mapping the species' distribution were selected in a previous study based on the same data

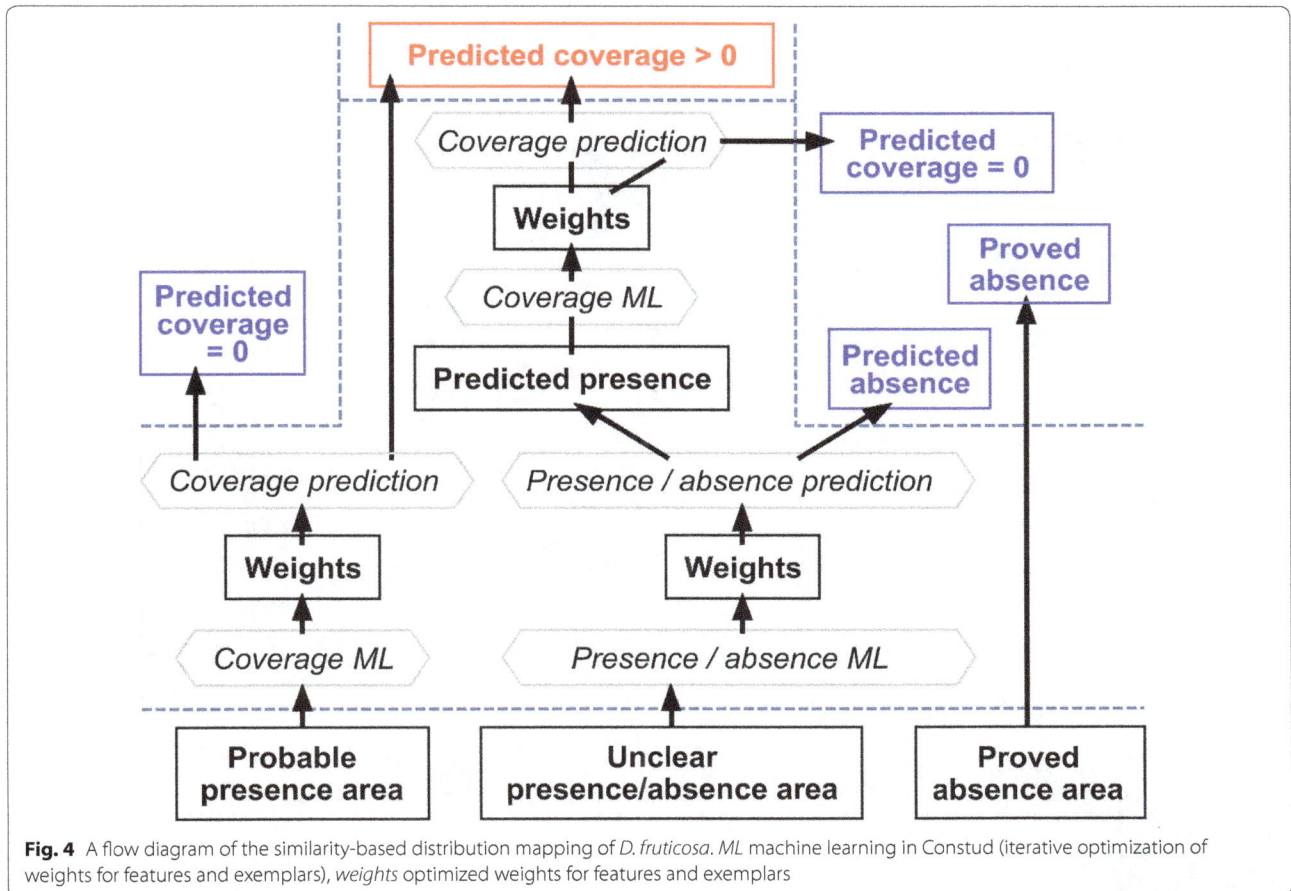

Fig. 4 A flow diagram of the similarity-based distribution mapping of *D. fruticosa*. *ML* machine learning in Constud (iterative optimization of weights for features and exemplars), *weights* optimized weights for features and exemplars

set [26]. These were: present land cover type (15 categories); historical land cover (6 categories); the most frequent (modal) historical land cover at a radius of 200 m; tussocks dominating or not within a 1 km radius; local soil type (25 categories); the modal soil type within a radius of 200 m; and features describing the spatial auto-covariation of the species' occurrence (the proportion of finds within radii of 100 and 1000 m, the mean coverage within 100 and 1000 m, and the reverse distance-weighted mean coverage).

The land cover and soil data for calculating these features were obtained from Land Board Estonia: the land cover data from the Estonian National Topographic Database, soil data from the Estonian 1: 10,000 soil map. Historical land cover types were digitized from a topographical map surveyed in the second half of the 1930s, the mean coverage and the proportion of presences in neighbourhood are calculated from observed coverages in locations retained after thinning. Site features used for machine learning were read from 10 × 10 m grid layers rasterized from vector format source data. Prediction maps were calculated as 100 × 100 m Idrisi (Clark Labs) raster format grids.

Methods

Dasiphora fruticosa distribution mapping consisted of six stages (Fig. 4). The weights for features and exemplars were separately calibrated using software Constud (see the next paragraph) for stages 3–5.

1. Selecting site features indicative to the presence/absence of the species.
2. Partitioning the study area to the species' probable presence region, proved absence region and unclear region (Fig. 5).
3. Calculating the expected coverage in the probable presence partition.
4. Calculating the expected presence/absence in the unclear partition.
5. Calculating the expected coverage at each predicted presence site in the initially unclear partition.
6. Overlaying the observed data to the map of predicted values.

The first and second points in this list refer to the stages of the wider research project and are reported in

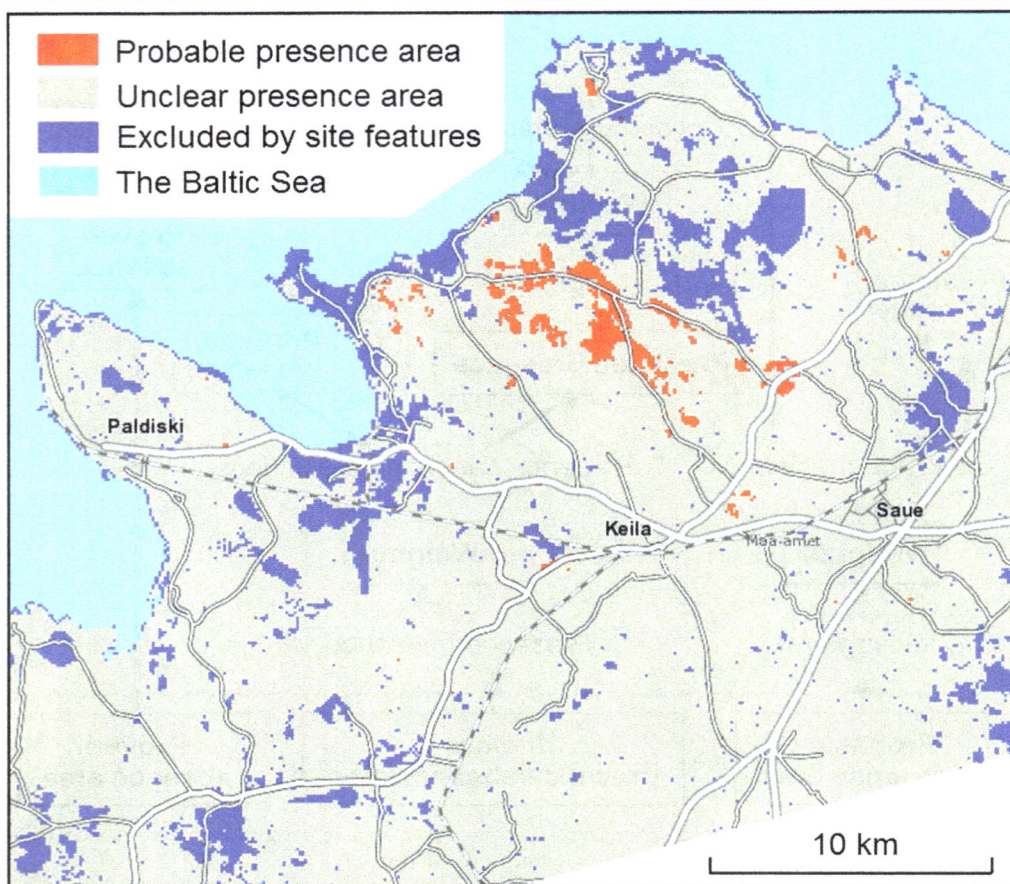

Fig. 5 Spatial partition of the study area on a background map from the Estonian Land Board

a separate publication [26] where boosted classification tree models were used to compare the value of 60 individual site features at thematically and spatially different levels of generalization as indicators of the species' presence or absence. The 60 site features were calculated from Estonian land cover database, soil map, a historical map, elevation data, from human population density, and from the species data in vicinity of every location.

The criteria for pre-classifying the study area were as follows. The proved absence partition (41.3 km^2): more than 1% of observations and no presences; the probable presence partition (11.4 km^2): values of at least two site features among the four most firm presence predictors indicate the species presence; unclear partition (766.3 km^2): most of the study area, which is meeting neither of these two criteria. The proved absence partition does not contain any currently known *D. fruticosa* find sites; the probable presence partition contains both presence and absence sites, but presence sites are more frequent. The unclear partition contains predominantly

absence records but includes sites potentially suitable for the species.

Calculated values (stages 3, 4 and 5 in the list above) were similarity based estimations. If the site features in a currently predicted site are similar to an exemplar site, then the species coverage in the predicted site is expected to be similar to the species coverage in the exemplar site. Details of the algorithm are given below. Finally the observed coverage records were overlaid to the layer of predicted coverage values assuming the field records to be more reliable than the calculated values.

Both spatial thinning and machine learning contributed to data reduction. Spatial thinning removes spatially redundant records and machine learning of weights removes redundant learning cases including of which does not reduce prediction error. In addition, only data from the same partition are included since a separate set of learning data was used in each partition (Fig. 6).

We used the software system Constud 3 [29] for machine learning of the optimal weights for features and

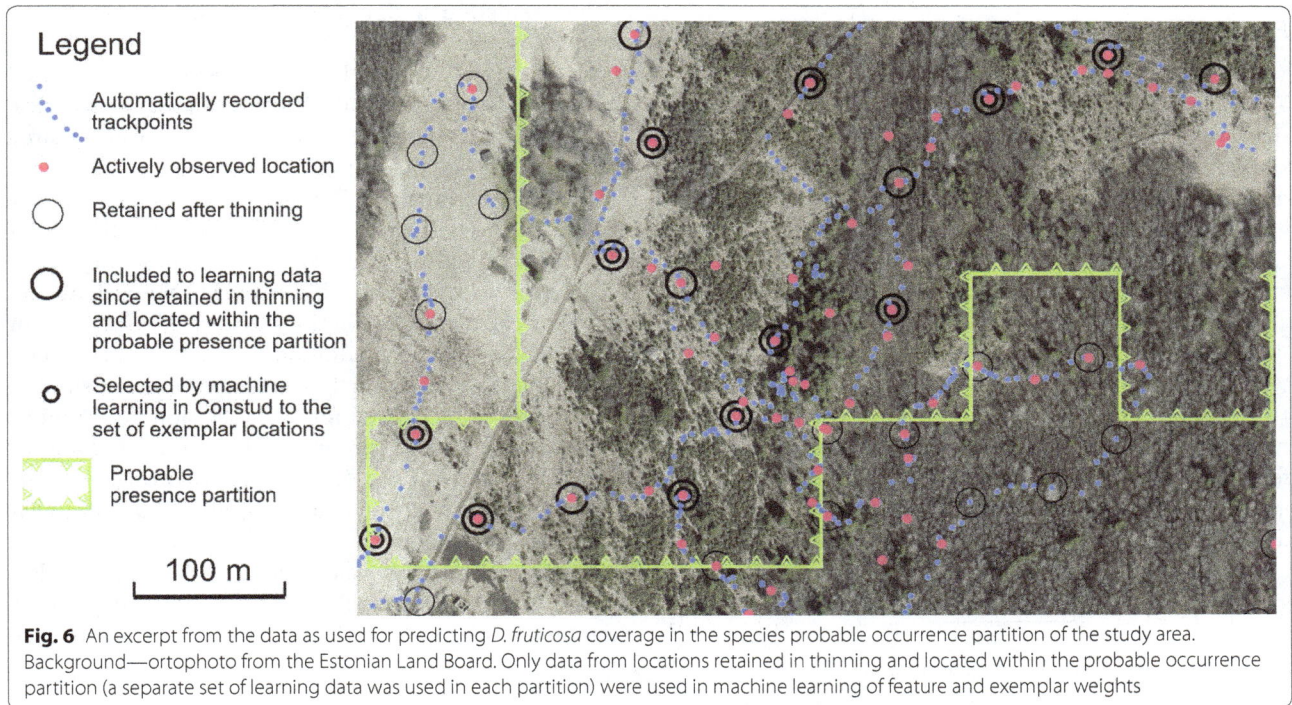

Fig. 6 An excerpt from the data as used for predicting *D. fruticosa* coverage in the species probable occurrence partition of the study area. Background—ortophoto from the Estonian Land Board. Only data from locations retained in thinning and located within the probable occurrence partition (a separate set of learning data was used in each partition) were used in machine learning of feature and exemplar weights

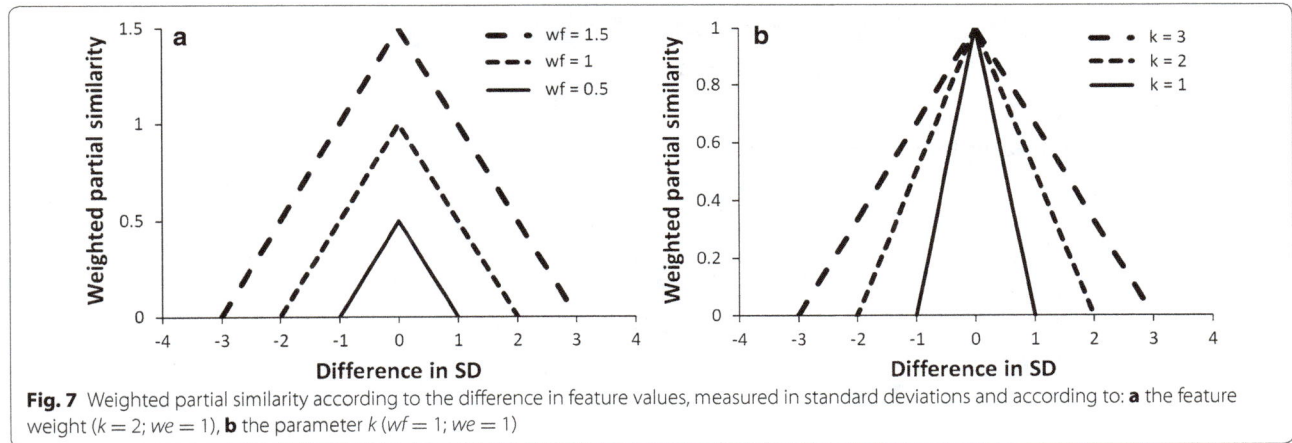

Fig. 7 Weighted partial similarity according to the difference in feature values, measured in standard deviations and according to: **a** the feature weight ($k = 2$; $we = 1$), **b** the parameter k ($wf = 1$; $we = 1$)

exemplars, and for the calculation of similarity-based maps for *D. fruticosa*'s occurrence and coverage. Constud as a software system for similarity-based predictions is described in [30]; recent changes compared to the previous version can be found at [31].

The central operation of the similarity-based reasoning is similarity metering. The Gower's [32] metric is commonly used for quantifying the distance between two objects. The Gower's metric uses range standardization to equalize the contribution from each numerical feature. The distance metric in Constud (1) differs from the Gower's metric, by using the sum of partial similarities *PS*

(Of, Ef) as weighted by feature weights (wf) and exemplar weights (we), and by replacing the range standardisation with $k \times SD$ for numerical features (where k is the sum of similarity searched for decision—the default value in Constud is 2—SD the standard deviation of the feature, and $|Fe - Fo|$ the difference in feature values) (1). The partial similarity is calculated between an observation O and exemplar E regarding only a single feature f. Negative partial similarity values are assigned a value of zero. Features with a higher weight have a wider accepted difference in values and have a larger share of the total similarity (Fig. 7).

$$PS(Of, Ef) = \text{wf} \times \text{wo} \times \left(1 - \frac{|Fe - Fo|}{k \times \text{SD} \times \text{wf} \times \text{we}}\right) \quad (1)$$

This formula (1) is applied for numerical features. When similarity between categories (Sc) is considered, the similarity of matching categories is equal to one and of different categories equal to zero (2); otherwise, category-specific similarity values can be assigned by the Constud user and stored in a special database table. We applied the first option in this study.

$$PS(Of, Ef) = Sc \times \text{wf} \times \text{we} \quad (2)$$

Total similarity (TS) between feature vectors is measured as the mean of partial similarities, weighted by the feature and exemplar weights (3). Zero-weight features, i.e. those unsuitable due to temporal limits or missing data, are skipped, as are zero weight exemplars.

$$TS(O, E) = \frac{\sum_f PS(Of, Ef)}{\sum_f (\text{wf} \times \text{we})} \quad (3)$$

The prediction fit for a binomial variable (presence/absence) is calculated in Constud as the True Skill Statistic ranging between −1 and +1, zero is the expected value in case of random decisions [33]; for a numerical variable, the objective function is the root mean squared error (RMSE) ranging from 0 to ∞.

Machine learning in Constud involves an iterative search for the best predictive weight sets for the exemplars records and the features to use in similarity-based recognition of the predictable variable. The weights are ready for calculating predictions, either as raster map or as a database object. In addition to the predicted values, Constud enables the user to calculate the mean similarity to the exemplars used for each decision; for nominal dependent variables, the similarity to a given category and the certainty of the decision can also be computed. A low similarity at any location indicates a lack of similar exemplars, and the probable need for additional data collection. A low certainty value means nearly equal similarity of a case to the exemplars of alternative categories; a certainty equal to one indicates that the exemplars used

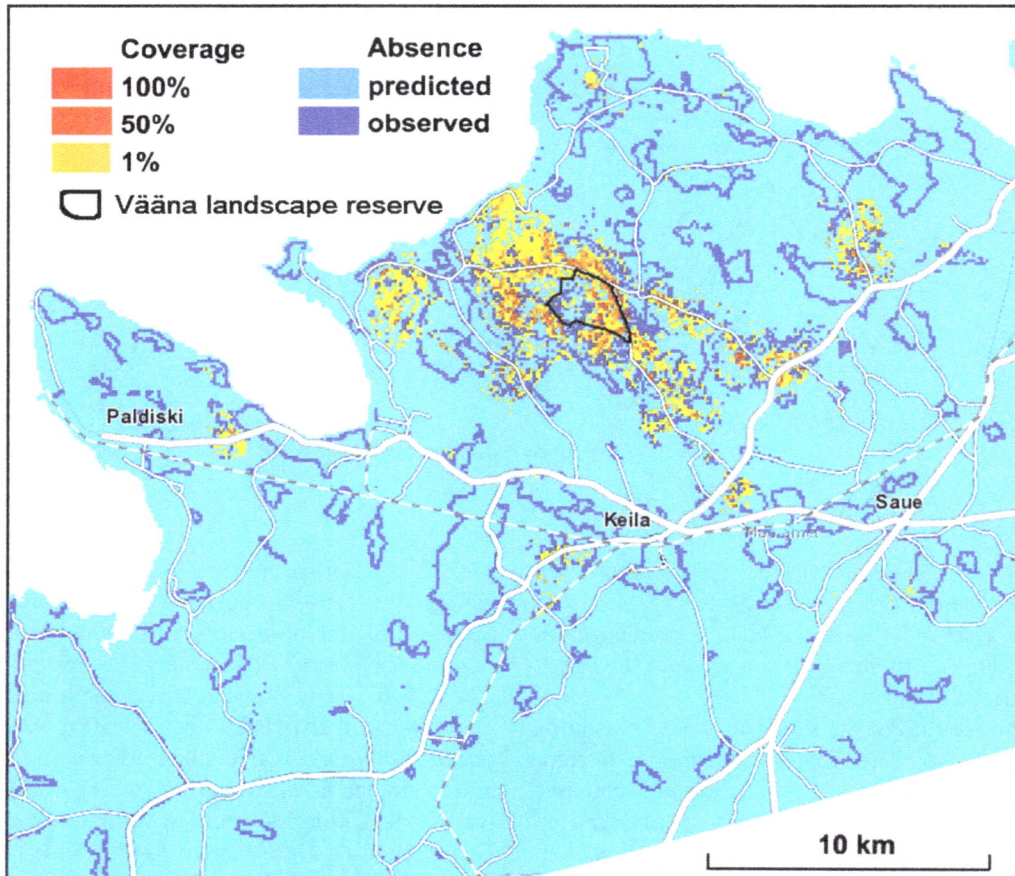

Fig. 8 Predicted *D. fruticosa* coverage in the study area on a background map from the Estonian Land Board

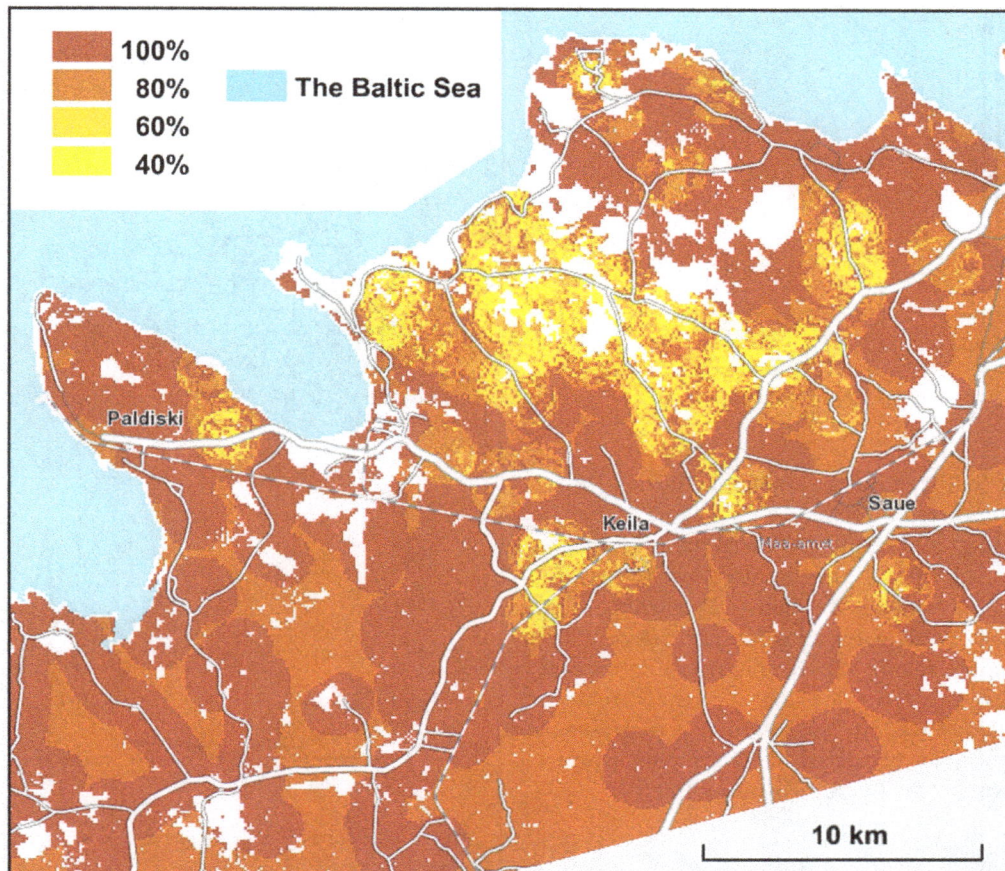

Fig. 9 The certainty of decision making in predicting *D. fruticosa* presence/absence in the unclear area on a background map from the Estonian Land Board

Table 1 Machine learned (ML) feature weights of the predictive sets selected in Constud

Feature	Num. of classes	Radius (m)	ML feature weights in predicting		
			Presence/absence in unclear partition	Coverage in	
				Predicted presence sites	Probable presence partition
Present land cover	15	0		0.8	1
Historical land cover	6	0		0.3	0.1
Historical land cover	6	200			0.4
Historical land cover	2	1000			0.1
Soil	25	0			1
Soil	25	200			0.3
Proportion of presences		100	2.9		0.1
Proportion of presences		1000	0.5	0.6	5.5
Mean coverage		100	0.2	2.0	0.2
Mean coverage		1000	0.3		0.6
Distance weighted mean coverage		10,000	1.1		1.8

The three ML sets of feature weights match the machine learning operations in Fig. 4

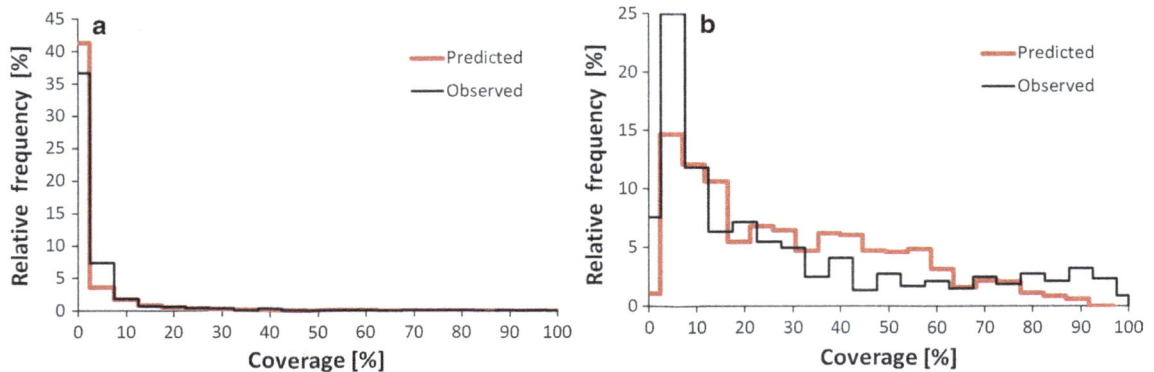

Fig. 10 The frequency of *D. fruticosa* coverage values: **a** in the unclear presence/absence partition; **b** in the probable presence partition

for predicting the dependent variable in this site represent only the predicted category.

Constud settings and results are stored in a SQL Server database, and in principal, can be modified by the user. The main initial parameters for Constud's learning of the *D. fruticosa* data were: a training sample size of 500; a validation sample size of 1000; the initial amount of similarity used to search for a decision = 5; the number of learning iterations was 200. The proportion of presences and absences was not equalized in the training samples. The grid interval for the output maps was 100 m.

Results

The main result was the mapping of estimated *D. fruticosa* coverage in the study area (Fig. 8). The recognition of *D. fruticosa* presence and absence sites within the hitherto unclear partition was not firm: the True Skill Statistic in the validation sample = 0.64. The certainty of decisions was low mainly around registered occurrence sites (Fig. 9); elsewhere the prediction was firmly the species absence.

The predictive set selected by the machine learning included only features that described autocovariation regarding the distribution of *D. fruticosa* (Table 1); i.e. soil and land cover did not support considerably the recognition of species presence sites, when the features describing autocovariation had values (the occurrence and coverage of the species in the neighbourhood was known). A high proportion of presences in the 100 m vicinity was the most reliable occurrence predictor, but was not a firm predictor, as there were always at least some absence records near observed presence sites.

The RMSE of coverage estimation for predicted presence sites in the unclear partition was 0.115 (with coverage as a continuous variable ranging from 0 to 1). Zero coverage dominated both the observed and predicted values (Fig. 10). In most cases, zero coverage was correctly

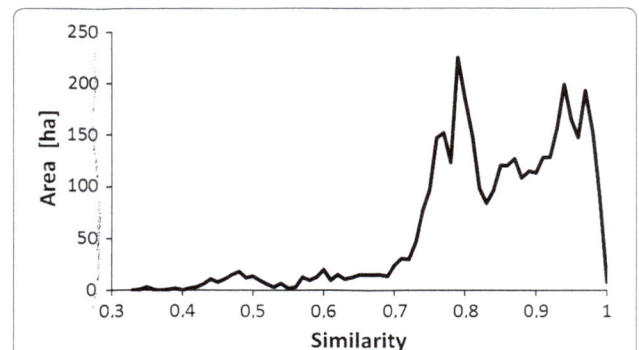

Fig. 11 The frequency of similarity values when predicting *D. fruticosa* coverage

predicted to be zero (specificity = 55.6%), which held the RMSE low.

The coverage estimation for the probable presence partition was much less reliable (RMSE = 0.244), despite all site features being included to the predictive set by machine learning in Constud. Species coverage in this partition was highly variable: there were 48 records among the thinned locations in the probable presence partition where *D. fruticosa* coverage reaches 0.9, together with 61 records of zero coverage. The mean similarity of predicted sites to exemplar sites used for *D. fruticosa* coverage mapping was mostly over 75% (Figs. 11, 12). Relatively low similarity values occurred on naturally rare or under-sampled land cover and soil types, e.g. on Floatic Histosol, which was represented by only two field records.

Discussion

A novel software system, Constud 3, developed for similarity-based predictions, was used to map a fragmentally distributed clonal shrub around one of its core

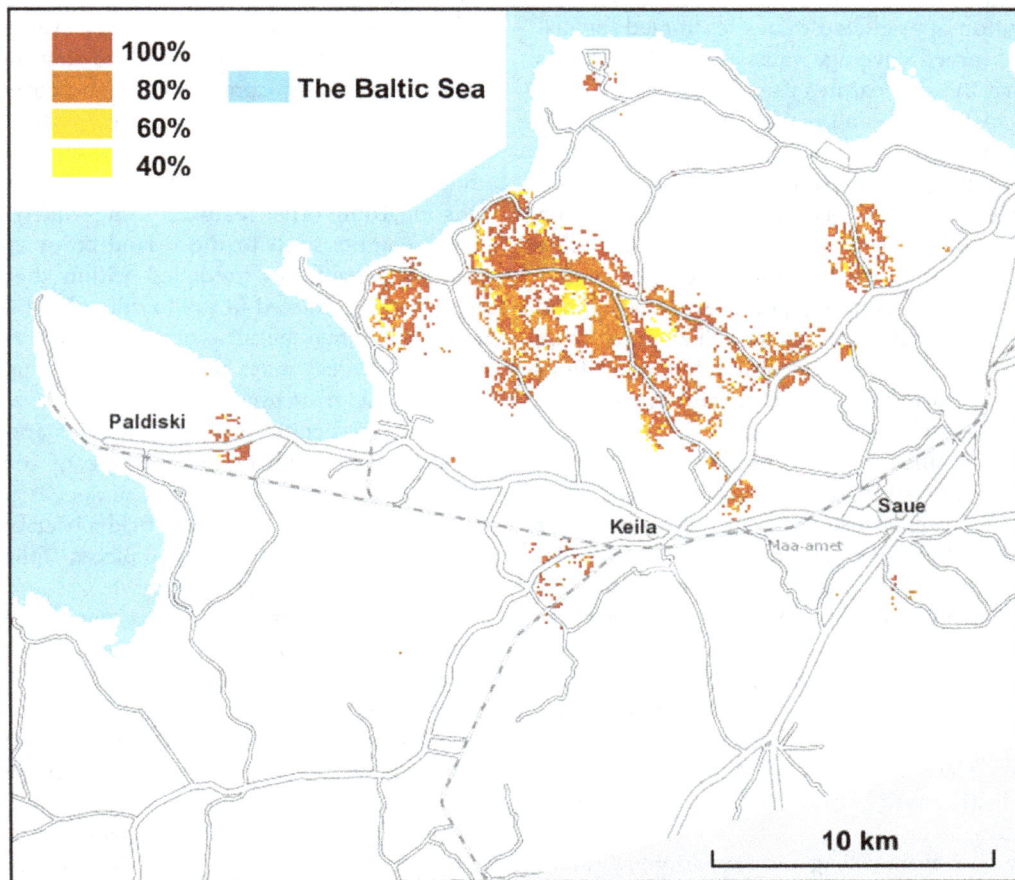

Fig. 12 The similarity of locations to the exemplar locations used for predicting *D. fruticosa* coverage on a background map from the Estonian Land Board

populations. A similarity-based approach to distribution mapping has rarely been used, and its implementation for abundance or plant coverage mapping is unknown to us. In their review, Elith and Leathwick [34], only denote similarity-based approaches in connection with species occurrence mapping, and not for numerical variables such as suitability or coverage. Predicting abundance is generally more complicated when compared to presence/absence, as a species' abundance is unstable and depends more on population processes than habitat properties [38–40]. This study and some other publications [35–37] have however highlighted the ability of similarity-based methods to predict numerical variables.

The predictive map of *D. fruticosa* coverage (Fig. 8), demonstrates that the main protected area, created for the species, extends only over the main population core. However, most of the other populations and part of the core population are situated outside of protected areas. The distribution map could be applied for improving conservation planning and management. The coverage

estimations could be used to find the most viable subpopulations, as well as to target management efforts, especially considering that dense *D. fruticosa* scrub shades out valuable herb layer of alvars, consequently resulting in a decrease in species richness [41].

In species distribution modelling approaches uncertainty generally increases near range edges, ecologically, due to increasing sensitivity to habitat conditions and, statistically, due to the procedure of converting probabilities from logistic regression into presence/absence [42]. In a more detailed scale of our study, the species actual presence or absence was remotely difficult to estimate (Fig. 9) in the population core area as the distribution is not continuous; the species absent locations are commonly near by the presence locations (Fig. 6).

In the case of coverage estimation, the RMSE was quite high—approximately equal to the mean), although the mean coverage of the species in the neighbourhood was included as predictive site feature. The high level of prediction error was not caused by excessive smoothing related to a high sum of similarity searched for decision, as the distribution of predicted values resembled the distribution of observed coverage values (Fig. 10b). It was more likely that the site features used did not adequately represent the spatial scale and real causes of variability in *D. fruticosa* coverage.

Spatial partitioning of a study area for separate models could be justified, if the area is diverse and each partition sufficiently covered by data. In most habitat suitability and species distribution studies, the study area is partitioned for stratified sampling [4]. Previously applied partitioning options for separate modelling have been based on site similarity [43], sampling effort (density of records) [44, 45], arbitrary geographical districts [46, 47], rings and quarters from the records distribution centroid [48], and random partitioning [49]. In the current study, as an innovation, separate predictive sets were used in spatial partitions formed according to the prior estimated expected presence/absence of the species. An additional option is to calibrate predictive sets separately for areas where one or another explanatory value is missing. For example, the soil data used in the current study contained void values, yet for similarity-based reasoning, data absence is not an obstacle, because the similarity level is calculated using existing site features. Splitting a study area complicates the predictive system but can produce more reliable results and is worth of comparative methodological studies.

The primary goal of modelling and similarity-based reasoning is to obtain predicted values for vague situations; predictions that confirm already well established knowledge are less useful. Therefore, data collection focused on possible, but unproven presence regions, and on less represented site types (rare land cover and soil categories in the current study), is less resource-consuming than performing spatially random or regular sampling. The latter would entail enormous time and resources spent in regions and habitat types unsuitable for the target species. Iterative adaptive observation effort is analogous to iterative boosting methods in machine learning [50], and to the construction of uncertainty or ignorance maps [51, 52]. This approach may reduce the formal goodness-of-fit of the predicted values in a learning sample, but is less inclined toward over fitting, i.e. it produces more reliable estimations outside of observed sites. However, a gradual redirection of observation effort to less represented site types, would lead to pseudoreplication, especially in the case of landscapes that are merely represented by single patches [53].

Equal and sufficient representation of all site types is difficult to achieve when using observations along transects or moving tracks. In the current study the number of thinned observation sites was as high as 9797, but some land cover and soil types were not represented by

a sufficient number (1% of observations according to [26]) to prove the species absence. For example, although *D. fruticosa* was never observed growing in a bog, peat extraction site, or quagmire; these land cover types were not considered to indicate the species absence as containing less observations than necessary. Still, in case of a high location similarity to the exemplar presence locations regarding other features, some—likely false—presence predictions fell to those land cover types. E.g. *D. fruticosa* presence is predicted within the Tõlinõmme bog, which is situated in the middle of the species main population range, because of the 'tussock area' symbols on the historical map, which were also used to mark abundant *D. fruticosa* occurrence on alvars. In addition, frequent occurrence of the species around the bog supported the predicted occurrence of the species. If a detailed distribution mapping project is planned as a continuous process, the next field observations would primarily be directed to such places, followed by an update of the knowledge base.

The essence of similarity-based mapping, if compared to species niche modelling, is that the aim is not to find the environmental variables that result in the occurrence or absence of a species, but just to predict the distribution as accurately as possible. Short range spatial autocovariates can be effective predictors of dependent variables (especially for plants spreading by stems), but are applicable only near existing data. For example, proportion presences within 100 m was an effective predictor for a particular site, but for most of the study area, this site feature was useless, because it only covered 5.5% of the unclear presence/absence partition, despite the enormous amount of fieldwork undertaken (700 km on foot). Autocovariates calculated at a larger radius or as distance weighted, covered a larger proportion of the study area, but were less reliable predictors. Additional direct observations could add the close-neighbourhood area of the observed sites. Methodological alternatives for including neighbourhood effects, such as Markov's chains, use the estimated and not the observed values of the target variable. The availability of estimated values is not limited by the extent of the neighbourhood, but these methods are computationally intensive and return unstable predictions [4].

Conclusions

We encourage the use of similarity-based reasoning for habitat suitability and species distribution mapping, and recommend using the predictive map of *D. fruticosa* for conservation planning. We also suggest using separate models in different spatial partitions of any larger study area; and call others specialists to use the same data in connection with different methods in comparative methodological distribution mapping studies. The coordinates of observed sites and the recorded *D. fruticosa* coverages are freely available from doi:10.13140/RG.2.1.4987.6724. Contact the authors or apply a licence from the Estonian Land Board for using the topographical data layers.

Abbreviations
Constud: a software system for similarity-based reasoning developed at the University of Tartu; GPS: Global Positioning System; ML: machine learning; in Constud software used in this project—iterative optimization of feature and case weights; RMSE: root mean squared error, the standard deviation of differences between observed and predicted values; SQL Server: a relational database management system developed by Microsoft.

Authors' contributions
KR designed the study, analysed the data, illustrated and wrote an initial version of the manuscript. LR participated in interpreting the analyses and writing the manuscript. Both authors contributed to the final version of the paper. Both authors read and approved the final manuscript.

Acknowledgements
The authors thank the Estonian Land Board and Statistics Estonia for supplying cartographical and statistical data, James Phillips for linguistic editing of the manuscript, and Mare Remm for participating in the field observations.

Competing interests
The authors declare that they have no competing interests.

Funding
The investigation was financially supported by the Estonian Ministry of Education and Research (Projects IUT 2-16 and IUT34-7). The funder played no role in designing the study, collecting, analysing the data, interpreting the results and writing the manuscript.

References
1. Guisan A, Zimmermann NE. Predictive habitat distribution models in ecology. Ecol Model. 2000;135:147–86.
2. Elith J, Graham CH, Anderson RP, Dudík M, Ferrier S, Guisan A, et al. Novel methods improve prediction of species' distributions from occurrence data. Ecography. 2006;29:129–51.
3. Austin MP. Species distribution models and ecological theory: a critical assessment and some possible new approaches. Ecol Model. 2007;200:1–19.
4. Franklin J. Mapping species distributions. Spatial inference and prediction. Cambridge: Cambridge University Press; 2009.
5. Remm K. Case-based predictions for species and habitat mapping. Ecol Model. 2004;177:259–81.
6. Guisan A, Thuiller W. Predicting species distributions: offering more than simple habitat models. Ecol Lett. 2005;8:993–1009.
7. Rodríguez JP, Brotons L, Bustamante J, Seoane J. The application of predictive modelling of species distribution to biodiversity conservation. Divers Distrib. 2007;13:243–51.
8. Aha DW. The omnipresence of case-based reasoning in science and application. Knowl Based Syst. 1998;11:261–73.
9. Chalom E, Asa E, Biton E. Measuring image similarity: an overview of some useful applications. IEEE Instrum Meas Mag. 2013;16:24–8.
10. Marling C, Montani S, Bichindaritz I, Funk P. Synergistic case-based reasoning in medical domains. Expert Syst Appl. 2014;41:249–59.
11. Choi D, Ko B, Kim H, Kim P. Text analysis for detecting terrorism-related articles on the web. J Netw Comput Appl. 2014;38:16–21.
12. Shokouhi SV, Skalle P, Aamodt A. An overview of case-based reasoning applications in drilling engineering. Artif Intell Rev. 2014;41:317–29.
13. Singh D, Ganju A, Singh A. Weather prediction using nearest-neighbour model. Curr Sci India. 2005;88:1283–9.
14. Jasiewicz J, Netzel P, Stepinski TF. Landscape similarity, retrieval, and machine mapping of physiographic units. Geomorphology. 2014;221:104–12.
15. Booth TH. A climatic analysis method for expert systems assisting tree species introductions. Agrofor Syst. 1990;10:33–45.
16. Carpenter G, Gillison AN, Winter J. DOMAIN: a flexible modelling procedure for mapping potential distributions of plants and animals. Biodivers Conserv. 1993;2:667–80.
17. Clark JD, Dunn JE, Smith KG. A multivariate model of female black bear habitat use for a geographic information system. J Wildlife Manag. 1993;57:519–26.
18. Skov F. Potential plant distribution mapping based on climatic similarity. Taxon. 2000;49:503–15.
19. de Siqueira MF, Durigan G, de Júnior MP, Peterson AT. Something from nothing: using landscape similarity and ecological niche modeling to find rare plant species. J Nat Conserv. 2009;17:25–32.
20. Remm K, Remm L. Similarity-based large-scale distribution mapping of orchids. Biodivers Conserv. 2009;18:1629–47.
21. Leht M, Reier Ü. Origin, chromosome number and reproduction biology of *Potentilla fruticosa* (Rosaceae) in Estonia and Latvia. Acta Bot Fenn. 1999;162:191–6.
22. Lonati M, Pascale M, Operti B, Lombardi G. Synecology, conservation status and IUCN assessment of *Potentilla fruticosa* L. in the Italian Alps. Acta Bot Gallica. 2014;161:159–73.
23. Elkington TT. Cytotaxonomic variation in *Potentilla fruticosa* L. New Phytol. 1969;68:151–60.
24. Pigott CD, Walters SM. On the interpretation of the discontinuous distributions shown by certain British species of open habitats. J Ecol. 1954;42:95–116.
25. Elkington TT. Woodell SRJ. *Potentilla fruticosa* L. (*Dasiphora fruticosa* (L.) Rydb.). J Ecol. 1963;51:769–81.
26. Remm K. Selecting site characteristics at different spatial and thematic scales for shrubby cinquefoil (*Potentilla fruticosa* L.) distribution mapping. For Stud. 2016;64:17–38.
27. Arold I. *Eesti Maastikud* (Estonian Landscapes). Tartu Ülikooli Kirjastus; 2005.
28. Remm K. Shrubby cinquefoil (*Dasiphora fruticosa*) cover records from a study site in the NW Estonia. 2016. doi:10.13140/RG.2.1.4987.6724.
29. Remm K, Remm M. Case-based estimation of the risk of enterobiasis. Artif Intell Med. 2008;43:167–77.
30. Remm K, Kelviste T. Constud Tutorial. University of Tartu; 2011.
31. http://digiarhiiv.ut.ee/Constud3/. Accessed 23 July 2016.
32. Gower JC. A general coefficient of similarity and some of its properties. Biometrics. 1971;27:857–71.
33. Allouche O, Tsoar A, Kadmon R. Assessing the accuracy of species distribution models: prevalence, kappa and the true skill statistic (TSS). J Appl Ecol. 2006;43:1223–32.
34. Elith J, Leathwick JR. Species distribution models: ecological explanation and prediction across space and time. Annu Rev Ecol Evol Syst. 2009;40:677–97.
35. Gayer G, Gilboa I, Lieberman O. Rule-based and case-based reasoning in housing prices. BE J Theor Econ 2007;7.
36. Liu F, Rossiter DG, Song X-D, Zhang G-L, Yang R-M, Zhao Y-G, Li D-C, Ju B. A similarity-based method for three-dimensional prediction of soil organic matter concentration. Geoderma. 2014;263:254–63.
37. Yang L, Huang C, Liu G, Liu J, Zhu A-X. Mapping soil salinity using a similarity-based prediction approach: a case study in Huanghe River Delta, China. Chin Geogr Sci. 2015;25:283–94.

38. van Horne B. Density as misleading indicator of habitat quality. J Wildl Manag. 1983;47:893–901.

39. Frescino TS, Edwards TC Jr, Moisen GG. Modelling spatially explicit forest structural attributes using generalized additive models. J Veg Sci. 2001;12:15–26.

40. Pearce J, Ferrier S. The practical value of modelling relative abundance of species for regional conservation planning: a case study. Biol Conserv. 2001;98:33–43.

41. Rejmánek M, Rosén E. The effects of colonizing shrubs (*Juniperus communis* and *Potentilla fruticosa*) on species richness in the grasslands of Stora Alvaret, Öland (Sweden). Acta Phytogeographica Suecica. 1988;76:67–72.

42. Hanspach J, Kühn I, Schweiger O, Pompe S, Klotz S. Geographical patterns in prediction errors of species distribution models. Global Ecol Biogeogr. 2011;20:779–88.

43. Estrada-Peña A, Thuiller W. An assessment of the effect of data partitioning on the performance of modelling algorithms for habitat suitability for ticks. Med Vet Entomol. 2008;22:248–57.

44. Fourcade Y, Engler JO, Besnard AG, Rödder D, Secondi J. Confronting expert-based and modelled distributions for species with uncertain conservation status: a case study from the Corncrake (*Crex crex*). Biol Conserv. 2013;167:161–71.

45. Fourcade Y, Engler JO, Rödder D, Secondi J. Mapping species distributions with MAXENT using a geographically biased sample of presence data: a performance assessment of methods for correcting sampling bias. PLoS ONE. 2014;9(5):e97122. doi:10.1371/journal.pone.0097122.

46. Murphy HT, Lovett-Doust J. Accounting for regional niche variation in habitat suitability models. Oikos. 2007;116:99–110.

47. Gonzalez SC, Soto-Centeno JA, Reed DL. Population distribution models: species distributions are better modeled using biologically relevant data partitions. BMC Ecol. 2011;11:20. doi:10.1186/1472-6785-11-20.

48. Osborne PE, Suárez-Seoane S. Should data be partitioned spatially before building large-scale distribution models? Ecol Model. 2002;157:249–59.

49. Phillips SJ, Anderson RP, Schapire RE. Maximum entropy modeling of species geographic distributions. Ecol Model. 2006;190:231–59.

50. Freund Y, Schapire RE. A decision-theoretic generalization of on-line learning and an application to boosting. J Comput Syst Sci. 1997;55:119–39.

51. Rocchini D, Hortal J, Lengyel S, Lobo JM, Jiménez-Valverde A, Ricotta C, et al. Accounting for uncertainty when mapping species distributions: the need for maps of ignorance. Progr Phys Geogr. 2011;35:211.

52. Beale CM, Lennon JJ. Incorporating uncertainty in predictive species distribution modelling. Phil Trans R Soc B. 2012;367:247–58.

53. Hurlbert SH. Pseudoreplication and the design of ecological field experiments. Ecol Monogr. 1984;54:187–211.

8

How anthropogenic changes may affect soil-borne parasite diversity? Plant-parasitic nematode communities associated with olive trees in Morocco as a case study

Nadine Ali[1,2*], Johannes Tavoillot[2], Guillaume Besnard[3], Bouchaib Khadari[4], Ewa Dmowska[5], Grażyna Winiszewska[5], Odile Fossati-Gaschignard[2], Mohammed Ater[6], Mohamed Aït Hamza[7], Abdelhamid El Mousadik[7], Aïcha El Oualkadi[8], Abdelmajid Moukhli[8], Laila Essalouh[4], Ahmed El Bakkali[9], Elodie Chapuis[2,10,11†] and Thierry Mateille[2†]

Abstract

Background: Plant-parasitic nematodes (PPN) are major crop pests. On olive (*Olea europaea*), they significantly contribute to economic losses in the top-ten olive producing countries in the world especially in nurseries and under cropping intensification. The diversity and the structure of PPN communities respond to environmental and anthropogenic forces. The olive tree is a good host plant model to understand the impact of such forces on PPN diversity since it grows according to different modalities (wild, feral and cultivated olives). A wide soil survey was conducted in several olive-growing regions in Morocco. The taxonomical and the functional diversity as well as the structures of PPN communities were described and then compared between non-cultivated (wild and feral forms) and cultivated (traditional and high-density olive cultivation) olives.

Results: A high diversity of PPN with the detection of 117 species and 47 genera was revealed. Some taxa were recorded for the first time on olive trees worldwide and new species were also identified. Anthropogenic factors (wild vs cultivated conditions) strongly impacted the PPN diversity and the functional composition of communities because the species richness, the local diversity and the evenness of communities significantly decreased and the abundance of nematodes significantly increased in high-density conditions. Furthermore, these conditions exhibited many more obligate and colonizer PPN and less persister PPN compared to non-cultivated conditions. Taxonomical structures of communities were also impacted: genera such as *Xiphinema* spp. and *Heterodera* spp. were dominant in wild olive, whereas harmful taxa such as *Meloidogyne* spp. were especially enhanced in high-density orchards.

Conclusions: Olive anthropogenic practices reduce the PPN diversity in communities and lead to changes of the community structures with the development of some damaging nematodes. The study underlined the PPN diversity as a relevant indicator to assess community pathogenicity. That could be taken into account in order to design control strategies based on community rearrangements and interactions between species instead of reducing the most pathogenic species.

Keywords: Anthropisation, Communities, Functional diversity, Morocco, Olive, Plant-parasitic nematodes, Taxonomical structures

*Correspondence: nadineali.tichrine.univ@gmail.com
†Elodie Chapuis and Thierry Mateille are co-leaders of the publication
[1] Plant Protection Department, Faculty of Agriculture, Tishreen University, PO Box 2233, Latakia, Syrian Arab Republic
Full list of author information is available at the end of the article

Background

A biological community refers to an assemblage of populations from different organisms living together in a habitat. This biological assemblage within a community could be described by several traits such as the number of species (richness), their relative abundance (evenness), the present species (taxonomical structure), the interactions among them as well as their temporal and spatial variation [1]. Species diversity is important for the stability of the community and consequently that of the ecosystems [2]. For instance, functional consequences on ecosystem processes are related to species richness and to species-specific traits. Moreover, species diversity can play a crucial role in ecosystems resilience and/or resistance to human disturbances and to environmental changes [1].

Soil communities have been described as the "poor man's tropical rainforest", because of the relatively high level of biodiversity and the large proportion of undescribed species, as well as the limited information available about their community structure and dynamics [3]. Human interventions in ecosystems such as land-use changes, invasive species and over-exploitation, lead to biodiversity loss and/or species extinction [4]. For example, in agrosystems, crop intensification greatly disturbs the soils, affecting composition and functions of their biota [5, 6].

Among soil biota, nematodes are ubiquitous soil inhabitants and among the most abundant and diversified biota [7]. They reflect several feeding behaviors that make it possible to allocate them to different trophic groups: bacterivores, fungivores, carnivores and plant feeders [8]. Due to the various life strategies of nematodes (r and K for colonizer and persister nematodes, respectively), their diversity and their co-existence in communities are closely related to short response time, to environmental changes and to disturbances in their habitats [9].

Plant-parasitic nematodes (PPN) are known to attack a wide range of crop plants (cereals, vegetables, tubers, fruits, flowers, etc.), causing annual crop losses estimated at billions of dollars in worldwide [10, 11]. On the olive tree (Olea europaea L.), PPN are able to reduce tree growth [12] and may be responsible for 5–10% yield losses [13]. Their impact is especially strengthened in nurseries and in intensive cultivation systems where irrigation conditions favor the development of roots and, as a result, nematode multiplication [14]. A high diversity of PPN on olive trees was reviewed worldwide [14, 15].

In Morocco, olive tree is a good example of ecological, botanical and genetic diversity. Spontaneous trees are distinguished under three different forms: (i) autochthonous wild trees, usually referred to as oleasters (O. europaea subsp. europaea var. sylvestris (Mill.) Lehr.), are common in coastal and mountainous regions [16]; (ii) the

Moroccan hexaploid olive subspecies O. europaea subsp. maroccana is endemic in the High Atlas Mountains [17]; (iii) feral forms are wild-looking olive trees that correspond either to abandoned cultivated olive trees or to olive trees grown from cultivated olive seeds spread by birds. Additionally, cultivated forms (O. europaea subsp. europaea var. europaea) are also widespread. Different olive cropping systems can be distinguished according to tree density [18]: traditional orchards (ca. 80–400 trees/ha) vs high-density orchards (up to 1800 trees/ha). However, these new intensive techniques, accompanied by the replacement of traditional low-intensive production with highly intensified and mechanized cultivation, including the use of herbicides to remove weeds, are expected to induce a possible degradation of the plant communities and their associated fauna [19]. As for olive propagation, it is generally performed from root cuttings that could be accompanied by soil transport and, consequently, by the spread of soil-borne parasites. Thus, PPN could be spread by soil transport or by unsanitized plant material (e.g. from uncertified nurseries). The local PPN populations in olive-growing areas could therefore have originated from historical mixtures made up of native (before olive introduction) and invasive (with root stocks from oleasters) communities. In this context, we hypothesize that PPN communities may have adapted to olive propagation processes and to cultivation practices. These anthropogenic forces could exist in Morocco where high-density cultivated areas have been extended and where ancestral or traditional cultivars have often been discarded in favor of a few highly productive varieties [20]. These new conditions of cultivation might have to face a resurgence of several pests, including PPN. To address these hypotheses, this study was undertaken in order to: (i) describe the species diversity of PPN communities associated with wild, feral and cultivated olives in Morocco where their diversity is completely unknown, and (ii) assess how anthropogenic forces (propagation and intensification practices) could impact the diversity and the structure of PPN communities by comparing them between different olive growing modalities.

Methods

Site description

Sampling of soil and olive leaves took place in Morocco from March to April 2012. Wild olive locations were as far as possible from current orchards. In contrast, feral olive locations were sampled within the proximity of cultivated olive stands or near main roads. The survey was conducted at 94 sites in several geographic regions all along a northeast-southwest 900-km long transect (Fig. 1; Table 1). The main regions sampled included: (i) the Souss region (15 sites), located on the southern side

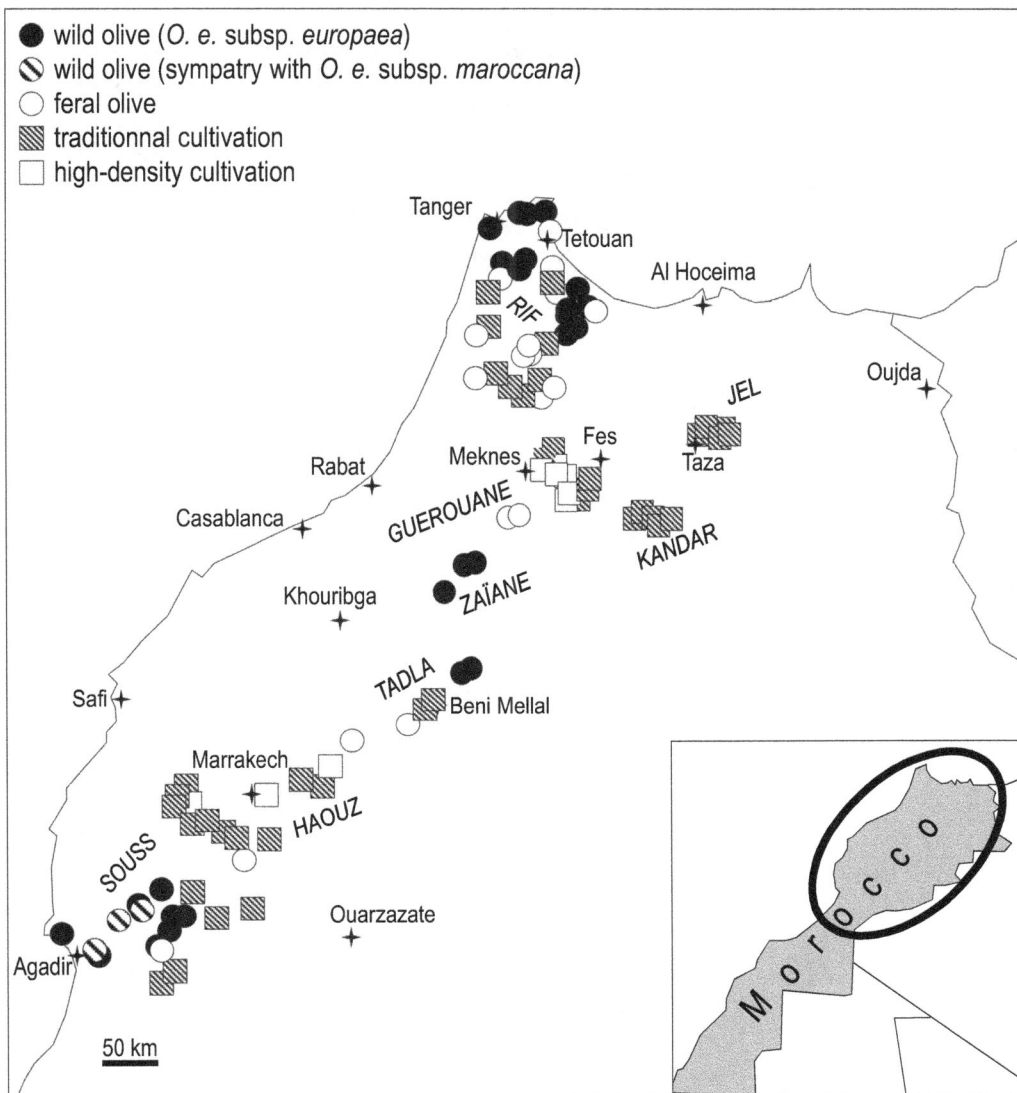

Fig. 1 Sites sampled in Morocco. Olive-growing modalities are given for each site

of the High Atlas Mountains near Agadir, where sampled trees were either wild (including trees of *O. europaea maroccana* in sympatry with *O. europaea* var. *sylvestris*), feral, or traditionally cultivated; (ii) the Haouz region (15 sites) located on the northern side of the High Atlas Mountains near Marrakech, where sampled trees were traditionally or high-density cultivated, or feral; (iii) the Tadla region (five sites) located along the northern side of the southern Middle Atlas Mountains near Beni Mellal, where sampled trees were either wild, feral, or traditionally cultivated; (iv) the Zaïane region (three wild olive sites), south of Meknes; (v) the Guerouane region (with traditionally or high-density cultivated sampled trees,

and less feral trees); (vi) the Kandar region (five sites) located in the northern Middle Atlas Mountains, south of Fes and the Jel plain situated to the east of Taza in eastern Morocco (five sites), where trees are traditionally cultivated; and (vii) both the Atlantic and Mediterranean slopes of the Rif mountains in the north (33 sites) where most of the sampled trees were wild or feral, and less traditionally cultivated.

Soil sampling
Considering that PPN spend all or almost all their life in the soil [21], the nematode sampling only included soil. A total of 213 samples were collected from the 94 sites.

Table 1 Location of the olive sampling sites surveyed in Morocco

Geographic region	City	Olive modality	No of sites	Latitude N (decimal°)	Longitude W (decimal°)
Souss	Tiguert	Wild	2	30.63	9.86
	Aourir	Wild	1	30.52	9.59
	Ouled Teïma	Wild	1	30.81	9.14
		Feral	1	30.42	9.02
		Traditional cultivation	1	30.42	9.02
	Taroudant	Wild	4	30.74	8.77
		Traditional cultivation	2	30.61	9.34
	Ouled Berhil	Traditional cultivation	1	30.65	8.18
	Aoulouz	Traditional cultivation	1	30.55	8.66
Haouz	El Kelaa Des Sraghna	Feral	1	32.15	7.26
		Traditional cultivation	1	31.37	7.95
	Tamellalt	Traditional cultivation	1	31.46	7.98
	Sidi Bou Othmane	High-density cultivation	1	31.70	7.69
	Marrakech	High-density cultivation	2	31.69	8.11
		Traditional cultivation	7	31.63	8.10
	Tahannaout	Traditional cultivation	1	31.57	7.97
	Asni	Feral	1	31.28	7.96
Tadla	Beni Mellal	Wild	2	32.58	5.98
		Traditional cultivation	1	32.21	6.83
	El Ksiba	Feral	2	32.32	6.39
Zaïane	Oulmes	Wild	2	33.32	6.07
	Oued Zem	Wild	1	33.33	6.00
Guerouane	El Hajeb	High-density cultivation	2	33.70	5.63
		Traditional cultivation	2	33.77	5.71
	Meknes	High-density cultivation	4	33.88	5.41
		Traditional cultivation	3	33.85	5.39
	Khemisset	Feral	2	33.63	5.83
Kandar	Sefrou	Traditional cultivation	5	33.87	4.88
Jel	Taza	Traditional cultivation	3	34.25	3.80
	Msoun	Traditional cultivation	2	34.26	3.74
Rif	Tanger	Wild	1	35.79	5.92
	Fnideq	Wild	2	35.78	5.37
	Tetouan	Wild	1	35.54	5.62
		Feral	1	34.79	5.77
	Asilah	Wild	4	35.07	5.33
		Traditional cultivation	1	35.05	5.35
	Chefchaouen	Wild	8	35.07	5.33
		Feral	3	35.07	5.32
		Traditional cultivation	2	35.38	5.37
	Bni Harchen	Wild	2	35.54	5.62
	Ouazzane	Wild	1	34.94	5.53
		Feral	2	34.79	5.77
		Traditional cultivation	5	34.79	5.77

This was done with a small spade under the foliage of each olive tree from the upper rhizosphere (the 15–20-cm deep layer inhabited by pleiotropic roots), in the close vicinity of active olive roots. This ensured that roots from weeds or other herbaceous plants were unlikely sampled. On cultivated olive (traditional and high-density cultivation), tillage and other human activities are frequent, which could lead to the homogenization of the PPN communities in an orchard. Each orchard was therefore considered as a repetition per modality. The sampling was carried out in each orchard along transects under four trees located at a distance of approximately 10 m. Five sub-samples were collected from each tree. These 20 sub-samples were thoroughly mixed to obtain a single representative sample per orchard. Contrary to cultivated orchards, heterogeneous PPN communities were expected in wild and feral olive trees because human interventions are scarce or absent. Each tree was thus taken as a repetition. Five sub-samples were also collected from each tree and then combined to form one 1-dm³ reference sample per tree.

Genetic characterization of the olive tree

In order to confirm the determination of olive-growing modalities, three olive branches corresponding to soil samples were collected to determine the chloroplast haplotype of each tree (according to [22]). All cultivated olive sampled trees only show the haplotype E1-1. Feral olive sampled trees show only E1-1 or mixtures with E2 and E3 haplotypes (i.e., E2-1, E2-2, E2-4 and E3-3, E3-4). E2 and E3 have been previously detected in Moroccan cultivars, but with frequencies below 5% [16]. Wild sampled trees show haplotypes characteristic of Moroccan-Iberian oleasters (i.e. E2-5, E2-6, E2-14, E3-4, E3-7, E3-8) and of *O. europaea maroccana* (M1-1, M1-2, M1-7).

Nematode extraction, identification and quantification

All of the nematode analyses were performed in the nematode quarantine area (French Government Agreement No 80622) of the Research Unit, "Centre de Biologie pour la Gestion des Populations" (Montpellier, France).

A 250-cm³ wet aliquot was taken from each soil sample for nematode extraction using the elutriation procedure [23]. PPN belonging to the Aphelenchida, Dorylaimida, Triplonchida and Tylenchida orders were enumerated in 5-cm³ counting chambers [24] and identified at the genus level based on dichotomous keys [25] and at the species level with genus-specific keys. The population levels were expressed per dm³ of fresh soil. Concerning specific identification, the nematode suspensions were preserved in mixture of formalin and glycerine [26], and then adult specimens were processed according to Seinhorst method [27] and mounted onto slides [28] for microscopic observation. Root-knot nematodes (*Meloidogyne* spp.) were identified at the species level by biochemical (esterase patterns) and molecular (SCAR markers and 28S rDNA D2-D3 expansion segments) approaches [29].

Analyses of nematode diversity

Several ecological indices were used:

a. Taxonomical diversity: (i) the total number of PPN in a community (N); (ii) the species richness (S); (iii) the Shannon–Wiener diversity index H' ($H' = -\sum p_i ln p_i$, where p_i is the proportion of individuals in each species (iii) that quantifies the local diversity or the heterogeneity of diversity (H' ranges from 0 to $ln(S)$); and (iv) the evenness ($E = H'/ln\ S$) that quantifies the regularity of species distribution within the community (E varies between 0 and 1).

b. Functional diversity: PPN species detected in communities were distributed into life-strategy groups according to the colonizer/persister value (cp-value) of the family to which they belong [30]. The diversity of the community was described by calculating: (i) the plant-parasitic index ($PPI = \sum cp_i n_i/N$), which quantifies the plant-feeding diversity of the communities; (ii) the relative mean abundance (%) of each cp-value class in a community calculated as follows: $Rcp_i = cp_i n_i/N$; (iii) the genus richness included in each cp-value class. PPN species were also assigned to the trophic groups according to their feeding habits [31, 32]: obligate plant feeders (OPF), facultative plant feeders (FPF) that alternatively feed on fungi, and fungal feeders (FF) that alternatively feed on plants. These trophic groups were also described according to (i) the relative mean abundance (%) of individuals within each of them, and (ii) the genus richness included in each [33].

c. The structure of PPN communities was designed at the genus level. The dominance of each nematode genus in the samples was first estimated by modeling the abundance (A) and the frequency (F) of each genus in the whole samples [34]. Afterwards, PPN community structures were described according to multivariate statistical analyses.

Data analyses

These diversity indices were calculated using the Vegan library [35]. In order to evaluate the impact of anthropogenic changes on biodiversity and community structures, different olive variables were defined according to olive-growing modalities: wild (WO), feral (FO), traditional or low-density cultivation (TR) and modern or high-density cultivation (HD), and according to olive irrigation conditions: irrigated or rainfed. The mean values of the different nematode diversity indices were compared according to olive propagation (wild vs cultivated) and to intensification practices (traditional vs high-density, irrigated vs rainfed). Principal Component Analysis (PCA) was carried out on nematode genera data in order to describe PPN community structures. To assess the impact of olive anthropogenic changes on taxonomical structures, a co-Inertia Analysis (CIA) was applied between olive-growing modality data (WO-FO-TR-HD) and PPN genera. The scarcest genera (with total abundance less than 1%) were then excluded from the dataset prior to running the analysis. These different multivariate analyses and graphs were performed using *ade4* library [36, 37]. All analyses were done using R version 3.3.2 [38]. The Wilcox (non-parametric) test was used for all pair-wise multiple comparisons. Differences obtained at levels of $P < 0.05$ were considered to be significant.

Results

PPN diversity associated with olive trees in Morocco

The PPN communities associated with olive trees in Morocco were highly diversified. A total of 117 species and 47 genera were identified. They belong to two families of Aphelenchida, to a family of Dorylaimida, to a family of Triplonchida and to 14 families of Tylenchida (Table 2).

At the family level, the Tylenchidae and Telotylenchidae were dispersed in all the regions sampled; they were the most diversified families, including 11, 9 genera in each, respectively. However, each genus was often represented by one or two species only (e.g. *Amplimerlinus, Bitylenchus, Tylenchus*). Most of these species were very rare as they were detected in one or two sites only (e.g. *Aglenchus agricola, Coslenchus gracilis* and *Paratrophurus loofi* in the Rif region). In contrast, the Hoplolaimidae family was represented by two genera only (*Helicotylenchus* and *Rotylenchus*), but the number of species identified in each genus was high (11 and 4 species, respectively), and they were distributed in all the regions, except in eastern Morocco (the Kandar and Jel regions). Longidoridae and Trichodoridae nematodes were detected mostly in the Rif region. Root-lesion nematodes (e.g. *Pratylenchus*) and Pin nematodes Paratylenchidae (e.g. *Paratylenchus*) were dispersed at all the sites surveyed. Four root-knot nematodes species were identified: *Meloidogyne arenaria* and *M. hapla* were detected in the Rif region, *M. javanica* was generally detected in southern Morocco (in the Souss and Haouz regions) and in the Guerouane and Tadla regions. *M. spartelensis* is a new species identified in the Rif region; another new species seems to occur in the Souss region (identification is in progress). Other families such as Criconematidae and Psilenchidae were detected in a few sites.

Among the 47 identified genera, *Filenchus, Helicotylenchus, Merlinius, Paratylenchus, Pratylenchus, Rotylenchus, Tylenchorhynchus* and *Xiphinema* were the most widespread in olive soils. Considering the species level, 11 *Helicotylenchus* species (Hoplolaimidae) were frequently collected in olive samples. Among them, *H. crassatus* was clearly the most dominant species (occurring in 58% of the samples). It was present in all regions except in the Jel and Kandar regions. *H. dihystera* and *H. varicaudatus* also occurred in 43 and 32% of the samples, respectively. In contrast, *H. exallus* and *H. minzi*, detected in the Guerouane region, and *H. pseudorobustus*, detected in the Haouz region, were scarcer. In addition, *Merlinius brevidens* (Telotylenchidae) and *Filenchus filiformis* (Tylenchidae) were also frequently recovered (51 and 40% of the samples, respectively).

Diversity of PPN communities according to anthropogenic changes

Diversity indices mean values were compared between to the four olive-growing modalities and between rainfed and irrigated olive samples.

(a) Taxonomical diversity

The total number of PPN (N) was up to two times higher on cultivated (HD & TR) than on non-cultivated olive (WO & FO). Similarly on irrigated olive, the total number of PPN was higher (Table 3). In contrast, the PPN communities were significantly richer in species (S), more diversified (H') and more homogenously distributed (E) in communities on WO and FO and on rainfed olive than on TR and HD and on irrigated olive.

(b) Functional diversity

The PPN identified were allocated in all the parasitic *cp*-values (*cp*-2 to *cp*-5 groups, Table 2). The WO and HD modalities revealed nematode communities with significantly higher plant-parasitic indices (PPI) than those in FO and in TR orchards (Table 4). This means that WO and HD olive areas had significantly more plant-feeding nematodes with higher *cp* values than other olive systems. The most opportunist/colonizer PPN (*cp*-2 and *cp*-3) dominated in all the communities (44 and 48%, respectively; Table 2). The overall abundance and occurrence of the persister nematodes (*cp*-4 and *cp*-5) was very low (4% for each *cp* class). Any effect was recorded on the *cp*-4 class. *Cp*-2 and *cp*-3 nematodes were more abundant in TR and HD, while *cp*-5 nematodes occurred more often in WO areas and were completely absent in HD orchards.

Concerning the trophic groups within communities, the OPF nematodes were the most dominant (62%), while the FPF and the FF nematodes were the least frequent (26 and 12%, respectively). FF nematodes were significantly more numerous in WO areas (Table 4). FPF and OPF nematodes were more abundant in TR and

Table 2 Plant-parasitic nematode taxa associated with olive trees in Morocco

Orders and families (cp value)	Species (trophic group)	Authors	Geographic regions														
			Rif							Jel		Kandar	Guerouane			Zaïane	
			Tanger	Fnideq	Tet-ouan	Asilah	Chef-chaouen	Bni Harchen	Ouaz-zane	Taza	Guercif	Sefrou	El Hajeb	Meknes	Khemis-set	Oulmes	Oued Zem
Aphelenchida																	
Aphelenchidae (2)	*Aphelenchus avenae* (F)	Bastian, 1865		+	+	+	+		+					+			
	A. isomerus (F)	Anderson and Hooper, 1980		+													
Aphelen-choididae (2)	*Aphelen-choides graminis* (F)	Baranovskaya and Haque, 1968		+													
	A. helicus (F)	Heyns, 1964				+											
	A. saprophilus (F)	Franklin, 1957		+													
	Aprutides guidetti (F)	Scognamiglio, 1974	+			+	+		+								
Dorylaimida																	
Longidoridae (5)	*Longidorus* sp. (OPF)	Micoletzky, 1922	+	+	+		+		+					+		+	
	Xiphinema pachtai-cum (OPF)	Tulaganov, 1938	+		+	+	+	+	+								
	X. turcicum (OPF)	Luc and Dal-masso, 1964	+														
	X. vuittenezi (OPF)	Luc et al. 1964														+	+
	Xiphinema sp. (OPF)	Cobb, 1913	+	+	+	+	+	+	+			+		+			
Triplonchida																	
Trichodoridae (4)	*Paratricho-dorus* sp. (OPF)	Siddiqi, 1974										+		+			
	Trichodorus sp. (OPF)	Cobb, 1913	+														

Table 2 continued

Orders and families (cp value)	Species (trophic group)	Authors	Geographic regions														
			Rif							Jel		Kandar	Guerouane			Zaïane	
			Tanger	Fnideq	Tet-ouan	Asilah	Chef-chaouen	Bni Harchen	Ouaz-zane	Taza	Guercif	Sefrou	El Hajeb	Meknes	Khemis-set	Oulmes	Oued Zem
Tylenchida																	
Anguinidae (2)	Ditylenchus emus (FPF)	Khan et al., 1969															
	D. equalis (FPF)	Heyns, 1964				+			+								
	Notho-tylenchus acutus (F)	Khan, 1965				+											
	N. adasi (F)	Syces, 1980						+									
	N. geraerti (F)	Kheiri, 1971	+				+		+								
	N. medians (F)	Thorne and Malek, 1968							+								
Criconemati-dae (3)	Ogma rhombos-quamatus (OPF)	Mehta and Raski, 1981	+														
	Criconema sp. (OPF)	Hofmänner and Menzel, 1914												+			
	Criconemella sp. (OPF)	De Grisse and Loof, 1965															
	Macropost-honia sp. (OPF)	De Man, 1880					+										
Dolichoridae (3)	Neodolicho-rhynchus microphas-mis (OPF)	Loof, 1960															
Heteroderi-dae (3)	Heterodera riparia (OPF)	Subbotin et al, 1997				+	+										
	Heterodera sp. (OPF)	Schmidt, 1871	+													+	+

Table 2 continued

Orders and families (cp value)	Species (trophic group)	Authors	Geographic regions														
			Rif							Jel		Kandar	Guerouane		Khemisset	Zaiane	
			Tanger	Fnideq	Tetouan	Asilah	Chefchaouen	Bni Harchen	Ouazzane	Taza	Guercif	Sefrou	El Hajeb	Meknes		Oulmes	Oued Zem
Hoplolaimidae (3)	*Helicotylenchus canadensis* (OPF)	Waseem, 1961		+	+				+								
	H. crassatus (OPF)	Anderson, 1973	+	+	+	+	+	+	+				+	+	+	+	+
	H. crenacauda (OPF)	Sher, 1966	+														
	H. digonicus (OPF)	Perry, 1959	+	+		+	+		+							+	+
	H. dihystera (OPF)	Cobb, 1893	+	+	+	+	+	+	+						+		
	H. exallus (OPF)	Sher, 1966													+		
	H. minzi (OPF)	Sher, 1966													+		
	H. pseudorobustus (OPF)	Steiner, 1914															
	H. tunisiensis (OPF)	Siddiqi, 1964							+								
	H. varicaudatus (OPF)	Yuen, 1964	+	+	+	+		+	+							+	+
	H. vulgaris (OPF)	Yuen, 1964		+		+	+		+								
	Helicotylenchus sp. (OPF)	Steiner, 1945			+	+	+		+	+	+		+	+			
	Rotylenchus buxophilus (OPF)	Golden, 1956															
	R. goodeyi (OPF)	Loof and Oostenbrink, 1958															
	R. pumilus (OPF)	Perry, 1959	+			+			+						+		
	R. robustus (OPF)	de Man, 1876				+			+								

Table 2 continued

Orders and families (cp value)	Species (trophic group)	Authors	Geographic regions														
			Rif							Jel		Kandar	Guerouane			Zaïane	
			Tanger	Fnideq	Tet-ouan	Asilah	Chef-chaouen	Bni Harchen	Ouaz-zane	Taza	Guercif	Sefrou	El Hajeb	Meknes	Khemis-set	Oulmes	Oued Zem
	Rotylenchus sp. (OPF)	Filipjev, 1936	+		+		+	+	+	+	+	+	+	+	+		+
Meloidogynidae (3)	*Meloidogyne arenaria* (OPF)	Neal, 1889	+														
	M. hapla (OPF)	Chitwood, 1949	+														
	M. spartelensis (OPF)	Ali et al. 2015	+														
	Meloidogyne sp2 (OPF)	Goeldi, 1892							+								
Paratylenchidae (2)	*Cacopaurus* sp. (OPF)	Thorne, 1943							+							+	
	Paratylenchus (Gracilacus) sp. (OPF)	Raski, 1962	+	+		+	+		+								
	Paratylenchus (P.) microdorus (OPF)	Andrássy, 1959							+								
	P. (P.) nanus (OPF)	Cobb, 1923	+														
	P. (P.) sheri (OPF)	Raski, 1973				+			+								
	P. (G.) straeleni (OPF)	de Coninck, 1931	+				+										
	P. (P.) vandenbrandei (OPF)	De Grisse, 1962	+				+										
	P. (P.) veruculatus (OPF)	Wu, 1962										+	+	+	+	+	
	Paratylenchus (Paratylenchus) sp. (OPF)	Micoletzky, 1922	+	+			+	+	+	+		+	+	+	+	+	+

Table 2 continued

Orders and families (cp value)	Species (trophic group)	Authors	Rif – Tanger	Fnideq	Tet-ouan	Asilah	Chef-chaouen	Bni Harchen	Ouaz-zane	Jel – Taza	Guercif	Kandar – Sefrou	Guerouane – El Hajeb	Meknes	Khemis-set	Zaïane – Oulmes	Oued Zem
Pratylenchidae (3)	Pratylenchoides hispaniensis (OPF)	Troccoli et al., 1997	+						+								
	P. laticauda (OPF)	Braun and Loof, 1967				+											
	Pratylenchoides sp. (OPF)	Winslow, 1958		+						+	+	+		+	+	+	+
	Pratylenchus crenatus (OPF)	Loof, 1960		+		+											
	P. mediterraneus (OPF)	Corbett, 1983							+								
	P. neglectus (OPF)	Rensch, 1924		+								+					
	P. pinguicaudatus (OPF)	Corbett, 1969				+											
	P. thornei (OPF)	Sher and Allen, 1953		+													
	Pratylenchus sp. (OPF)	Filipjev, 1936	+	+	+	+	+	+	+	+	+	+	+	+	+	+	
	Zygotylenchus guevarai (OPF)	Tobar Jiménez, 1963		+	+	+	+		+	+	+	+	+	+		+	
Psilenchidae (2)	Psilenchus aestuarius (FPF)	Andrássy, 1962															
	P. hilarulus (FPF)	de Man, 1921			+												
Rotylenchulidae (3)	Rotylenchulus sp. (OPF)	Linford and Oliveira, 1940	+			+	+		+	+	+	+				+	+
Telotylenchidae (3)	Amplimerlinius globigerus (OPF)	Siddiqi, 1979				+	+	+	+					+	+		

Table 2 continued

Orders and families (cp value)	Species (trophic group)	Authors	Geographic regions														
			Rif							Jel		Kandar	Guerouane			Zaïane	
			Tanger	Fnideq	Tet-ouan	Asilah	Chef-chaouen	Bni Harchen	Ouaz-zane	Taza	Guercif	Sefrou	El Hajeb	Meknes	Khemis-set	Oulmes	Oued Zem
	A. interme-dius (OPF)	Bravo, 1976							+								
	A. paraglo-bigerus (OPF)	Castillo et al., 1990															+
	Bitylenchus aerolatus (OPF)	Tobar Jiménez, 1970															
	Merlinius breviders (OPF)	Allen, 1955	+	+	+	+	+		+					+	+	+	
	M. micro-dorus (OPF)	Geraert, 1966						+									
	M. nothus (OPF)	Allen, 1955	+	+			+	+									
	Merlinius sp. (OPF)	Siddiqi, 1970								+		+					
	Nagelus obscurus (OPF)	Allen, 1955		+													
	Paratrophu-rus loofi (OPF)	Arias, 1970				+											
	Scutylenchus lenorus (OPF)	Brown, 1956															
	S. mamillatus (OPF)	Tobar-Jiménez, 1966															
	S. tessellatus (OPF)	Goodey, 1952															
	Telotylenchus avaricus (OPF)	Kleynhans, 1975															
	T. paaloofi (OPF)	Tikyani and Khera, 1970															
	T. ventralis (OPF)	Loof, 1963							+								

Table 2 continued

Orders and families (cp value)	Species (trophic group)	Authors	Rif							Jel		Kandar	Guerouane		Khemisset	Zäiane	
			Tanger	Fnideq	Tetouan	Asilah	Chefchaouen	Bni Harchen	Ouazzane	Taza	Guercif	Sefrou	El Hajeb	Meknes	Khemisset	Oulmes	Oued Zem
	Trophurus sculptus (OPF)	Loof, 1956		+		+	+										
	Tylenchorhynchus clarus (OPF)	Allen, 1955	+		+	+		+	+								
	T. crassicaudatus (OPF)	Williams, 1960						+									
	Tylenchorhynchus sp. (OPF)	Cobb, 1913			+					+	+	+	+	+			+
Tylenchidae (2)	*Aglenchus agricola* (FPF)	de Man, 1884						+									
	Basiria flandriensis (FPF)	Gerraert, 1968				+											
	B. graminophila (FPF)	Siddiqi, 1959	+			+	+								+		
	B. tumida (FPF)	Colbran, 1960	+					+									
	Boleodorus clavicaudatus (F)	Thorne, 1941					+										
	B. thylactus (F)	Thorne, 1941		+		+	+										
	B. volutus (F)	Lima and Siddiqi, 1963				+		+									
	Coslenchus gracilis (FPF)	Andrássy, 1982															
	Discotylenchus sp. (FPF)	Siddiqi, 1980															
	Filenchus andrassyi (FPF)	Szczygiel, 1969														+	

Table 2 continued

Orders and families (cp value) / Species (trophic group)	Authors	Geographic regions														
		Rif							Jel		Kandar	Guerouane			Zaïane	
		Tanger	Fnideq	Tetouan	Asilah	Chefchaouen	Bni Harchen	Ouazzane	Taza	Guercif	Sefrou	El Hajeb	Meknes	Khemisset	Oulmes	Oued Zem
F. baloghi (FPF)	Andrássy, 1958						+								+	+
F. filiformis (FPF)	Bütschli, 1873	+	+	+	+	+	+	+				+	+	+		
F. hamatus (FPF)	Thorne and Malek, 1968		+					+								
F. misellus (FPF)	Andrássy, 1958		+		+	+								+	+	+
F. sandneri (FPF)	Wasilewska, 1965	+														
Filenchus sp. (FPF)	Andrássy, 1954			+		+	+	+	+	+	+	+	+			
Irantylenchus vicinus (FPF)	Szczygiel, 1970				+											
Malenchus acarayensis (FPF)	Andrássy, 1968															
M. andrassyi (FPF)	Merny, 1970															
M. exiguus (FPF)	Massey, 1969					+										
Malenchus sp. (FPF)	Andrássy, 1968															
Miculenchus salvus (FPF)	Andrássy, 1959															
Ottolenchus discrepans (FPF)	Andrássy, 1954															
O. facultativus (FPF)	Szczygiel, 1970				+									+		
Tylenchus elegans (FPF)	De Man, 1876				+											
Tylenchus sp. (FPF)	Bastian, 1865															

Table 2 continued

Orders and families (cp value)	Species (trophic group)	Authors	Tadla		Haouz				Tahnaout	Asni	Souss					
			Beni Mellal	El Ksiba	El Kelaa Des Sraghna	Tamellalt	Sidi Bou Othmane	Mar-rakech			Tiguert	Aourir	Ouled Teima	Taroudant	Ouled Berhil	Aoulouz
Aphelenchida																
Aphelenchidae (2)	*Aphelenchus avenae* (F)	Bastian, 1865	+	+				+			+		+	+		+
	A. isomerus (F)	Anderson and Hooper, 1980														
Aphelenchoididae (2)	*Aphelenchoides graminis* (F)	Baranovskaya and Haque, 1968														
	A. helicus (F)	Heyns, 1964						+								
	A. saprophilus (F)	Franklin, 1957		+							+					
	Aprutides guidetti (F)	Scognamiglio, 1974														
Dorylaimida																
Longidoridae (5)	*Longidorus* sp. (OPF)	Micoletzky, 1922	+	+												
	Xiphinema pachtaicum (OPF)	Tulaganov, 1938									+		+			
	X. turcicum (OPF)	Luc and Dalmasso, 1964									+					
	X. vuittenezi (OPF)	Luc et al. 1964														
	Xiphinema sp. (OPF)	Cobb, 1913	+	+	+					+	+	+	+	+	+	

Table 2 continued

Orders and families (cp value)	Species (trophic group)	Authors	Geographic regions													
			Tadla		Haouz				Tahnaout	Asni	Souss					
			Beni Mellal	El Ksiba	El Kelaa Des Sraghna	Tamellalt	Sidi Bou Othmane	Mar-rakech			Tiguert	Aourir	Ouled Teima	Taroudant	Ouled Berhil	Aoulouz
Triplonchida																
Trichodoridae (4)	*Paratricho-dorus* sp. (OPF)	Siddiqi, 1974						+								
	Tricho-dorus sp. (OPF)	Cobb, 1913		+												
Tylenchida																
Anguini-dae (2)	*Ditylen-chus emus* (FPF)	Khan et al. 1969									+					
	D. equalis (FPF)	Heyns, 1964									+					
	Nothoty-lenchus acutus (F)	Khan, 1965														
	N. adasi (F)	Syces, 1980	+													
	N. geraerti (F)	Kheiri, 1971									+		+			
	N. medians (F)	Thorne and Malek, 1968														
Criconema-tidae (3)	*Ogma rhom-bosqua-matus* (OPF)	Mehta and Raski, 1981														
	Criconema sp. (OPF)	Hofmän-ner and Menzel, 1914			+											
	Cricone-mella sp. (OPF)	De Grisse and Loof, 1965														
	Macropo-sthonia sp. (OPF)	De Man, 1880														

Table 2 continued

Orders and families (cp value)	Species (trophic group)	Authors	Tadla		Haouz					Souss						
			Beni Mellal	El Ksiba	El Kelaa Des Sraghna	Tamellalt	Sidi Bou Othmane	Mar-rakech	Tahnaout	Asni	Tiguert	Aourir	Ouled Teima	Taroudant	Ouled Berhil	Aoulouz
Dolichoridae (3)	*Neodolichorhynchus microphasmis* (OPF)	Loof, 1960											+			
Heteroderidae (3)	*Heterodera riparia* (OPF)	Subbotin et al, 1997														
	Heterodera sp. (OPF)	Schmidt, 1871		+	+						+		+	+		
Hoplolaimidae (3)	*Helicotylenchus canadensis* (OPF)	Waseem, 1961								+						
	H. crassatus (OPF)	Anderson, 1973	+	+	+			+	+	+	+		+	+		+
	H. crenacauda (OPF)	Sher, 1966	+		+			+								
	H. digonicus (OPF)	Perry, 1959	+	+								+				
	H. dihystera (OPF)	Cobb, 1893		+	+	+	+	+		+		+	+			
	H. exallus (OPF)	Sher, 1966														
	H. minzi (OPF)	Sher, 1966														
	H. pseudorobustus (OPF)	Steiner, 1914					+	+								
	H. tunisiensis (OPF)	Siddiqi, 1964														
	H. varicaudatus (OPF)	Yuen, 1964		+				+		+		+				+

Table 2 continued

Orders and families (*cp* value)	Species (trophic group)	Authors	Geographic regions														
			Tadla		Haouz						Souss						
			Beni Mellal	El Ksiba	El Kelaa Des Sraghna	Tamellalt	Sidi Bou Othmane	Mar-rakech	Tahnaout	Asni	Tiguert	Aourir	Ouled Teima	Taroudant	Ouled Berhil	Aoulouz
	H. vulgaris (OPF)	Yuen, 1964	+		+			+								+
	Helicoty-lenchus sp. (OPF)	Steiner, 1945		+	+	+		+	+		+		+	+	+	
	Rotylen-chus bux-ophilus (OPF)	Golden, 1956						+								
	R. goodeyi (OPF)	Loof and Oosten-brink, 1958									+					
	R. pumilus (OPF)	Perry, 1959		+				+								
	R. robustus (OPF)	de Man, 1876							+							
	Rotylen-chus sp. (OPF)	Filipjev, 1936	+	+	+			+	+	+	+	+	+	+	+	+
Meloidogy-nidae (3)	*Meloi-dogyne arenaria* (OPF)	Neal, 1889														
	M. hapla (OPF)	Chitwood, 1949														
	M. spart-elensis (OPF)	Ali et al, 2015														
	Meloido-gyne sp2 (OPF)	Goeldi, 1892			+			+			+					

Table 2 continued

Orders and families (cp value)	Species (trophic group)	Authors	Geographic regions														
			Tadla		Haouz						Souss					
			Beni Mellal	El Ksiba	El Kelaa Des Sraghna	Tamellalt	Sidi Bou Othmane	Mar-rakech	Tahnaout	Asni	Tiguert	Aourir	Ouled Teima	Taroudant	Ouled Berhil	Aoulouz
Para-tylenchi-dae (2)	*Cacopau-rus* sp. (OPF)	Thorne, 1943														
	Paratylen-chus (Gracila-cus) sp. (OPF)	Raski, 1962										+		+		
	Paratylen-chus (P.) micro-dorus (OPF)	Andrássy, 1959									+					
	P. (P.) nanus (OPF)	Cobb, 1923														
	P. (P.) sheri (OPF)	Raski, 1973														
	P. (G.) straeleni (OPF)	de Coninck, 1931														
	P. (P.) vanden-brandei (OPF)	De Grisse, 1962														
	P. (P.) verrucu-latus (OPF)	Wu, 1962						+								
	Paratylen-chus (Para-tylen-chus) sp. (OPF)	Micoletzky, 1922	+	+	+	+	+	+	+	+	+	+	+	+	+	+

Table 2 continued

Orders and families (cp value)	Species (trophic group)	Authors	Geographic regions													
			Tadla		Haouz						Souss					
			Beni Mellal	El Ksiba	El Kelaa Des Sraghna	Tamellalt	Sidi Bou Othmane	Mar-rakech	Tahnaout	Asni	Tiguert	Aourir	Ouled Teima	Taroudant	Ouled Berhil	Aoulouz
Pratylenchi-dae (3)	Pratylenchoides hispaniensis (OPF)	Troccoli et al. 1997												+		
	P. laticauda (OPF)	Braun and Loof, 1967												+		
	Pratylenchoides sp. (OPF)	Winslow, 1958	+	+				+			+	+	+			
	Pratylenchus crenatus (OPF)	Loof, 1960														
	P. mediterraneus (OPF)	Corbett, 1983														
	P. neglectus (OPF)	Rensch, 1924		+												
	P. pinguicaudatus (OPF)	Corbett, 1969	+					+								
	P. thornei (OPF)	Sher and Allen, 1953	+	+	+		+	+	+	+		+	+			
	Pratylenchus sp. (OPF)	Filipjev, 1936														+
	Zygotylenchus guevarai (OPF)	Tobar Jiménez, 1963			+						+			+		

Table 2 continued

Orders and families (cp value)	Species (trophic group)	Authors	Geographic regions													
			Tadla		Haouz						Souss					
			Beni Mellal	El Ksiba	El Kelaa Des Sraghna	Tamellalt	Sidi Bou Othmane	Marrakech	Tahnaout	Asni	Tiguert	Aourir	Ouled Teima	Taroudant	Ouled Berhil	Aoulouz
Psilenchidae (2)	Psilenchus aestuarius (FPF)	Andrássy, 1962			+											
	P. hilarulus (FPF)	de Man, 1921														
Rotylenchulidae (3)	Rotylenchulus sp. (OPF)	Linford and Oliveira, 1940	+	+	+			+			+	+		+		
Telotylenchidae (3)	Amplimerlinius globigerus (OPF)	Siddiqi, 1979						+								
	A. intermedius (OPF)	Bravo, 1976														
	A. paraglobigerus (OPF)	Castillo et al. 1990														
	Bitylenchus aerolatus (OPF)	Tobar Jiménez, 1970						+								
	Merlinius brevidens (OPF)	Allen, 1955	+		+		+	+		+	+	+	+	+		+
	M. microdorus (OPF)	Geraert, 1966					+									
	M. nothus (OPF)	Allen, 1955											+			
	Merlinius sp. (OPF)	Siddiqi, 1970		+	+			+								
	Nagelus obscurus (OPF)	Allen, 1955		+												

Table 2 continued

Geographic region groups: **Tadla** (Beni Mellal, El Ksiba); **Haouz** (El Kelaa Des Sraghna, Tamellalt, Sidi Bou Othmane, Mar‑rakech); Tahnaout, Asni; **Souss** (Tiguert, Aourir, Ouled Teima, Taroudant, Ouled Berhil, Aoulouz).

Orders and families (cp value) / Species (trophic group)	Authors	Beni Mellal	El Ksiba	El Kelaa Des Sraghna	Tamellalt	Sidi Bou Othmane	Mar-rakech	Tahnaout	Asni	Tiguert	Aourir	Ouled Teima	Taroudant	Ouled Berhil	Aoulouz
Paratrophurus loofi (OPF)	Arias, 1970														
Scutylenchus lenorus (OPF)	Brown, 1956								+		+	+			
S. mamillatus (OPF)	Tobar-Jiménez, 1966									+					
S. tessellatus (OPF)	Goodey, 1952								+						
Telotylenchus avaricus (OPF)	Kleynhans, 1975											+			
T. paalooti (OPF)	Tikyani and Khera, 1970	+							+			+			
T. ventralis (OPF)	Loof, 1963						+			+		+			
Trophurus sculptus (OPF)	Loof, 1956														
Tylenchorhynchus clarus (OPF)	Allen, 1955			+	+	+	+	+		+	+	+		+	
T. crassicaudatus (OPF)	Williams, 1960										+		+		
Tylenchorhynchus sp. (OPF)	Cobb, 1913	+	+	+	+		+	+		+		+	+	+	

Table 2 continued

Orders and families (cp value)	Species (trophic group)	Authors	Geographic regions													
			Tadla		Haouz					Asni	Souss			Taroudant	Ouled Berhil	Aoulouz
			Beni Mellal	El Ksiba	El Kelaa Des Sraghna	Tamellalt	Sidi Bou Othmane	Marrakech	Tahnaout		Tiguert	Aourir	Ouled Teima			
Tylenchidae (2)	Aglenchus agricola (FPF)	de Man, 1884														
	Basiria flandriensis (FPF)	Gerraert, 1968														
	B. graminophila (FPF)	Siddiqi, 1959														+
	B. tumida (FPF)	Colbran, 1960	+								+					
	Boleodorus clavicaudatus (F)	Thorne, 1941														
	B. thylactus (F)	Thorne, 1941	+	+				+			+		+			+
	B. volutus (F)	Lima and Siddiqi, 1963														
	Coslenchus gracilis (FPF)	Andrássy, 1982														
	Discotylenchus sp. (FPF)	Siddiqi, 1980	+													
	Filenchus andrassyi (FPF)	Szczygiel, 1969														
	F. baloghi (FPF)	Andrássy, 1958														
	F. filiformis (FPF)	Bütschli, 1873	+		+	+	+	+	+	+	+					+
	F. hamatus (FPF)	Thorne and Malek, 1968			+	+		+	+	+						+

Table 2 continued

Orders and families (cp value)	Species (trophic group)	Authors	Geographic regions													
			Tadla		Haouz						Souss					
			Beni Mellal	El Ksiba	El Kelaa Des Sraghna	Tamellalt	Sidi Bou Othmane	Mar-rakech	Tahnaout	Asni	Tiguert	Aourir	Ouled Teima	Taroudant	Ouled Berhil	Aoulouz
	F. misellus (FPF)	Andrássy, 1958									+					+
	F. sandneri (FPF)	Wasilewska, 1965									+					+
	Filenchus sp. (FPF)	Andrássy, 1954	+	+	+	+		+	+	+	+	+	+	+	+	+
	Irantylenchus vicinus (FPF)	Szczygiel, 1970			+											
	Malenchus acarayensis (FPF)	Andrássy, 1968									+					
	M. andrassyi	Merny, 1970														+
	M. exiguus (FPF)	Massey, 1969														+
	Malenchus sp. (FPF)	Andrássy, 1968														+
	Miculenchus salvus (FPF)	Andrássy, 1959						+								
	Ottolenchus discrepans (FPF)	Andrássy, 1954						+								
	O. facultativus (FPF)	Szczygiel, 1970										+				
	Tylenchus elegans (FPF)	De Man, 1876								+						
	Tylenchus sp. (FPF)	Bastian, 1865			+											

Trophic groups: *FF* fungal feeders, *FPF* facultative plant feeders, *OPF* obligate plant feeders

Table 3 Taxonomical diversity indices in PPN communities associated with olive (mean values) according to olive-growing modalities and water supply

Olive variables	Nb of samples	N	S	H′	E
Growing modality					
WO	88	2227 b	10.31 a	1.55 a	0.68 a
FO	75	2751 b	9.51 a	1.58 a	0.69 a
TR	40	4369 a	7.50 b	1.24 b	0.58 a
HD	10	4352 a	6.90 b	1.04 b	0.50 b
Water supply					
Rainfed	171	2512 b	9.87 a	1.56 a	0.69 a
Irrigated	42	4365 a	7.36 b	1.19 b	0.56 b

The letters (a–c) indicate significant differences among the variables measured according to ANOVA and Wilcoxon tests. $P < 0.05$

WO wild olive, *FO* feral olive, *TR* traditional cultivation, *HD* high-density cultivation, *N* total number of PPN/dm³ of soil, *S* species richness, *H′* local diversity, *E* evenness

HD orchards,w respectively. The ratio between FPF and OPF nematodes was unbalanced in favor of OPF in HD orchards, and in favor of FPF in TR orchards and in FO areas. The rainfed-irrigation modalities did not have any effect on the trophic groups.

The *cp*-2, *cp*-3, FPF and OPF functional groups were represented by the highest number of genera (44, 48, 26 and 62%, respectively). Comparing this richness in each group between olive-growing modalities only, the PPN communities detected in WO and FO demonstrated higher richness and diversity compared to those detected in TR and HD (Table 5).

(c) Community patterns

Community structure was described at the genus level. Modeling the dominance of each genus in the samples (Fig. 2a), 83% of the genera were classified as less frequent

(F < 30%) according to the model and 35% as occasional (F < 5%). A total of 62.5% of the nematode genera were classified as highly abundant according to the abundance threshold defined by the model (A = 200 nematodes/ dm³ of soil). Eight genera were classified as dominant (F ≥ 30% and A ≥ 10,000 nematodes/dm³ of soil): *Filenchus* and *Helicotylenchus* (F > 80%); and *Rotylenchus*, *Merlinius*, *Paratylenchus*, *Xiphinema*, *Pratylenchus* and *Tylenchorhynchus* (40 < F < 70%). Six other highly abundant genera were less frequent, including root-knot nematodes (*Meloidogyne* spp., F = 12.2%) and cyst nematodes (*Heterodera* spp., F = 10%). No genus was found to be frequent and in low abundance.

As shown by the PCA loading plot of the nematode taxa (Fig. 2b), Hoplolaimidae nematodes (*Helicotylenchus* and *Rotylenchus*), and *Paratylenchus*, *Filenchus* and *Pratylenchus* genera to a lesser extent, were correlated to the PC1 axis (negative values). The PC2 axis indicated contrasted positions for *Tylenchorhynchus* spp. (negative values), opposed to *Boleodorus*, *Xiphinema*, *Nothotylenchus*, *Merlinius*, *Rotylenchulus*, *Meloidogyne*, *Heterodera* and *Telotylenchus* (positive values).

Correspondences between PPN community patterns and olive-growing modalities

Considering olive-growing modalities, the loading plot of the Co-Inertia Analysis (CIA) analysis between nematode and olive data (Fig. 3) indicated an important contribution of the anthropogenic gradient (WO-FO-TR-HD) to the CIA1 axis. The CIA2 axis was essentially correlated with the feral growing modality (FO, positive values) and with the wild olive (WO, negative values). Regarding the projection of the nematode genera in the loading plot (Fig. 3), the analysis indicated that the genera *Merlinius*, *Xiphinema*, *Heterodera*, *Nothotylenchus*, *Rotylenchulus* and *Boleodorus* were correlated with WO. In contrast,

Table 4 Functional diversity in PPN communities on olive (mean values) according to olive-growing modalities and water supply

Olive variables	PPI	Rcp-2	Rcp-3	Rcp-4	Rcp-5	FF	FPF	OPF
Growing modality								
WO	2.65 a	45.58 a	48.89 b	0.08	5.45 a	8.69 a	32.63 b	58.68 ab
FO	2.57 ab	46.19 a	52.19 b	0.03	1.59 b	3.62 b	39.35 ab	57.03 ab
TR	2.49 b	52.62 a	46.68 b	0.04	0.66 b	3.69 b	46.89 a	49.42 b
HD	2.74 a	25.96 b	73.71 a	0.33	0.00 b	0.12 b	25.13 b	74.76 a
Water supply								
Rainfed	2.61	45.91	50.69	0.05	3.35 a	5.93	36.28	57.78
Irrigated	2.55	46.27	53.11	0.11	0.50 b	2.84	41.71	55.45

The letters (a–c) indicate significant differences among the variables measured according to ANOVA and Wilcoxon tests. $P < 0.05$

WO wild olive, *FO* feral olive, *TR* traditional cultivation, *HD* high-density cultivation, *PPI* plant parasitic index, relative mean abundance (%) of each *cp*-value (R*cp*-i) and of each trophic group (*FF* fungal feeders, *FPF* facultative plant feeders, *OPF* obligate plant feeders)

Table 5 Genus richness of PPN within each functional group according to olive-growing modalities

Olive-growing modalities	cp-2	cp-3	cp-4	cp-5	FF	FPF	OPF
WO	24	23	1	3	5	14	32
FO	13	14	1	2	3	8	19
TR	11	11	1	2	4	6	15
HD	5	11	2	0	1	3	14

WO wild olive, *FO* feral olive, *TR* traditional cultivation, *HD* high-density cultivation, *cp-2* to *cp-5* *cp*-values, *FF* fungal feeders, *FPF* facultative plant feeders, *OPF* obligate plant feeders

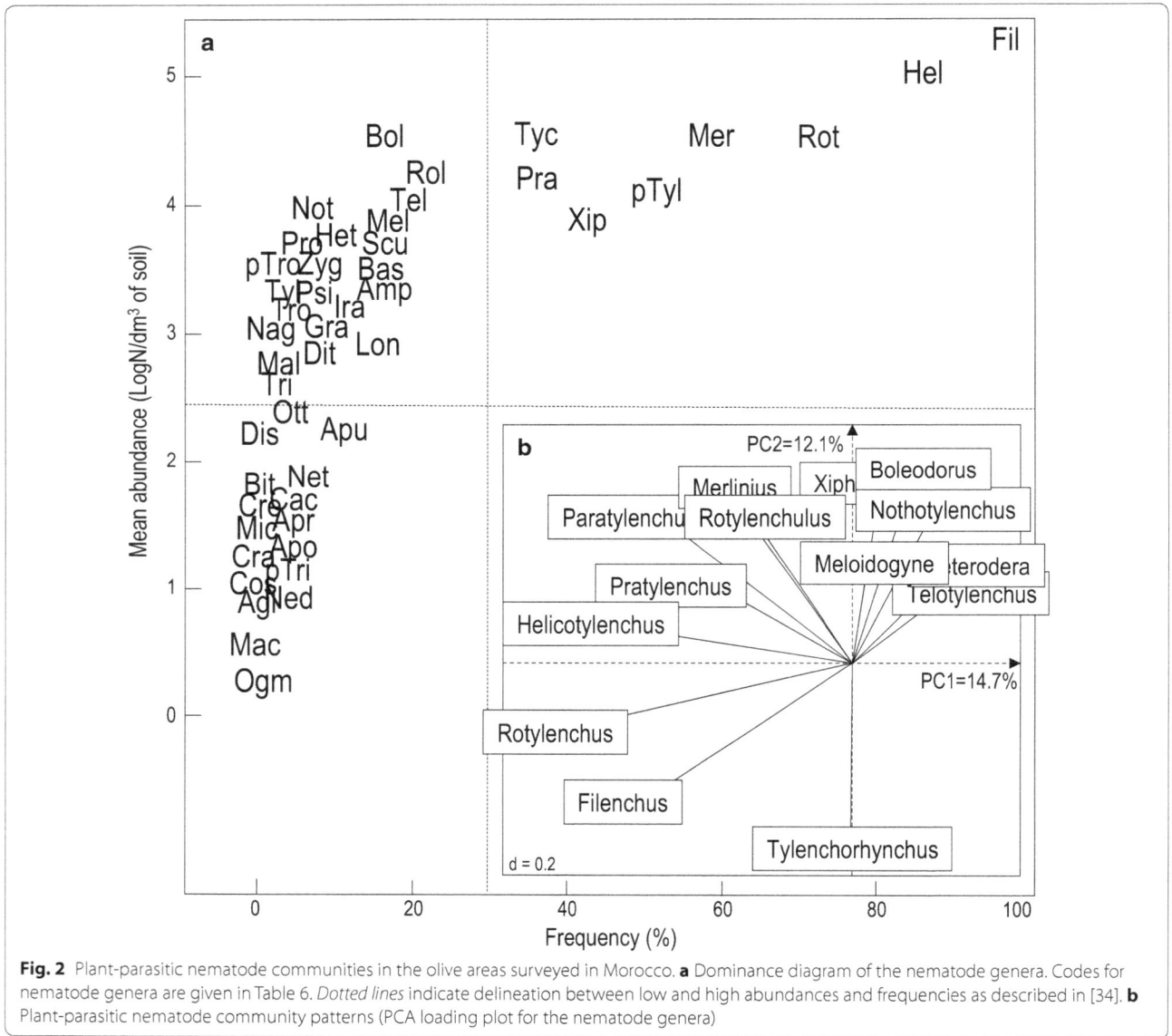

Fig. 2 Plant-parasitic nematode communities in the olive areas surveyed in Morocco. **a** Dominance diagram of the nematode genera. Codes for nematode genera are given in Table 6. *Dotted lines* indicate delineation between low and high abundances and frequencies as described in [34]. **b** Plant-parasitic nematode community patterns (PCA loading plot for the nematode genera)

Meloidogyne and *Tylenchorhynchus* were enhanced by cultivation practices (especially HD). The other nematode genera (*Filenchus, Pratylenchus*) were more closely related to TR, while *Telotylenchus, Helicotylenchus,* *Rotylenchus* and *Paratylenchus* were more closely related to FO. The mean comparisons of nematode abundances between the modality groups arranged according to their CIA1 eigenvalues (HD, TR and WO + FO) (Fig. 3)

Table 6 Nematodes genera and their corresponding codes

PPN genus	Code	PPN genus	Code	PPN genus	Code
Aglenchus	Agl	Helicotylenchus	Hel	Paratrophorus	pTro
Amplimerlinius	Amp	Heterodera	Het	Paratylenchus	pTyl
Aphelenchoides	Apo	Irantylenchus	Ira	Pratylenchoides	Pro
Aphelenchus	Apu	Longidorus	Lon	Pratylenchus	Pra
Aprutides	Apr	Macroposthenia	Mac	Psilenchus	Psi
Basiria	Bas	Malenchus	Mal	Rotylenchulus	Rol
Bitylenchus	Bit	Meloidogyne	Mel	Rotylenchus	Rot
Boleodorus	Bol	Merlinius	Mer	Scutylenchus	Scu
Cacopaurus	Cac	Miculenchus	Mic	Telotylenchus	Tel
Coslenchus	Cos	Nagelus	Nag	Trichodorus	Tri
Criconema	Cra	Neodolichorhynchus	Ned	Trophurus	Tro
Criconemella	Cre	Neotylenchus	Net	Tylenchorhynchus	Tyc
Discotylenchus	Dis	Nothotylenchus	Not	Tylenchus	Tyl
Ditylenchus	Dit	Ogma	Ogm	Xiphinema	Xip
Filenchus	Fil	Ottolenchus	Ott	Zygotylenchus	Zyg
Gracilacus	Gra	Paratrichodorus	pTri		

confirmed that *Meloidogyne* and *Tylenchorhynchus* nematodes were significantly more abundant in HD orchards compared to TR orchards or to WO + FO. Some significant differences were also detected between traditional and non-cultivated olive orchards. However, other nematodes such as *Merlinius*, *Xiphinema* and *Heterodera* were found to be significantly more abundant in WO + FO compared to cultivated olive conditions (HD, TR).

Discussion

Biodiversity is an essential ecological phenomenon because it represents a complex set of interacting ecological, evolutionary, biogeographical and physical processes [39]. Native biodiversity is being lost at a rapid rate owing to anthropogenic causes, including habitat destruction, pollution and the spread of non-native species [1, 40]. In this context, the main focus of this study was to understand how human activities (e.g. agricultural practices) in ecosystems could impact the diversity of PPN communities. The Mediterranean olive tree is particularly suitable for this study because it concerns ancient ecosystems with post-glacial refugia [16], many spots of Oleaster and many cases of feral olive. It also offers a large range of varieties, cultivated traditionally or at high-density, as present in Morocco.

PPN diversity associated with olive trees in Morocco

The PPN fauna and their distribution was totally unknown in Morocco before this study, except for a few

reports on some nematodes such as root-knot nematodes *Meloidogyne morocciensis* [41] and cereal cyst nematodes [42]. This study clearly highlights a high taxonomical diversity of PPN communities where 117 species belonging to 47 genera were recorded. In addition, the study adds taxa (seven genera and 60 species) that were recorded for the first time in association with olive trees worldwide. The dominance pattern was also revealed by PCA analyses that demonstrated that the nematode dataset was mainly structured by the most frequent and abundant genera, and by less frequent but abundant nematodes to a lesser extent. The communities observed were mainly dominated by *Filenchus* and *Helicotylenchus* genera, and other nematodes such as *Rotylenchus*, *Merlinius*, *Paratylenchus*, *Xiphinema*, *Pratylenchus* and *Tylenchorhynchus*. Some of them have been previously reported as widespread on olive trees worldwide [15]. High population levels of some nematode genera such as root-knot nematodes (*Meloidogyne* spp.) and cyst nematodes (*Heterodera* spp.), considered as very dangerous soil-borne plant pests were also recorded [43].

The taxonomical diversity of PPN analyzed in Morocco is the greatest when compared to other surveys on olive trees that documented 223 species worldwide (reported in [14, 15, 44–47]). This high diversity and the detection of new taxa could be essentially explained by: (i) a large sampling effort (213 soil samples corresponding to 363 trees sampled), conducted along a long transect (about 900 km) covering a wide range of olive-growing regions in Morocco; and (ii) a large proportion of samples collected in wild and feral olive areas (163 samples). These olive habitats could be considered as reservoirs of high diversity where a part remains unknown [48]. As evidence, a new root-knot nematode species, *Meloidogyne spartelensis*, was detected on wild olive in Northern Morocco [49]. However, other species could not be detected because they may occur only under unidentifiable life stages (e.g. juveniles), or their development may be linked to other periods of the year or to specific microhabitats [50]. As an example, no *Rotylenchulus* could be identified at the species level because all individuals were in the juvenile stage.

Impact of anthropogenic changes on the PPN communities associated with olive trees in Morocco

Taxonomical diversity indices were revealed impacted by olive propagation practices (from wild to cultivated olive): a high PPN richness was found in non-cultivated olive areas (wild and feral), with an equal distribution of species within communities (high evenness), contrary to what was observed in cultivated orchards (traditional and high-density). Nematode abundance was also significantly higher in orchards. A main conclusion also arose

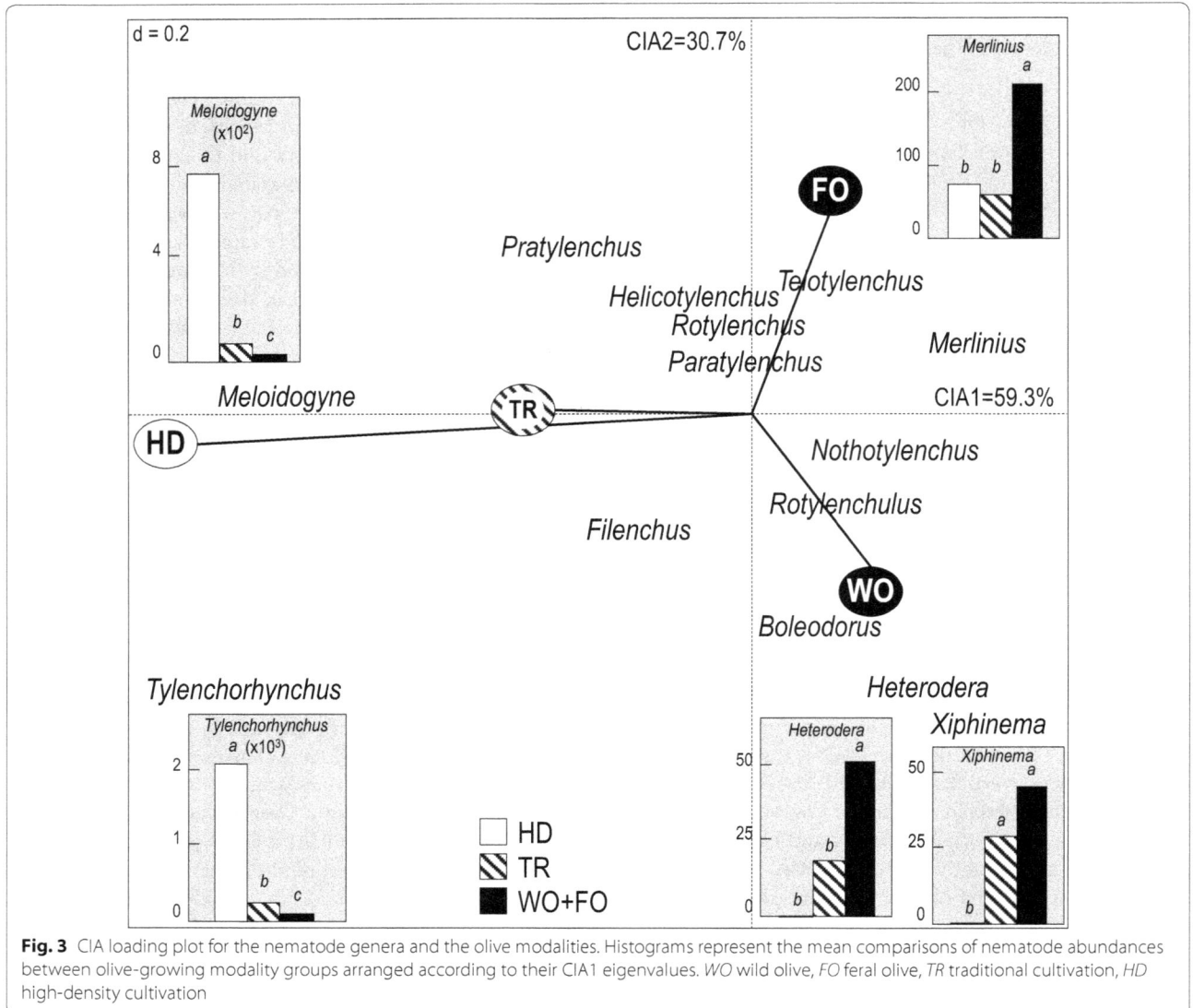

Fig. 3 CIA loading plot for the nematode genera and the olive modalities. Histograms represent the mean comparisons of nematode abundances between olive-growing modality groups arranged according to their CIA1 eigenvalues. *WO* wild olive, *FO* feral olive, *TR* traditional cultivation, *HD* high-density cultivation

in this study that showed that PPN are abundant in cultivated conditions while richness, local diversity and evenness are low, and vice versa in non-cultivated conditions. In other words, a high PPN species diversity within a community may prevent the multiplication of the species as a potential effect of trade-off interactions between nematode species and/or between them and other soil microorganisms [51, 52].

The study also highlighted the impact of anthropogenic practices on the functional diversity in communities: persisters and fungal-feeders were more diverse and numerous in wild olive conditions, whereas colonizers were frequently present under high-density conditions. Colonizer nematodes were represented by fewer genera, confirming imbalance between the high relative abundance and the low-genus richness and vice versa. Moreover, cp-5 nematodes were particularly related to wild olive, and totally absent under high-density olive cultivation conditions. This is consistent with other studies that demonstrated that cultivation intensification usually does not reduce the number of nematode trophic groups, but may change the composition of these groups [53]. The taxonomical structures of the communities were also distinguished between wild and cultivated olive: genera such as *Xiphinema* and *Heterodera* were detected in relation to natural ecosystems (wild olive), while others (e.g. *Meloidogyne* and *Tylenchorhynchus*) were favoured in cultivated areas. Dominant taxa such as *Helicotylenchus*, *Rotylenchus* and *Filenchus* did not appear to be impacted, which could explain their high dominance in the samples.

The taxonomical biodiversity indices were affected by the intensification level of farming systems between low or high tree-density orchards. The genus richness was usually higher in traditional than in high-density orchards. However, the intensification practices also impacted the functional diversity, as abundant cp-2 and FPF nematodes were found in traditional orchards, while cp-3 and OPF were more abundant in high-density olive orchards. The taxonomical structures of communities were also affected by olive cultivation intensification: genera such as *Meloidogyne* spp. and *Tylenchorhynchus* spp. were dominant in high-density orchards, whereas the traditional orchards were more favorable for the development of other genera such as *Pratylenchus* spp.

This study suggests that the PPN communities associated with non-cultivated olives (wild and feral) are not disturbed as a consequence of low or no human intervention in these ecosystems. This is consistent with other ecological observations that show that lowly-disturbed ecosystems generally host more diverse communities of soil organisms, as demonstrated for earthworms [54], for PPN [55] and for other soil biota communities [5, 6]. That is completely reversed in cultivated areas where the PPN communities were characterized by high abundances and low PPN's diversity. It is usually assumed that cropping systems are disturbed by human activities via agricultural practices (e.g. crop intensification, irrigation, tillage). These anthropogenic practices lead to species decline, as it has already been demonstrated on bees, birds and plants species [56], and soil biota [5, 6] including nematodes [57]. The decrease of nematodes diversity with increasing human activities can be attributed to several constraints such as physical disturbances, changes in quantity and quality of organic matter being returned to the soil and to the increase in the number of specific plant-feeding nematodes that are favoured by the selected crops [58].

These impacts on communities could be related to the biological characteristics of nematodes, leading them to respond differently to disturbances in their environment. These conditions induce favourable environments for PPN multiplication, especially irrigation, which enhances the development of roots [14]. This was consistent with others observations in southern Morocco [59]. This could explain the high abundance of colonizer species and, consequently, the high pathogenicity (PPI value) of the communities recorded in these cropping conditions. Moreover, agricultural practices applied in olive are very likely to select and multiply the most competitive and harmful PPN species such as *Meloidogyne* spp. in high-density orchards. That could also explain the absence of persister species in these conditions, since they are very sensitive to environmental disturbances. That agrees with

previous studies [53] showing that the greater cp-value nematodes are usually associated with low stress and undisturbed environments [9].

Conclusion

Anthropogenic changes such as propagation and intensification practices greatly impact the diversity of PPN communities associated with olive trees. Cultural practices (from wild to cultivated ecosystems or cropping intensification) could lead to community rearrangements in favour of highly pathogenic species defined as major agricultural pests [60]. In this vein, intensive production systems (high-yield varieties, irrigation, fertilization, etc.) induce environmental conditions suitable for the development of soil-borne diseases caused directly or indirectly (e.g. *Verticillium* wilt) by nematodes [14], such as root-knot (Meloidogynidae) and root-lesion (Pratylenchidae) nematodes. These groups of nematodes are known to affect olive production worldwide [15] and to be among the most frequent nematodes in nurseries [61]. Considering that the dispersal of PPN over long distances is passive (via contaminated irrigation, infected planting material or the dispersion of infested soil, etc. [30]), olive tree protection relies first on the use of healthy plant material (rootstocks) transplanted in a soil free of these parasites. The first step in avoiding PPN therefore starts in nurseries from where they could be introduced into olive orchards. This study also underlined PPN diversity and community structures as relevant indicators to assess resilient strategies in olive cropping systems. Further investigations should therefore focus on community rearrangements and on interactions between species co-existence mechanisms in order to develop diversity conservation or restoration (resilience) strategies [60] instead of reducing the most pathogenic species.

Abbreviations

A: abundance; ade4: analyse de données écologiques version 4; ANOVA: analysis of variation; CIA: co-inertia analysis; cp-value: a functional diversity index assigned to families of soil nematodes, which are categorized into a 1-5 colonizer-persister series; E: evenness; E1, E2, E3, M1, M2, M7: olive chloroplast lines; E1-1, E2-1, E3-4, M1-1, etc.: olive chloroplast haplotypes; F: frequency; FF: fungal feeders; FO: feral olive; FPF: facultative plant feeders; H': Shannon–Wiener diversity index; HD: high-density or modern cultivation; *In*: natural logarithm; N: total number of nematodes in a community; OPF: obligate plant feeders; PCA: principal component analysis; p_i: proportion of individuals in each species *i*; PPI: plant-parasitic index; PPN: plant-parasitic nematodes; *Rcp*: relative mean abundance (%) of each cp-value class in a community; S: species richness; SCAR: sequence characterized amplified region; TR: traditional or low-density cultivation; WO: wild olive.

Authors' contributions

NA, GB, BK, EC and TM designed the sampling device; NA, JT, GB, BK, MA, MAH, AEM, AEO, AEB, AM, EC and TM acquired the field data; GB, BK, LE and AEB processed the olive genotyping; NA, JT and TM processed the nematode extraction from soils; NA, JT, ED, GW and TM carried out the morphological characterization of the nematode genera and species; NA and JT carried out the biochemical and molecular characterization of the root-knot nematode species; NA, JT, OFG and TM analyzed the data; NA, GB, OFG, EC and TM drafted the manuscript. All authors read and approved the final manuscript.

Author details
[1] Plant Protection Department, Faculty of Agriculture, Tishreen University, PO Box 2233, Latakia, Syrian Arab Republic. [2] IRD, UMR CBGP, 755 Avenue du Campus Agropolis, CS30016, 34988 Montferrier-sur-Lez Cedex, France. [3] CNRS, UMR EDB, Université Toulouse III Paul Sabatier, Bâtiment 4R1, 118 Route de Narbonne, 31062 Toulouse Cedex 9, France. [4] UMR AGAP, SUPAGRO, Campus CIRAD, TAA-108/03, Avenue Agropolis, 34398 Montpellier Cedex 5, France. [5] Museum and Institute of Zoology PAS, Wilcza 64, 00-679 Warsaw, Poland. [6] Faculté des Sciences et Techniques, Université Abdelmalek Essaadi, BP 2062, 93030 Tétouan, Morocco. [7] Laboratoire LBVRN, Faculté des Sciences d'Agadir, Université Ibn Zohr, BP 8106, 80000 Agadir, Morocco. [8] INRA, CRRA, BP 513, 40000 Marrakech, Morocco. [9] INRA, UMR APCRPG, BP 578, 50000 Meknes, Morocco. [10] IRD, UMR IPME (IRD/Université de Montpellier/CIRAD), 911 Avenue Agropolis, BP 64501, 34394 Montpellier Cedex 5, France. [11] UMR PVBMT, 3P-CIRAD, 7 chemin de l'Irat, Ligne paradis, 97410 Saint Pierre, Réunion.

Acknowledgements
We would like to thank Simon Benateau (AgroCampus Ouest, Rennes, France) for his help in data analysis.

Competing interests
The authors declare that they have no competing interests.

Funding
This work was supported by a PhD grant from Tishreen University (Latakia, Syrian Arabic Republic). It was also funded by the PESTOLIVE project: Contribution of olive history for the management of soil-borne parasites in the Mediterranean Basin from EU and non-EU Mediterranean countries (ARIMNet action KBBE 219262) and by the LABEX entitled TULIP supported by the Agence Nationale de la Recherche (ANR-10-LABX-0041).

References
1. Chapin FS III, Zavaleta ES, Eviner VT, Naylor RL, Vitousek PM, Reynolds HL, Hooper DU, Lavorel S, Sala OE, Hobbie SE, Mack MC. Consequences of changing biodiversity. Nature. 2000;405:234–42.
2. Hooper DU, Chapin FS III, Ewel JJ, Hector A, Inchausti P, Lavorel S, Lawton JH, Lodge DM, Loreau M, Naeem S, Schmid B, Setälä H, Symstad AJ, Vandermeer J, Wardle DA. Effects of biodiversity on ecosystem functioning: a consensus of current knowledge. Ecol Monogr. 2005;75:3–35.
3. Giller PS. The diversity of soil communities, the 'poor man's tropical rainforest'. Biodivers Conserv. 1996;5:135–68.
4. Vackár D, ten Brink B, Loh J, Baillie JE, Reyers B. Review of multispecies indices for monitoring human impacts on biodiversity. Ecol Indic. 2012;17:58–67.
5. Postma-Blaauw MB, de Goede RGM, Bloem J, Faber JH, Brussaard L. Soil biota community structure and abundance under agricultural intensification and extensification. Ecology. 2010;91:460–73.
6. Postma-Blaauw MB, de Goede RGM, Bloem J, Faber JH, Brussaard L. Agricultural intensification and de-intensification differentially affect taxonomic diversity of predatory mites, earthworms, enchytraeids, nematodes and bacteria. Appl Soil Ecol. 2012;57:39–49.
7. Pen-Mouratov S, Rakhimbaev M, Steinberger Y. Seasonal and spatial variation in nematode communities in a Negev desert ecosystem. J Nematol. 2003;35:157–66.
8. Yeates GW, Bongers T, De Goede RG, Freckman DW, Georgieva SS. Feeding habits in soil nematode families and genera-an outline for soil ecologists. J Nematol. 1993;25:315–31.
9. Bongers T, Ferris H. Nematode community structure as a bioindicator in environmental monitoring. Trends Ecol Evol. 1999;14:224–8.
10. Djian-Caporalino C, Védie H, Arrufat A. Gestion des nématodes à galles: lutte conventionnelle et luttes alternatives. L'atout des plantes pièges. Phytoma. 2009;624:21–5.
11. Nicol JM, Turner SJ, Coyne DL, Den Nijs L, Hockland S, Maafi ZT. Current nematode threats to world agriculture. In: Genomics and molecular genetics of plant-nematode interactions. Springer: Netherlands; 2011. p. 21–43.
12. Nico AI, Jiménez-Díaz RM, Castillo P. Host suitability of the olive cultivars Arbequina and Picual for plant-parasitic nematodes. J Nematol. 2003;35:29–34.
13. Koenning SR, Overstreet C, Noling JW, Donald PA, Becker JO, Fortnum BA. Survey of crop losses in response to phytoparasitic nematodes in the United States for 1994. J Nematol. 1999;31:587–618.
14. Castillo P, Nico AI, Navas-Cortés JA, Landa BB, Jiménez-Díaz RM, Vovlas N. Plant-parasitic nematodes attacking olive trees and their management. Plant Dis. 2010;94:148–62.
15. Ali N, Chapuis E, Tavoillot J, Mateille T. Plant-parasitic nematodes associated with olive tree (Olea europaea L.) with a focus on the Mediterranean Basin: a review. CR Biol. 2014;337:423–42.
16. Besnard G, Khadari B, Navascues M, Fernandez-Mazuecos M, El Bakkali A, Arrigo N, Baali-Cherif D, Brunini-Bronzini de Caraffa V, Santoni S, Vargas P, Savolainen V. The complex history of the olive tree: from late quaternary diversification of Mediterranean lineages to primary domestication in the northern Levant. Proc Biol Sci. 2013;280:20122833.
17. Médail F, Quézel P, Besnard G, Khadari B. Systematics, ecology and phylogeographic significance of Olea europaea L. subsp. maroccana (Greuter & Burdet) P. Vargas et al., a relictual olive tree in south-west Morocco. Bot J Linn Soc. 2001;137:249–66.
18. El Mouhtadi I, Agouzzal M, Guy F. L'olivier au Maroc. OCL. 2014;21:D203.
19. Allen HD, Randall RE, Amable GS, Devereux BJ. The impact of changing olive cultivation practices on the ground flora of olive groves in the Messara and Psiloritis regions, Crete, Greece. Land Degrad Dev. 2006;17:249–73.
20. Khadari B, Charafi J, Moukhli A, Ater M. Substantial genetic diversity in cultivated Moroccan olive despite a single major cultivar: a paradoxical situation evidenced by the use of SSR loci. Tree Genet Genome. 2008;4:213–21.
21. Cade P, Thioulouse J. Identification of soil factors that relate to plant parasitic nematode communities on tomato and yam in the French West Indies. Appl Soil Ecol. 1998;8:35–49.
22. Besnard G, Hernandez P, Khadari B, Dorado G, Savolainen V. Genomic profiling of plastid DNA variation in the Mediterranean olive tree. BMC Plant Biol. 2011;11:80.
23. Seinhorst JW. Modifications of the elutriation method for extracting nematodes from soil. Nematologica. 1962;8:117–28.
24. Merny G, Luc M. Les techniques d'évaluation des populations dans le sol. In: Lamotte M, Boulière F, editors. Problèmes d'écologie: l'échantillonnage des peuplements animaux dans les milieux terrestres, Masson. Paris; 1969. p. 257–92.
25. Mai WF, Mullin PG. Plant-parasitic nematodes. A pictorial key to genera, 5th ed. New-York: Cornell University Press; 1996.
26. De Grisse AT. Redescription ou modifications de quelques techniques utilisées dans l'étude des nématodes phytoparasites. Meded Fakulteit Landbouw Gent. 1969;34:351–69.
27. van Benzoijen J. Methods and techniques for nematology. Wageningen: Wageningen University; 2006.
28. Cobb NA. Notes on nemas Intra vitam color reactions in nemas. Contrib Sci Nematol. 1917;5:120–4.
29. Ali N, Tavoillot J, Chapuis E, Mateille T. Trend to explain the distribution of root-knot nematodes Meloidogyne spp. associated with olive trees in Morocco. Agric Ecosyst Environ. 2016;225:22–32.
30. Bongers T. The maturity index: an ecological measure of environmental disturbance based on nematode species composition. Oecologia. 1990;83:14–9.
31. Bongers T, Bongers M. Functional diversity of nematodes. Appl Soil Ecol. 1998;10:239–51.
32. Wasilewska L. Changes in the structure of the soil nematode community over long-term secondary grassland succession in drained fen peat. Appl Soil Ecol. 2006;32:165–79.

33. Neher D, Bongers T, Ferris H. Computation of nematode community indices. In: Society of nematologists workshop, vol. 2. Estes Park, Colorado; 2004. p. 1–33.

34. Fortuner R, Merny G. Les nématodes parasites des racines associés au riz en Basse-Casamance (Sénégal) et en Gambie. Cah ORSTOM Ser Biol. 1973;21:3–20.

35. Oksanen J, Blanchet FG, Friendly M, Kindt R, Legendre P, McGlinn D, Minchin PR, O'Hara RB, Simpson GL, Solymos P, Stevens MHH, Szoecs E, Wagner H. Vegan: community ecology package, version 2.3–5. 2016. http://CRAN.R-project.org/package=vegan. Accessed 7 Sept 2016.

36. Chessel D, Dufour AB, Thioulouse J. The ade4 package -I- one-table methods. R News. 2004;4:5–10.

37. Dray S, Dufour AB. The ade4 package: implementing the duality diagram for ecologists. J Stat Softw. 2007;22:1–20.

38. R Core Team. R: a language and environment for statistical computing. Vienna: R Foundation for Statistical Computing; 2016. http://www.Rproject.org/.

39. Huston MA. Hidden treatments in ecological experiments: re-evaluating the ecosystem function of biodiversity. Oecologia. 1997;110:449–60.

40. Barnosky AD, Matzke N, Tomiya S, Wogan GO, Swartz B, Quental TB, Marshall C, McGuire JL, Lindsey EL, Maguire KC, Mersey B. Has the Earth's sixth mass extinction already arrived? Nature. 2011;471:51–7.

41. Rammah A, Hirschmann H. *Meloidogyne morocciensis* n. sp. (Meloidogyninae), a root-knot nematode from Morocco. J Nematol. 1990;22:279.

42. Mokrini F, Andaloussi FA, Alaoui Y, Troccoli A. Importance and distribution of the main cereal nematodes in Morocco. In: Cereal cyst nematodes: status, research and outlook. Proceedings of the first workshop of the international cereal cyst nematode initiative, Antalya, Turkey, 21–23 October; 2009. p. 45–50.

43. Jones JT, Haegeman A, Danchin EG, Gaur HS, Helder J, Jones MG, Kikuchi T, Manzanilla-López R, Palomares-Rius JE, Wesemael WM, Perry RN. Top 10 plant-parasitic nematodes in molecular plant pathology. Mol Plant Pathol. 2013;14:946–61.

44. Hashim Z. Distribution, pathogenicity and control of nematodes associated with olive. Fundam Appl Nematol. 1982;5:169–81.

45. Lamberti F, Vovlas N. Plant-parasitic nematodes associated with olive. EPPO Bull. 1993;23:481–8.

46. Sasanelli N. Olive-nematodes and their control. In: Ciancio A, Mukerji KG, editors. Integrated management of fruit crops nematodes. Berlin: Springer; 2009. p. 275–315.

47. Palomares-Rius JE, Castillo P, Montes-Borrego M, Navas-Cortés JA, Landa BB. Soil properties and olive cultivar determine the structure and diversity of plant-parasitic nematode communities infesting olive orchards soils in southern Spain. PLoS ONE. 2015;10:e0116890.

48. Christensen M, Emborg J. Biodiversity in natural versus managed forest in Denmark. For Ecol Manag. 1996;85:47–51.

49. Ali N, Tavoillot J, Mateille T, Chapuis E, Besnard G, El Bakkali A, Cantalapiedra-Navarrete C, Liébanas G, Castillo P, Palomares-Rius JE. A new root-knot nematode *Meloidogyne spartelensis* n. sp. (Nematoda: Meloidogynidae) in northern Morocco. Eur J Plant Pathol. 2015;143:25–42.

50. Hodda M, Stewart E, FitzGibbon F, Reid I, Longstaff BC, Packer I. Nematodes: useful indicators of soil conditions. Canberra: RIRDC (Rural Industries Research and Development Company); 1999.

51. Kneitel JM, Chase JM. Trade-offs in community ecology: linking spatial scales and species coexistence. Ecol Lett. 2004;7:69–80.

52. Piskiewicz AM, Duyts H, Van der Putten WH. Multiple species-specific controls of root-feeding nematodes in natural soils. Soil Biol Biochem. 2008;40:2729–35.

53. Van Eekeren N, Bommelé L, Bloem J, Schouten T, Rutgers M, de Goede R, Reheul D, Brussaard L. Soil biological quality after 36 years of ley-arable cropping, permanent grassland and permanent arable cropping. Appl Soil Ecol. 2008;40:432–46.

54. Fragoso C, Brown GG, Patrón JC, Blanchart E, Lavelle P, Pashanasi B, Senapati B, Kumar T. Agricultural intensification, soil biodiversity and agroecosystem function in the tropics: the role of earthworms. Appl Soil Ecol. 1997;6:17–35.

55. Cadet P. Gestion écologique des nématodes phytoparasites tropicaux. Cah Agric. 1998;7:187–94.

56. Yamaura Y, Royle JA, Shimada N, Asanuma S, Sato T, Taki H, Makino SI. Biodiversity of man-made open habitats in an underused country: a class of multispecies abundance models for count data. Biodivers Conserv. 2012;21:1365–80.

57. Pan FJ, Xu YL, McLaughlin NB, Xue AG, Yu Q, Han XZ, Liu W, Zhan L, Zhao D, Li CJ. Response of soil nematode community structure and diversity to long-term land use in the black soil region in China. Ecol Res. 2012;27:701–14.

58. Kimenju JW, Karanja NK, Mutua GK, Rimberia BM, Wachira PM. Nematode community structure as influenced by land use and intensity of cultivation. Trop Subtrop Agroecosyst. 2009;11:353–60.

59. Aït-Hamza M, Ferji Z, Ali N, Tavoillot J, Chapuis E, El Oualkadi A, Moukhli A, Khadari B, Boubaker H, Lakhtar H, Roussos S, Mateille T, El Mousadik A. Plant-parasitic nematodes associated with olive in southern Morocco. Int J Agric Biol. 2015;17:719–26.

60. Mateille T, Cadet P, Fargette M. Control and management of plant-parasitic nematode communities in a soil conservation approach. In: Ciancio A, Mukerji KG, editors. Integrated management and biocontrol of vegetable and grain crops nematodes. Dordrecht: Springer; 2008. p. 79–97.

61. Nico AI, Rapoport HF, Jiménez-Díaz RM, Castillo P. Incidence and population density of plant-parasitic nematodes associated with olive planting stocks at nurseries in southern Spain. Plant Dis. 2002;86:1075–9.

9

High dietary quality of non-toxic cyanobacteria for a benthic grazer and its implications for the control of cyanobacterial biofilms

Sophie Groendahl[1]* ⓘ and Patrick Fink[1,2]

Abstract

Background: Mass occurrences of cyanobacteria frequently cause detrimental effects to the functioning of aquatic ecosystems. Consequently, attempts haven been made to control cyanobacterial blooms through naturally co-occurring herbivores. Control of cyanobacteria through herbivores often appears to be constrained by their low dietary quality, rather than by the possession of toxins, as also non-toxic cyanobacteria are hardly consumed by many herbivores. It was thus hypothesized that the consumption of non-toxic cyanobacteria may be improved when complemented with other high quality prey. We conducted a laboratory experiment in which we fed the herbivorous freshwater gastropod *Lymnaea stagnalis* single non-toxic cyanobacterial and unialgal diets or a mixed diet to test if diet-mixing may enable these herbivores to control non-toxic cyanobacterial mass abundances.

Results: The treatments where *L. stagnalis* were fed non-toxic cyanobacteria and a mixed diet provided a significantly higher shell and soft-body growth rate than the average of all single algal, but not the non-toxic cyanobacterial diets. However, the increase in growth provided by the non-toxic cyanobacteria diets could not be related to typical determinants of dietary quality such as toxicity, nutrient stoichiometry or essential fatty acid content.

Conclusions: These results strongly contradict previous research which describes non-toxic cyanobacteria as a low quality food resource for freshwater herbivores in general. Our findings thus have strong implications to gastropod-cyanobacteria relationships and suggest that freshwater gastropods may be able to control mass occurrences of benthic non-toxic cyanobacteria, frequently observed in eutrophied water bodies worldwide.

Keywords: Herbivore, Balanced diet hypothesis, Nutrients, Nitrogen, Phosphorus, Compensatory feeding, Benthic algae, *Lymnaea stagnalis*, Fatty acids

Background

Cyanobacteria are a common component in the diets of herbivores in freshwater ecosystems. Cyanobacteria often occur in eutrophied water bodies and represent a low-quality food source for consumer species caused by a variety of factors such as the possession of toxins [1, 2], feeding inhibitors [3], unsuitable morphology [1] and the lack of essential dietary lipids [4]. In particular, their

low amounts of the essential polyunsaturated fatty acids (PUFA) omega 3 and omega 6 [5] and sterols, strongly constrain the fitness of herbivores on cyanobacterial diets [4, 6]. Within a cyanobacterial species, individual strains can be toxic or non-toxic. Although less frequently studied, non-toxic cyanobacteria are known to reduce growth [7] and reproduction [8], at a similar magnitude as toxin-bearing cyanobacteria, to cladocerans and copepods. Surveys conducted in different parts of the world showed that up to 75% of cyanobacterial blooms can be non-toxic [9–11]. Non-toxic cyanobacteria may consequently impact ecosystems, trophic cascades and

*Correspondence: sgroenda@uni-koeln.de
[1] Cologne Biocenter, Workgroup Aquatic Chemical Ecology, University of Cologne, Zuelpicher Strasse 47b, 50674 Koeln, Germany
Full list of author information is available at the end of the article

geochemical cycles [12]. Interestingly, in a meta-analysis by Wilson et al. [1], it was found that cyanobacterial toxins were actually less important with respect to their negative effects on consumer fitness than cyanobacterial cell morphology. Wilson et al. [1] therefore concluded that the role of cyanobacterial toxins in the determination of food quality may be less important than widely assumed and suggested that future research should focus more on nutritional deficiencies, morphology, and the toxicity of undescribed cyanobacterial compounds as mediators of the poor food quality of cyanobacteria.

Due to the multiple threats that cyanobacteria may pose to ecosystems, various attempts have been made to control cyanobacterial mass abundances ('blooms') through herbivory [13, 14]. However, this requires that the herbivores are able to efficiently utilize cyanobacteria as a food resource. While pure cyanobacteria may be a low quality food resource, a mixed diet with cyanobacterial and eukaryotic components may be easier to assimilate for many herbivores. For example, a severe sterol limitation of the planktonic herbivore *Daphnia* is assumed to occur only if cyanobacteria make up more than 80% of phytoplankton biomass [15]. Moreover, the growth of *Dreissena polymorpha* was strongly reduced while feeding upon a pure cyanobacterial diet deficient in polyunsaturated fatty acids (PUFAs) in comparison to a mixed diet rich in PUFAs [16]. By consuming a mixed diet, herbivores may obtain all nutrients required for growth. This is called the balanced diet hypothesis [17, 18] and has been described for numerous herbivores for example insects [19], snails [20] and fish [21]. Diet mixing may enable grazers to feed upon cyanobacteria without any significant decrease in fitness, and thus reduce cyanobacterial blooms. In a study by DeMott and Müller-Navarra [22], *Daphnia* feeding upon non-toxic cyanobacteria did not display any increase in growth, but when supplemented with a green alga a significant increase in growth was observed. Additionally, rotifers were found to grow better on a diet consisting of a mixture between green algae and cyanobacteria than on either single algal diet [23]. Herbivores may also compensate for a low quality diet through compensatory feeding [24–26]. This is a strategy in which herbivores increase their consumption rate as dietary nutrient concentrations decrease in order to maintain a sufficient uptake of the limiting nutrient(s). Compensatory feeding may thereby increase the fitness of herbivores, but it may also be associated with costs [27]. With respect to macronutrients, herbivores maintain a rather strict homeostasis [28, 29], for instance they need to maintain their body's elemental composition by excreting excess nutrients. This requires energy and results in a reduction of fitness [30, 31]. Moreover, when consuming food in higher quantities, the dosage of potential toxins in the diet can increase [32].

Effects of cyanobacterial diets on freshwater gastropods have rarely been studied. While feeding upon benthic biofilms gastropods rarely only encounter cyanobacteria, but mixtures of various microalgae, bacteria and protozoa embedded in a mucopolysaccharide matrix [33]. Gastropods can represent up to 60% of the total biomass of macroinvertebrates in freshwater ecosystems [34] and they play a key role in the top-down control of benthic primary production. For instance, *Lymnaea stagnalis* (L.), is a benthic herbivore [34] and important grazer in freshwater habitats [35]. It is often found in small eutrophic water bodies [36] where cyanobacteria are extremely common. It is thus plausible to assume that *L. stagnalis* has evolved strategies to cope with cyanobacterial presence in its diet. *L. stagnalis* detects its food via semiochemicals [37], but prey selection on the level of individual food items (e.g. algal cells) is not possible due to the rather unspecific ingestion mode via the gastropod radula [35]. Moreover, it is a common model organism in experimental ecology [37–39].

The aim of this study was to test whether gastropods may be able to feed upon non-toxic cyanobacteria without any significant decrease in fitness through the benefits of diet-mixing. Further, we aimed to investigate which factors are responsible for the typically observed low food quality of non-toxic cyanobacteria to freshwater herbivores [40, 41]. We thus hypothesized (1) that a pure diet consisting of non-toxic cyanobacteria will decrease the fitness of freshwater gastropods, and that (2) diet-mixing allows these gastropods to feed upon non-toxic cyanobacteria without any significant decrease in fitness. To test our hypotheses, we conducted a laboratory experiment using juveniles of the great pond snail *L. stagnalis* which we fed with either single, non-toxic cyanobacterial and algal diets or a mixture of all (six) primary producer species. The aquatic primary producers chosen for this experiment belonged to the three most important groups of organisms in freshwater biofilms: cyanobacteria, chlorophytes and diatoms.

Methods

Cultivation

We randomly selected six species of primary producers, two chlorophytes (*Aphanochaete repens* and *Klebsormidium flaccidum*), two cyanobacteria (*Cylindrospermum* sp. *and Lyngbya halophila*), and two diatoms (*Navicula* sp. *and Nitzschia communis*) from the Culture Collection of Algae at Cologne (CCAC, see Table 1). The size and structure of the primary producer cells were determined by microscopy to test whether the morphology of the cells may impact the fitness of *L. stagnalis* (Table 1).

Table 1 The six benthic primary producers used in the experiment together with their cell shape, average biovolume and origin

Species	Origin/strain	Shape	Average biovolume (μm^3)
Aphanochaete repens	CCAC/M2227	Sphere	350
Klebsormidium flaccidum	CCAC/2007 B	Cylinder	265
Cylindrospermum sp.	CCAC/1160 B	Cylinder	85
Lyngbya halophila	CCAC/1164 B	Cylinder, two half spheres	60
Navicula sp.	CCAC/1772 B	Prism on eliptic base	185
Nitzschia communis	CCAC/1762 B	Prism on eliptic base	410

The cell-specific biovolumes calculated on basis of the geometric shapes according to [45]. The shell morphology of the primary producer species were estimated as it may impact the ingestion by herbivores

All six primary producer strains were cultivated under continuous aeration in 8 L of cyanophyceae medium [42] for cyanobacteria and chlorophytes or in diatom medium [43] for diatoms, respectively. All cultures were kept in a climatized chamber at 20 ± 1 °C at a light intensity of 150 μmol photons s^{-1} m^{-2} as described elsewhere [44]. After one (chlorophytes and cyanobacteria) or two months (for diatoms) of exponential growth, the primary producers were harvested by centrifugation at 4500×g and the resulting pellets were freeze-dried [44]. Juvenile *L. stagnalis*, originating from a pond in Appeldorn, Germany, were raised in aquaria in a climatized chamber at 18 ± 1 °C with a light–dark period of 16:8 h and fed ad libitum with Tetra Wafer Mix™ fish food pellets (Tetra, Melle, Germany) prior to the experiment [44].

Growth experiment

A total of 64 juvenile *L. stagnalis* with a shell height (defined as the distance from the apex to the lower edge of the aperture) of 2.0 ± 0.2 mm were selected. Out of these, eight had their shells removed under a dissecting microscope and were subsequently freeze-dried for the determination of their initial soft body dry mass (dm) using a microbalance (Mettler UTM2, Giessen, Germany). The experiment took place in a climatized chamber at 20 ± 1 °C. The snails were individually kept in square polyethylene containers (length = 11 cm) with each 100 ml aged and aerated tap water and fed on a daily basis. The primary producers were rehydrated in 1 ml water each and then added through a hollow glass cylinder (d = 2.3 cm, h = 2.5 cm) that was submerged halfway in the water in the center of the snails' container. After approximately 1 h when the primary producers had sedimented to the bottom of the container, the glass cylinders were removed. This yielded one clear resource patch and prevented selective feeding by the snails in the mixed primary producer treatment. Water was exchanged daily and the containers were replaced with clean ones every other day. The experiment consisted of seven treatments

with eight replicates each, we thus used 56 containers, each containing one *L. stagnalis* individual. The seven treatments consisted of snails fed with a mixture of all six primary producer species or with one of the six primary producer species. To ensure that the snails could feed ad libitum, the amount of added food was increased during the 33 days experimental duration using previously estimated shell size specific ingestion rates [44]. The shell height of the snails was measured every three days as described above. After 33 days, the soft body dry mass of the remaining 54 snails (single incidents of death had occurred in the treatments with *A. repens* and *K. flaccidum* as diet organism during the experiment) was determined as described above. Since the juvenile growth of *L. stagnalis* can be assumed to be exponential [44], the somatic growth rate of the snails was estimated using the following equation:

$$g = \frac{\ln(m_{end}) - \ln(m_{start})}{time[d]}$$

where m_{start} is the initial dry mass of snails and m_{end} is the dry mass of the snails at the end of the experiment (day 33) over time (d), yielding the somatic growth rate (g).

Ingestion rate

On the final day of the experiment, the snails' ingestion rates were determined in the same setup used to determine the somatic growth rates of the snails. Additionally, three control units per dietary treatment were set up without snails. After 19 h, the snails and their fecal pellets were separately removed from the containers and the remaining primary producers were filtered onto precombusted glass fiber filters (GF/F, d = 25 mm, VWR GmbH, Darmstadt, Germany) and dried at 60 °C for 24 h. The amount of ingested food was determined by subtracting the primary producer dry mass remaining in the snails' containers after 19 h from the primary producer dry mass in the consumer-free controls.

Elemental analysis

To determine the C:N ratios of the primary producers, approximately 1 mg of each freeze-dried primary producer culture, including a mixture of all six primary producers in equal dry mass were packed into tin capsules (HekaTech, Wegberg, Germany) and subsequently analyzed using a Thermo Flash EA 2000 elemental Analyser (Schwerte, Germany). For the analysis of particulate phosphorus, approximately 1 mg of freeze-dried primary producers were transferred into a solution of potassium peroxodisulphate and 1.5% sodium hydroxide. The solution was subsequently autoclaved for 1 h at 120 °C and the soluble reactive phosphorus was analyzed using the molybdate–ascorbic acid method [46]. Both analyses were replicated fivefold per food treatment. The same method was applied to separately determine the molar C:N:P ratios of the 54 snails. When the mass of individual gastropod samples did not reach the required 1 mg dry mass, samples of several individuals from the same treatment were pooled randomly to reach sufficient sample masses.

Fatty acid analysis

Cultures of all primary producers were harvested in the exponential growth phase and filtered in triplicate onto precombusted GF/F filters to assess their fatty acid contents. Filters were placed into 5 mL CH_2Cl_2/MeOH (2:1 v/v) to extract total lipids. The samples were incubated over night at 4 °C whereafter 10 μg of methyl heptadecanoate (C17:0 ME) and 5 μg methyl tricosanoate (C23:0 ME), were added as internal standards. Subsequently, the samples were homogenized in an ultrasonic bath for 1 min and then centrifuged at $4500{\times}g$ for 5 min. Afterwards, the supernatants were dried at 40 °C under a gentle stream of nitrogen. Hydrolysis of lipids and subsequent methylation of fatty acids were achieved by adding 5 mL of 3 N methanolic HCl (Supelco) to the sample and then incubating the sample for 20 min at 70 °C to yield fatty acid methyl esters (FAMEs). The FAMEs were extracted using 6 mL isohexane. The hexane phase was again dried at 40 °C under a gentle stream of nitrogen. Finally, all samples were dissolved in 100 μL isohexane. The samples were then subjected to gas chromatographic analyses on a 6890 N GC System (Agilent Technologies, Waldbronn, Germany) equipped with a DB-225 capillary column (30 m, 0.25 mm i.d., 0.25 μm film thickness, J&W Scientific, Folsom, CA, USA) and a flame ionization detector (FID). The conditions of the GC were as follows: injector and FID temperatures were set to 220 °C; the initial oven temperature were 60 °C for 1 min, followed by a 2 min temperature ramp to 180 °C, then the temperature was increased to 200 °C over a time period of 12.9 min followed by a final 20.6 min temperature increase to

220 °C; helium (5.0 purity) with a flow rate of 1.5 ml min was used as carrier gas. FAMEs were identified by comparison of retention times with those of reference compounds and quantified using the internal standard and previously established calibration functions for each individual FAME. For more details we refer the reader to [47].

Toxin analysis

The two cyanobacterial species used in the experiment (*L. halophila* and *Cylindrospermum* sp.) are known to sometimes contain the toxins cylindrospermopsin and lyngbyatoxin-a. We therefore used high-resolution LC–MS to screen for the cyanobacterial toxins cylindrospermopsin and lyngbyatoxin-a in the cyanobacterial cultures to ensure that the toxins were not produced by the cyanobacterial strains. To screen the two cyanobacteria for the toxins cylindrospermopsin and lyngbyatoxin-a, a crude extract from the freeze-dried samples (500 mg dry mass) of the two cyanobacteria (*Cylindrospermum* sp. and *L. halophila*) were prepared using 50 mL of 100% methanol (HPLC Grade, VWR). The extracts were incubated for 1 h on a rotary shaker and subsequently centrifuged at $4500{\times}g$ for 5 min. The supernatant was then evaporated and the residue dissolved in 5 mL of 100% methanol. Three 5 μL subsamples each were analyzed on an Accela Ultra high pressure liquid chromatography (UPLC) system (Thermo Fisher) coupled with an exactive orbitrap mass spectrometer (Thermo Fisher) with electrospray ionization (ESI) in positive and negative ionization mode. As stationary phase, a Nucleosil C18 column (2 × 125 mm length, pore size 100 Å, particle size 3 μm; Macherey–Nagel, Düren, Germany) was used with a gradient of acetonitrile (ACN) and ultrapure water, each containing 0.05% trifluoroacetic acid (TFA). The column temperature was set to 30 °C and the flow rate to 300 μL/min. The solvent gradient started with 0 min: 38% ACN; 2 min: 40% ACN; 12 min: 50% ACN; 12.5 min: 100% ACN; 15 min: 100% ACN; 15.5 min: 38% ACN; 17 min: 38% ACN. Mass spectrometry was carried out at 1 scan s^{-1} from 150 to 1500 Da in positive and from 120 to 1500 Da in negative ionization mode, respectively (spray voltage 4.5 kV pos./4.3 kV neg. capillary temperature 325 °C, sheath and aux gas nitrogen set to 40 pos./35 neg. and 15 pos./5 neg. respectively). For the identification of potential cyanobacterial toxins, we used extracted ion chromatograms for the respective specific masses of the different compounds (Cylindrospermopsin pos. 416.12345 Da/neg. 414.10889 Da, Lyngbyatoxin-a pos. 438.31150 Da/neg. 436.29695 Da) granting a maximum mass deviation of 3 parts per million (ppm). Electrospray ionization resulted in adduct ions with one positive/one negative charge for the two compounds. We used the

Xcalibur software package (Thermo Fisher) for qualitative analysis.

Statistics

The gastropods' shell height increase over time was analysed via repeated measures ANOVA in R (v. 3.3.1) [48], followed by Bonferroni correction. All other statistical tests were performed in SigmaPlot (v. 11, SysStat). The data were checked for normal distribution using the Shapiro–Wilk's test and for homoscedasticity using Levene's test. When the data fulfilled the criteria for a parametric test, one-way ANOVAs was performed followed by Tukey's HSD. When the data did not fulfil the criteria for an ANOVA, a Kruskal–Wallis test was performed followed by Dunn's post hoc test. Linear regressions were performed to test for relationships between the algal C:N ratio and the ingestion rate of the snails, the algal C:N ratio and the somatic growth rates of the snails, the molar C:N, C:P, N:P ratios of the algae and the molar C:N, C:P, N:P ratios of the snails, and the biovolume of the algae and the somatic growth rate of the snails. An exponential growth, single, 2 parameter regression was performed to test for the relationship between snail dry mass and shell height.

Results

Lymnaea stagnalis fed the cyanobacteria or a mixed diet grew faster than on any of the pure eukaryotic algae, except for the comparison with *K. flaccidum* (repeated measures ANOVA, F = 26.6, df = 6, P < 0.001, Fig. 1a). The somatic growth rate of *L. stagnalis* was significantly higher in the treatments where *L. halophila* and the

mixed treatment was offered compared to the treatments in which *L. stagnalis* was provided with diatoms (Fig. 1b).

The molar C:N ratio (one-way ANOVA, F = 100.00, df = 6, P < 0.001, Fig. 2a) and the molar C:P ratio (one-way ANOVA, F = 59.16, df = 6, P < 0.001, Fig. 2c) of the cyanobacteria were significantly lower than that of the algae. The N:P ratio of *L. halophila* were significantly lower than the N:P ratio of *A. repens* and *N. communis* (Kruskal–Wallis, H = 31.24, df = 6, P < 0.001, Fig. 2e). The C:N ratios of the snails varied significantly between the diet treatments (one-way ANOVA, F = 14.01, df = 5, P < 0.001, Fig. 2b). *L. stagnalis* feeding upon *L. halophila* had a significantly lower C:N ratio compared to all other treatments (Fig. 2b), except for *A. repens* and *N. communis*. While C:N ratios were lower in *L. stagnalis* compared to their diets (Fig. 2a, b), the C:N:P ratios of the snails and their dietary organisms were not significantly correlated (see Additional file 1).

The food consumption of *L. stagnalis* was significantly higher when offered *Navicula* sp. compared to the treatment in which the snails fed on *Cylindrospermum* sp. (Kruskal–Wallis, H = 19.88, df = 6, P = 0.003, Fig. 3). However, no significant differences of the ingestion rates between any other treatments were found (Fig. 3). Moreover, the mean ingestion rate of the snails in each treatment increased linearly with the dietary C:N ratio (y = −0.0241 + (0.00683×), R^2 = 0.61, df = 6, P = 0.039, N = 7; Fig. 4a), but the mean somatic growth rate and the dietary C:N ratio did not correlate (linear regression, x = 8.458 − (8.825y), R^2 = 0.18, df = 6, P = 0.34, N = 7; Fig. 4b). The mean C:N ratios of the snails in each treatment were lower than the mean C:N ratios of any of the

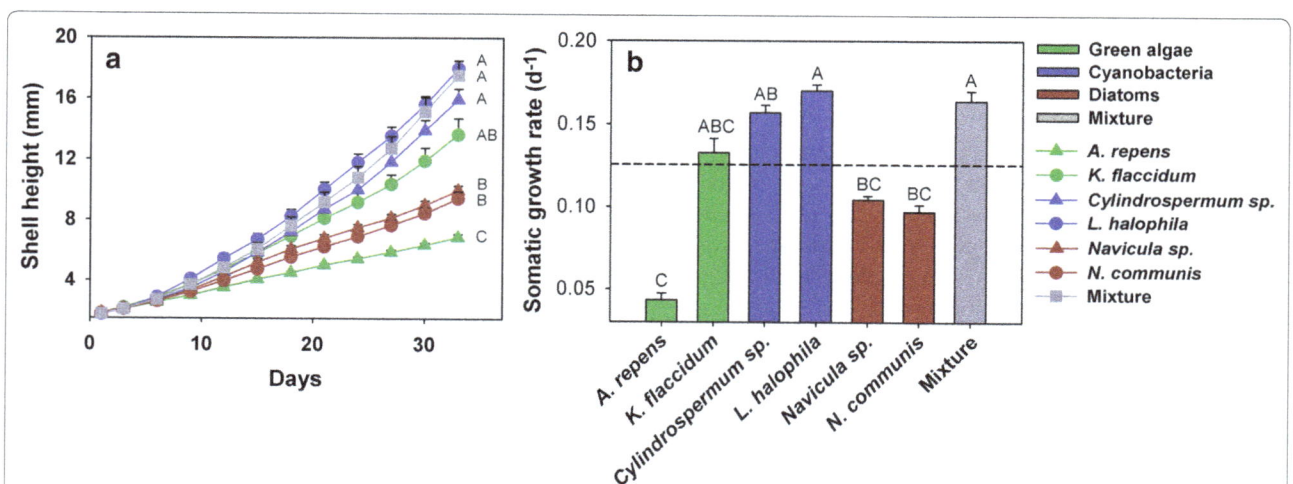

Fig. 1 Increase of shell height (**a**) over time and somatic growth rate (**b**) of *L. stagnalis*. The snails were fed a diet consisting of single primary producer species or a mixture of all six species (Mixture) in equal biomass (mean + 1 SE; N = 7–8). The *dashed line* indicates the average of all single algal treatments; means which were found to be significantly different in Tukey post hoc comparisons are labelled with *different letters*

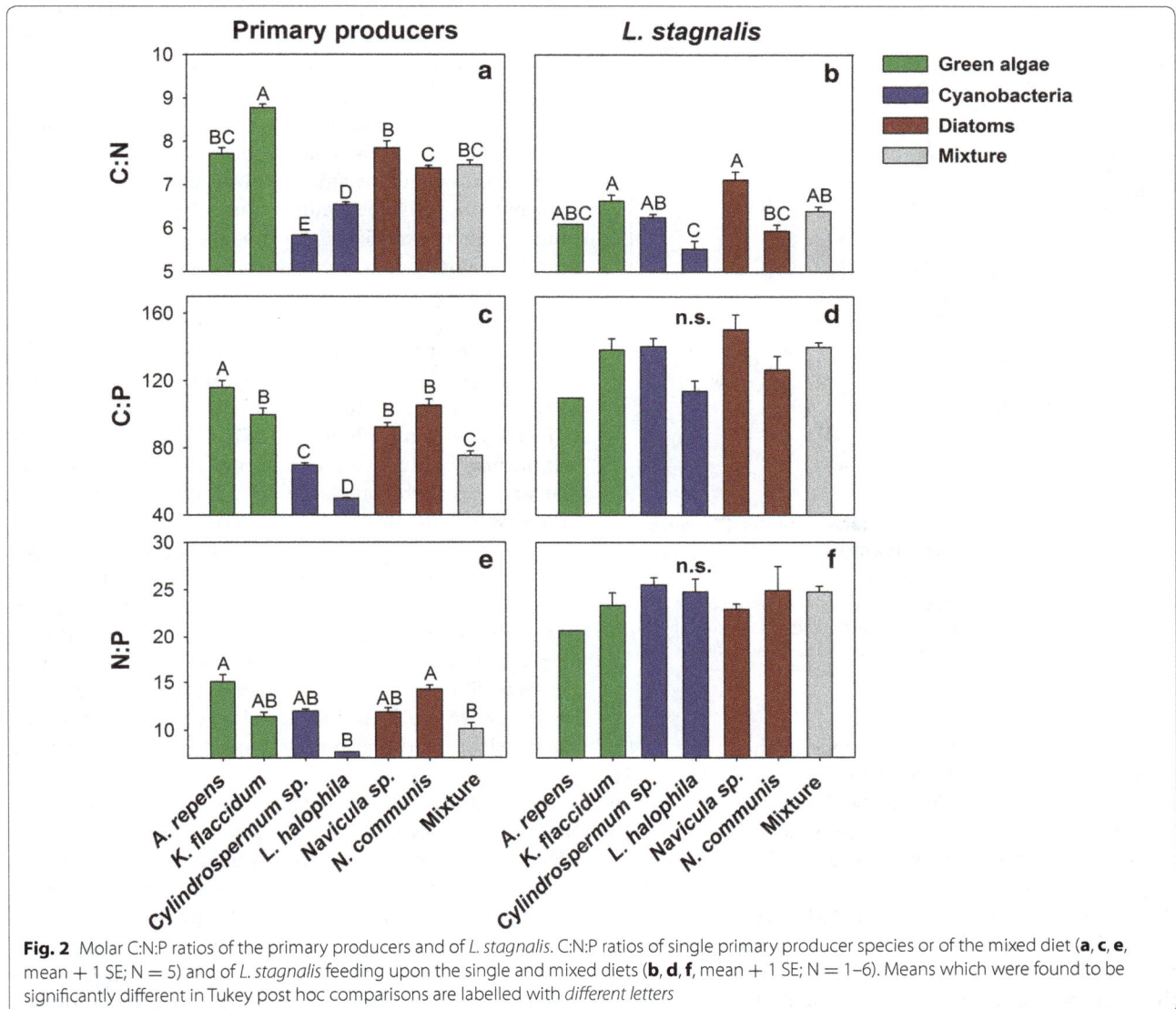

Fig. 2 Molar C:N:P ratios of the primary producers and of *L. stagnalis*. C:N:P ratios of single primary producer species or of the mixed diet (**a**, **c**, **e**, mean + 1 SE; N = 5) and of *L. stagnalis* feeding upon the single and mixed diets (**b**, **d**, **f**, mean + 1 SE; N = 1–6). Means which were found to be significantly different in Tukey post hoc comparisons are labelled with *different letters*

diets (N = 7; Fig. 4c). Additionally, no significant relationships between somatic growth rate and the dietary organisms' cell sizes were found (see Additional file 2).

The fatty acid concentration of the primary producers differed, *A. repens* and *Cylindrospermum* sp. were particularly rich in α-linolenic acid (C 18:3 n − 3) whereas *N. communis* was rich in eicosapentaenoic acid (C20:5 n − 3, see Additional file 3). *A. repens* and the cyanobacteria contained the highest absolute amount of palmitic acid (C 16:0, see Additional file 3), while the *A. repens* and *N. communis* contained the highest total amounts of fatty acids and PUFAs (see Additional file 4).

We did not detect the cyanobacterial toxin cylindrospermopsin in the cyanobacterium *Cylindrospermum* sp. nor lyngbyatoxin-a in *L. halophila* via high-resolution LC–MS.

Discussion

Contrary to our hypothesis, a mixed diet containing nontoxic cyanobacteria did not provide a higher growth rate for *L. stagnalis* compared to pure non-toxic cyanobacterial diets. Surprisingly, the single non-toxic cyanobacteria provided growth rates higher than or equal to the diet consisting of mixed pro- and eukaryotic primary producers. Non-toxic cyanobacteria—at least the two strains investigated here—may therefore be considered a high quality resource for *L. stagnalis*. Negative effects of non-toxic cyanobacteria on the fitness of animals have frequently been reported [7, 8]. However, a screening of cyanobacterial strains demonstrated that some cyanobacterial strains can have a high nutritional value [49]. The importance of the supply ratio or stoichiometry of carbon (C), nitrogen (N) and phosphorus (P) is well

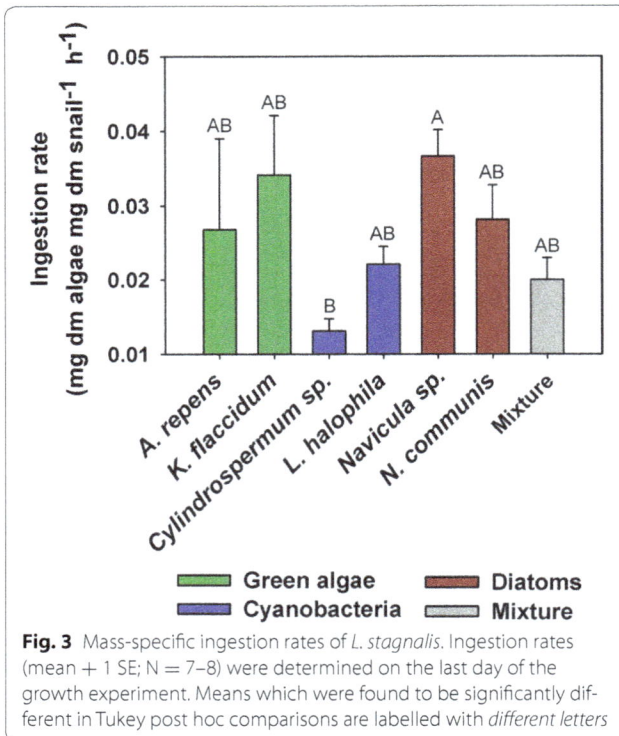

Fig. 3 Mass-specific ingestion rates of *L. stagnalis*. Ingestion rates (mean + 1 SE; N = 7–8) were determined on the last day of the growth experiment. Means which were found to be significantly different in Tukey post hoc comparisons are labelled with *different letters*

studied, as the balance or imbalance in the molar C:N:P supply ratio has been linked to herbivore growth [50, 51], fecundity [52, 53], developmental times [52] and survival rates [52]. While nitrogen is mainly needed for protein synthesis [29], phosphorus is required for the synthesis of phospholipids and nucleic acids [54]. We found that the cyanobacterial species exhibited slightly lower C:N and C:P ratios than the green algae. On the other hand, the overall differences in nutrient stoichiometry between the diet organisms were not particularly pronounced and the nutrient ratios typically below those observed for *L. stagnalis*, which makes a direct nitrogen or phosphorus limitation of snail growth on the green algal diets unlikely. Furthermore, we did not find a significant correlation between the C:N ratio of the algae and the somatic growth rate of the snails, suggesting that the snails were not nitrogen limited. Also, the C:P ratios and the C:N ratios of the snails and the primary producers did not correlate.

The ingestion rate of the snails increased linearly with the C:N ratio of the primary producers, suggesting that compensatory feeding occurred. However, the somatic growth rate of the snails did not correlate with the C:N ratio of the primary producers. It is possible that compensatory feeding by *L. stagnalis* decreased a reduction in growth caused by nitrogen limitation in our experiment. In a previous study, it was found that the freshwater snail

Radix ovata displayed compensatory feeding on low nutrient diets [24]. This dampened the differences in fitness compared to the treatments in which *R. ovata* was fed a nutrient-rich diet.

The lack of essential fatty acids and sterols in cyanobacteria is frequently held responsible for the reduction in growth of pelagic herbivorous zooplankton [4, 15], but also of filter-feeding clams *Corbicula* [40] and mussels *Dreissena* [55]. Even though the diatom diets in our experiment contained much more PUFAs than the cyanobacteria, *L. stagnalis* grew better on both cyanobacteria compared to any of the diatoms. This supports previous findings that lymnaeid gastropods appear to be less susceptible to dietary PUFA limitations than other freshwater invertebrates [56].

Difference in morphology [1, 57, 58] and cell size [44] can strongly influence the ingestion of algae and cyanobacteria by herbivores. We found that snails feeding upon two filamentous cyanobacterial species grew best. Similar results have been found for the lymnaeid species *Radix peregra* which ingests filamentous green algae better than diatoms [58]. Grazing by *Lymnaea elodes* for instance increased the abundance of small coccoid cells of green algae and cyanobacteria at the expense of larger diatoms [59], suggesting a preference for larger algal cells. However, we could not find any clear correlation between algal cell size and the somatic growth rate of *L. stagnalis* in this study.

If the fitness of *L. stagnalis* is increased by the consumption of non-toxic cyanobacteria rather than green algae and diatoms, *L. stagnalis* might have the ability to decrease non-toxic cyanobacterial abundances. In fact, it has been found that snails have the potential to reduce cyanobacterial blooms. In a study by Armitage and Fong [60], primary producers were subjected to nutrient enrichment which led to an increase in cyanobacterial blooms by up to 200%. When snails were allowed to feed upon the primary producers only cyanobacteria decreased in biomass [60]. The snails did not avoid consumption of the toxic cyanobacteria, which indicated that they perceived the cyanobacteria as a suitable food resource, even though cyanobacteria could be linked to an increased mortality of the snails [60]. Toxic cyanobacteria are likely to influence competitive interactions by consumer species favoring the most tolerant ones [61]. Previous studies found that crustacean zooplankton of the genus *Daphnia* are able to locally adapt to environments where cyanobacteria occur in high abundances [62–64]. Similarly physical adaptations in freshwater mussels to cyanobacterial toxins have been found [65]. As *L. stagnalis* often occur in eutrophicated water bodies [36], habitats where cyanobacteria are known to occur, it is possible that—similar to *zooplankton*

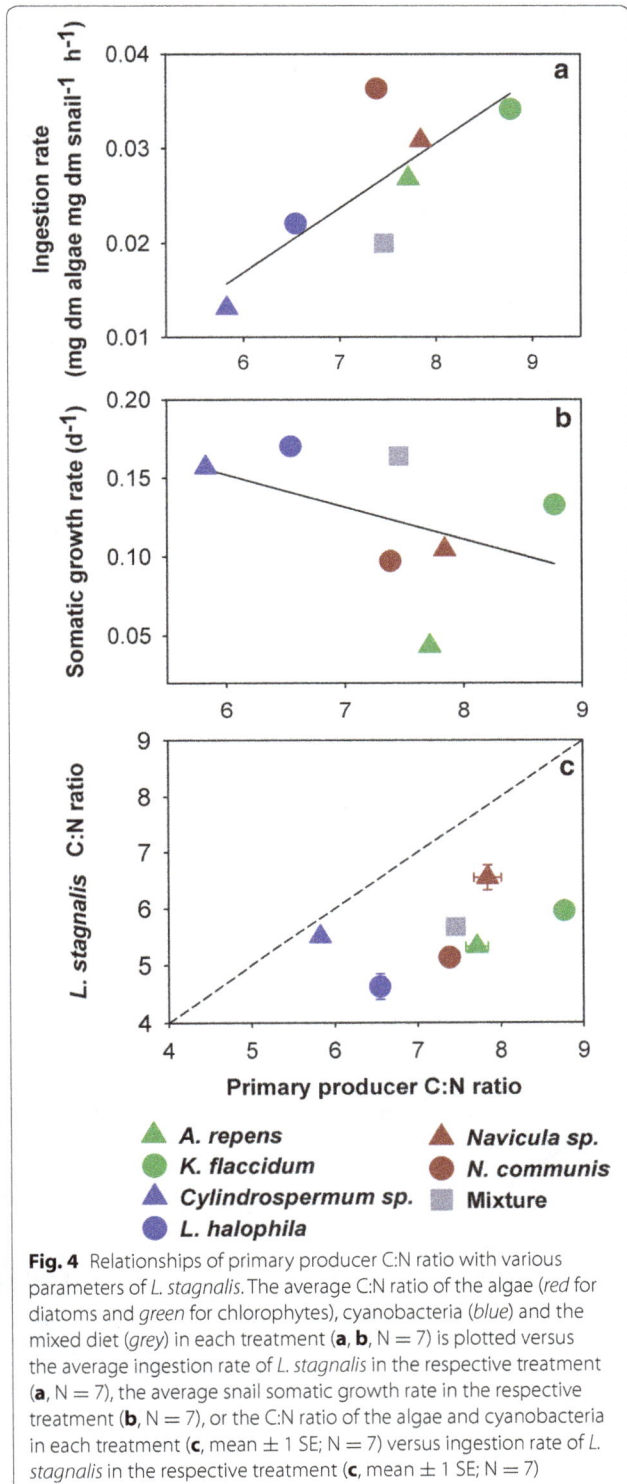

Fig. 4 Relationships of primary producer C:N ratio with various parameters of *L. stagnalis*. The average C:N ratio of the algae (*red* for diatoms and *green* for chlorophytes), cyanobacteria (*blue*) and the mixed diet (*grey*) in each treatment (**a**, **b**, N = 7) is plotted versus the average ingestion rate of *L. stagnalis* in the respective treatment (**a**, N = 7), the average snail somatic growth rate in the respective treatment (**b**, N = 7), or the C:N ratio of the algae and cyanobacteria in each treatment (**c**, mean ± 1 SE; N = 7) versus ingestion rate of *L. stagnalis* in the respective treatment (**c**, mean ± 1 SE; N = 7)

we did not investigate the effects of all diet combinations possible on the fitness of *L. stagnalis*, however, the experimental set-up still it provided insights into gastropod-cyanobacteria relationships.

Conclusions

Efforts have been made in order to control cyanobacterial abundances by manipulating the biomass of herbivores and thereby increasing the top–down grazing pressure on cyanobacteria. However, most studies conducted used toxic cyanobacterial species; therefore, the knowledge of non-toxic cyanobacteria-grazer relationships remains limited. We hypothesized that the growth rate of *L. stagnalis* should be significantly reduced when feeding upon non-toxic cyanobacteria, we found however, the opposite pattern. Non-toxic cyanobacteria and the mixed diet provided the best growth rates for the snails. *L. stagnalis* might thus be a good biological control agent for non-toxic cyanobacterial mass occurrences. These results hence have considerable repercussions to how the dietary quality of non-toxic cyanobacteria for gastropods is perceived.

Additional files

Additional file 1. Relationship between the C:N:P ratios (mean ± SE) of the primary producers and *L. stagnalis*. Nonsignificant linear regressions for C:N (A, y = 3.006 + (0.335 x), R^2 = 0.27, df = 6, P = 0.23), C:P (B, y = 134.678 - (0.0388 x), R^2 < 0.005, df = 6, P = 0.90), and N:P (C, y = 27.849 - (0.342 x), R^2 = 0.26, df = 6, P = 0.25).

Additional file 2. Relationship between primary producer biovolume and somatic growth rate of *L. stagnalis*. Not statistically significant linear regression, y = 0.176 - (0.0000260 x), R^2 = 0.63, df = 5, P = 0.06.

Additional file 3. A table of the fatty acid composition of the primary producers used as a food resource for *L. stagnalis*. Values given are means ± 1 SE of N = 3 replicates analyzed via gas chromatography of fatty acid methyl esters (n.d. = not detected), the standard errors are given in parentheses.

Additional file 4. Total fatty acid and polyunsaturated fatty acid (PUFA) concentration of the primary producers. Values given are means ± 1 SE of N = 3 replicates analyzed via gas chromatography of fatty acid methyl esters, the standard errors are given in parentheses.

Authors' contributions
SG and PF designed the research. SG conducted the laboratory work, performed the statistical analyses and drafted the manuscript. PF contributed to manuscript writing. Both authors read and approved the final manuscript.

Author details
[1] Cologne Biocenter, Workgroup Aquatic Chemical Ecology, University of Cologne, Zuelpicher Strasse 47b, 50674 Koeln, Germany. [2] Present Address: Institute for Zoomorphology and Cell Biology, Heinrich-Heine University of Duesseldorf, Universitaetsstrasse 1, 40225 Duesseldorf, Germany.

Acknowledgements
We thank Sofia Albrecht for help with the experiment, Barbara Melkonian at the Culture Collection of Algae at the University of Cologne (CCAC) for providing us with the cyanobacterial and algal cultures, Katja Preuss for assistance with the GC analysis of fatty acids and Christian Burberg for help with the LC–MS screening for cyanobacterial toxins.

and mussels—Lymnaea have evolved to coexist with cyanobacteria.

We feed *L. stagnalis* with single primary producer species or a mixture of all species. Due to time constraints

Competing interests
The authors declare that they have no competing interests.

Funding
This study was supported by the Deutsche Forschungsgemeinschaft (DFG), Grant FI 1548/5-1 to PF.

References
1. Wilson AE, Sarnelle O, Tillmanns AR. Effects of cyanobacterial toxicity and morphology on the population growth of freshwater zooplankton: meta-analyses of laboratory experiments. Limnol Oceanogr. 2006;51(4):1915–24.
2. Carmichael WW. Health effects of toxin-producing cyanobacteria: "The CyanoHABs". Hum Ecol Risk Assess. 2001;7(5):1393–407.
3. Schwarzenberger A, Zitt A, Kroth P, Mueller S, Von Elert E. Gene expression and activity of digestive proteases in *Daphnia*: effects of cyanobacterial protease inhibitors. BMC Physiol. 2010;10:6.
4. Martin-Creuzburg D, von Elert E. Good food versus bad food: the role of sterols and polyunsaturated fatty acids in determining growth and reproduction of *Daphnia magna*. Aquat Ecol. 2009;43(4):943–50.
5. Sargent J, Bell J, Bell M, Henderson R, Tocher D. Requirement criteria for essential fatty acids. J Appl Ichthyol. 1995;11(3–4):183–98.
6. Wacker A, Becher P, von Elert E. Food quality effects of unsaturated fatty acids on larvae of the zebra mussel *Dreissena polymorpha*. Limnol Oceanogr. 2002;47(4):1242–8.
7. FerrÃo-Filho AS, Azevedo SM, DeMott WR. Effects of toxic and non-toxic cyanobacteria on the life history of tropical and temperate cladocerans. Freshw Biol. 2000;45(1):1–19.
8. Koski M, Engström J, Viitasalo M. Reproduction and survival of the calanoid copepod Eurytemora affinis fed with toxic and non-toxic cyanobacteria. Mar Ecol Prog Ser. 1999;186:187–97.
9. Lindholm T, Vesterkvist P, Spoof L, Lundberg-Niinistö C, Meriluoto J. Microcystin occurrence in lakes in Åland, SW Finland. Hydrobiologia. 2003;505(1):129–38.
10. Sivonen K. Cyanobacterial toxins and toxin production. Phycologia. 1996;35(6S):12–24.
11. Vezie C, Brient L, Sivonen K, Bertru G, Lefeuvre JC, Salkinoja-Salonen M. Variation of microcystin content of cyanobacterial blooms and isolated strains in Lake Grand-Lieu (France). Microb Ecol. 1998;35(2):126–35.
12. Sukenik A, Quesada A, Salmaso N. Global expansion of toxic and non-toxic cyanobacteria: effect on ecosystem functioning. Biodivers Conserv. 2015;24(4):889–908.
13. Xie P, Liu J. Practical success of biomanipulation using filter-feeding fish to control cyanobacteria blooms: a synthesis of decades of research and application in a subtropical hypereutrophic lake. Sci World J. 2001;1:337–56.
14. Matveev V, Matveeva L, Jones GJ. Study of the ability of *Daphnia carinata* King to control phytoplankton and resist cyanobacterial toxicity: implications for biomanipulation in Australia. Mar Freshw Res. 1994;45(5):889–904.
15. von Elert E, Martin-Creuzburg D, Le Coz JR. Absence of sterols constrains carbon transfer between cyanobacteria and a freshwater herbivore (*Daphnia galeata*). Proc R Soc Lond B: Biol Sci. 2003;270(1520):1209–14.
16. Wacker A, von Elert E. Strong influences of larval diet history on subsequent post–settlement growth in the freshwater mollusc *Dreissena polymorpha*. Proc R Soc Lond B: Biol Sci. 2002;269(1505):2113–9.
17. Pulliam HR. Diet optimization with nutrient constraints. Am Nat. 1975;109(970):765–8.
18. Westoby M. What are the biological bases of varied diets? Am Nat. 1978;112(985):627–31.
19. Unsicker SB, Oswald A, Koehler G, Weisser WW. Complementarity effects through dietary mixing enhance the performance of a generalist insect herbivore. Oecologia. 2008;156(2):313–24.
20. Watanabe JM. Food preference, food quality and diets of three herbivorous gastropods (*Trochidae*: *Tegula*) in a temperate kelp forest habitat. Oecologia. 1984;62(1):47–52.
21. Lobel PS, Ogden JC. Foraging by the herbivorous parrotfish *Sparisoma radians*. Mar Biol. 1981;64(2):173–83.
22. DeMott W, Müller-Navarra D. The importance of highly unsaturated fatty acids in zooplankton nutrition: evidence from experiments with *Daphnia*, a cyanobacterium and lipid emulsions. Freshw Biol. 1997;38(3):649–64.
23. Alva-Martínez AF, Fernández R, Sarma S, Nandini S. Effect of mixed toxic diets (*Microcystis* and *Chlorella*) on the rotifers Brachionus calyciflorus and Brachionus havanaensis cultured alone and together. Limnol-Ecol Manag Inland Waters. 2009;39(4):302–5.
24. Fink P, Von Elert E. Physiological responses to stoichiometric constraints: nutrient limitation and compensatory feeding in a freshwater snail. Oikos. 2006;115(3):484–94.
25. Cruz-Rivera E, Hay ME. Can quantity replace quality? Food choice, compensatory feeding, and fitness of marine mesograzers. Ecology. 2000;81(1):201–19.
26. Berner D, Blanckenhorn WU, Körner C. Grasshoppers cope with low host plant quality by compensatory feeding and food selection: N limitation challenged. Oikos. 2005;111(3):525–33.
27. Zehnder CB, Hunter MD. More is not necessarily better: the impact of limiting and excessive nutrients on herbivore population growth rates. Ecol Entomol. 2009;34(4):535–43.
28. Persson J, Fink P, Goto A, Hood JM, Jonas J, Kato S. To be or not to be what you eat: regulation of stoichiometric homeostasis among autotrophs and heterotrophs. Oikos. 2010;119(5):741–51.
29. Sterner RW, Elser JJ. Ecological stoichiometry: the biology of elements from molecules to the biosphere. Princeton: Princeton University Press; 2002.
30. Darchambeau F, Faerøvig PJ, Hessen DO. How *Daphnia* copes with excess carbon in its food. Oecologia. 2003;136(3):336–46.
31. Suzuki-Ohno Y, Kawata M, Urabe J. Optimal feeding under stoichiometric constraints: a model of compensatory feeding with functional response. Oikos. 2012;121(4):569–78.
32. Slansky F, Wheeler G. Caterpillars' compensatory feeding response to diluted nutrients leads to toxic allelochemical dose. Entomol Exp Appl. 1992;65(2):171–86.
33. Anderson M. Variations in biofilms colonizing artificial surfaces: seasonal effects and effects of grazers. J Mar Biol Assoc UK. 1995;75(03):705–14.
34. Habdija I, Lajtner J, Belinic I. The contribution of gastropod biomass in macrobenthic communities of a karstic river. Int Rev Gesamt Hydrobiol. 1995;80(1):103–10.
35. Dillon R. The ecology of freshwater molluscs, vol. 1. 1st ed. Cambridge: Cambridge University Press; 2000.
36. Clarke AH. Gastropods as indicators of trophic lake stages. Nautilus. 1979;94:4.
37. Moelzner J, Fink P. The smell of good food: volatile infochemicals as resource quality indicators. J Anim Ecol. 2014;83(5):1007–14.
38. Bakker ES, Dobrescu I, Straile D, Holmgren M. Testing the stress gradient hypothesis in herbivore communities: facilitation peaks at intermediate nutrient levels. Ecology. 2013;94(8):1776–84.
39. Dalesman S, Rundle SD, Bilton DT, Cotton PA. Phylogenetic relatedness and ecological interactions determine antipredator behavior. Ecology. 2007;88(10):2462–7.
40. Basen T, Martin-Creuzburg D, Rothhaupt KO. Role of essential lipids in determining food quality for the invasive freshwater clam *Corbicula fluminea*. J N Am Benthol Soc. 2011;30(3):653–64.
41. von Elert E. Determination of limiting polyunsaturated fatty acids in *Daphnia galeata* using a new method to enrich food algae with single fatty acids. Limnol Oceanogr. 2002;47(6):1764–73.
42. von Elert E, Jüttner F. Phosphorus limitation and not light controls the extracellular release of allelopathic compounds by *Trichormus doliolum* (cyanobacteria). Limnol Oceanogr. 1997;42(8):1796–802.
43. Wendel T. Lipoxygenase-katalysierte VOC-Bildung: Untersuchungen an Seewasser und Laborkulturen von Diatomeen. Shaker; 1994.
44. Groendahl S, Fink P. The effect of diet mixing on a nonselective herbivore. PLoS ONE. 2016;11(7):e0158924.
45. Hillebrand H, Dürselen CD, Kirschtel D, Pollingher U, Zohary T. Biovolume calculation for pelagic and benthic microalgae. J Phycol. 1999;35(2):403–24.
46. Greenberg A, Trussel R, Clesceri L. Standard method for the examination of water and wastewater. Washington: American Public Health Association (APHA); 1985.
47. Fink P. Invasion of quality: high amounts of essential fatty acids in the invasive Ponto-Caspian mysid Limnomysis benedeni. J Plankton Res. 2013: fbt029.

48. Team RC. R: a language and environment for statistical computing. Vienna; 2013.

49. Becker E, Venkataraman L. Production and utilization of the blue-green alga *Spirulina* in India. Biomass. 1984;4(2):105–25.

50. Hessen DO. Nutrient element limitation of zooplankton production. Am Nat. 1992;149:799–814.

51. Elser JJ, Fagan WF, Denno RF, Dobberfuhl DR, Folarin A, Huberty A, Interlandi S, Kilham SS, McCauley E, Schulz KL, et al. Nutritional constraints in terrestrial and freshwater food webs. Nature. 2000;408(6812):578–80.

52. Huberty AF, Denno RF. Consequences of nitrogen and phosphorus limitation for the performance of two planthoppers with divergent life-history strategies. Oecologia. 2006;149(3):444–55.

53. Kilham S, Kreeger D, Goulden C, Lynn S. Effects of algal food quality on fecundity and population growth rates of *Daphnia*. Freshw Biol. 1997;38(3):639–47.

54. Weider LJ, Elser JJ, Crease TJ, Mateos M, Cotner JB, Markow TA. The functional significance of ribosomal (r) DNA variation: impacts on the evolutionary ecology of organisms. Annu Rev Ecol Evol Syst. 2005;36:219–42.

55. Wacker A, Von Elert E. Food quality controls reproduction of the zebra mussel (*Dreissena polymorpha*). Oecologia. 2003;135(3):332–8.

56. Fink P. Food quality and food choice in freshwater gastropods: field and laboratory investigations on a key component of littoral food webs. Berlin: Logos-Verlag; 2006.

57. Lodge D. Selective grazing on periphyton: a determinant of freshwater gastropod microdistributions. Freshw Biol. 1986;16(6):831–41.

58. Calow P. Studies on the natural diet of *Lymnaea pereger obtusa* (Kobelt) and its possible ecological implications. J Molluscan Stud. 1970;39(2–3):203–15.

59. Cuker B. E m: Grazing and nutrient interactions in controlling the activity and composition of the epilithic algal community of an arctic lake. Limnol Oceanogr. 1983;28(1):133–41.

60. Armitage AR, Fong P. Upward cascading effects of nutrients: shifts in a benthic microalgal community and a negative herbivore response. Oecologia. 2004;139(4):560–7.

61. Gerard C, Poullain V, Lance E, Acou A, Brient L, Carpentier A. Influence of toxic cyanobacteria on community structure and microcystin accumulation of freshwater molluscs. Environ Pollut. 2009;157(2):609–17.

62. Schwarzenberger A, D'Hondt S, Vyverman W, von Elert E. Seasonal succession of cyanobacterial protease inhibitors and *Daphnia magna* genotypes in a eutrophic Swedish lake. Aquat Sci. 2013;75(3):433–45.

63. Gustafsson S, Rengefors K, Hansson L-A. Increased consumer fitness following transfer of toxin tolerance to offspring via maternal effects. Ecology. 2005;86(10):2561–7.

64. Hairston N Jr, Holtmeier C, Lampert W, Weider L, Post D, Fischer J, Caceres C, Fox J, Gaedke U. Natural selection for grazer resistance to toxic cyanobacteria: evolution of phenotypic plasticity? Evolution. 2001;55(11):2203–14.

65. Burmester V, Nimptsch J, Wiegand C. Adaptation of freshwater mussels to cyanobacterial toxins: Response of the biotransformation and antioxidant enzymes. Ecotoxicol Environ Saf. 2012;78:296–309.

10

Suppression of reproductive characteristics of the invasive plant Mikania micrantha by sweet potato competition

Shicai Shen[1], Gaofeng Xu[1], David Roy Clements[2*], Guimei Jin[1], Shufang Liu[1], Yanxian Yang[1], Aidong Chen[1], Fudou Zhang[1*] and Hisashi Kato-Noguchi[3]

Abstract

Background: As a means of biologically controlling *Mikania micrantha* H.B.K. in Yunnan, China, the influence of sweet potato [*Ipomoea batatas* (L.) Lam.] on its reproductive characteristics was studied. The trial utilized a de Wit replacement series incorporating six ratios of sweet potato and *M. micrantha* plants in 25 m^2 plots over 2 years.

Results: Budding of *M. micrantha* occurred at the end of September; flowering and fruiting occurred from October to February. Flowering phenology of *M. micrantha* was delayed ($P < 0.05$), duration of flowering and fruiting was reduced ($P < 0.05$) and duration of bud formation was increased ($P < 0.05$) with increasing proportions of sweet potato. Reproductive allocation, reproductive investment and reproductive index of *M. micrantha* were significantly reduced ($P < 0.05$) with increasing sweet potato densities. Apidae bees, and Calliphoridae or Syrphidae flies were the most abundant visitors to *M. micrantha* flowers. Overall flower visits decreased ($P < 0.05$) as sweet potato increased. Thus the mechanism by which sweet potato suppressed sexual reproduction in *M. micrantha* was essentially two-fold: causing a delay in flowering phenology and reducing pollinator visits. The number, biomass, length, set rate, germination rate, and 1000-grain dry weight of *M. micrantha* seeds were suppressed ($P < 0.05$) by sweet potato competition. With proportional increases in sweet potato, sexual and asexual seedling populations of *M. micrantha* were significantly reduced ($P < 0.05$). The mortality of both seedling types increased ($P < 0.05$) with proportional increases in sweet potato.

Conclusions: These results suggest that sweet potato significantly suppresses the reproductive ability of the invasive species *M. micrantha*, and is a promising alternative to traditional biological control and other methods of control. Planting sweet potato in conjunction with other control methods could provide a comprehensive strategy for managing *M. micrantha*. The scenario of controlling *M. micrantha* by utilizing a crop with a similar growth form may provide a useful model for similar management strategies in other systems.

Keywords: Biological control, *Mikania micrantha*, Sweet potato, Competition, Flowering and fruiting phenology, Reproductive suppression

Background

Plant invasions have received attention on a global scale. Invasive plants have caused great economic harm, biodiversity loss, environmental problems, and even human and animal health issues [1, 2]. Currently, methods to control invasive plants have been widely investigated and practiced, including mechanical [3–5], chemical [6, 7] and biological control [8–11]. However, mechanical measures may potentially accelerate invasions [3], chemical measures can be detrimental to non-target species and environment health [12–14], and traditional biological control measures via introduction

---
*Correspondence: clements@twu.ca; fdzh@vip.sina.com
[1] Agricultural Environment and Resource Research Institute, Yunnan Academy of Agricultural Sciences, Kunming 650205, Yunnan, China
[2] Biology Department, Trinity Western University, 7600 Glover Road, Langley, BC V2Y1Y1, Canada
Full list of author information is available at the end of the article

of pathogens, parasites, and predators against invaders are expensive and pose risks to ecosystem integrity [15, 16]. Thus, as a potential alternative to traditional biological control which generally employs insects or pathogens, replacement control relies on growth characteristics of one or more plants to suppress exotic plants, simultaneously reducing damage caused by the invasive species and improving local natural ecosystem health by reducing the potential for invasive plants to spread beyond agricultural fields. Adoption of this alternative method has received considerable attention in recent years [17–21].

Mikania micrantha H.B.K. is a rapidly-growing perennial creeping vine belonging to the family Asteraceae native to Central and South America [22]. The vine has been listed among the top 100 worst invasive species [22, 23] and as one of the top 10 worst weeds in the world [22]. *M. micrantha* is present in tropical Asia, parts of Papua New Guinea, Indian Ocean islands, Pacific Ocean Islands, and Florida in the US [22, 24, 25]. It has colonized a broad range of farming systems and forest lands, banks of streams and rivers, roadsides and railway tracks, pastures, and open disturbed areas [22], leading to serious economic loss, biodiversity loss and negative environmental impacts [7, 18, 22, 26, 27].

To explore ecological methods for managing *M. micrantha*, biological control measures through replacement control with high value species (e.g., local food, native species and/or cash crops) have been investigated [18, 28–31]. In 2006 and 2007, sweet potato [*Ipomoea batatas* (L.) Lam.: Convolvulaceae], an important locally grown cash crop native to the American tropics, was observed to inhibit *M. micrantha* growth in Longchuan County of Yunnan Province, China in sweet potato fields where *M. micrantha* occurred [30]. Subsequent studies examined the effects of a local crop, sweet potato, on *M. micrantha* growth and soil nutrients [18]. Sweet potato exhibited greater competitive ability than *M. micrantha*, with plant height, branch, leaf, stem node, adventitious root, and biomass of *M. micrantha* suppressed significantly; furthermore sweet potato also demonstrated higher levels of nutrient uptake than *M. micrantha*. Moreover, flowering of *M. micrantha* was significantly suppressed in mixed culture with sweet potato and with decreasing proportions of *M. micrantha*, the competitiveness of sweet potato increased at a rate exceeding what would be predicted by the increase in relative density [18]. However, no literature is available on the effects of sweet potato competition on the entire suite of reproductive characteristics of *M. micrantha*.

Building on our former studies [18, 30, 31], the present research examined how sweet potato suppressed reproductive characteristics of *M. micrantha* in Yunnan Province, China. Previous reports did not refer to impacts on reproductive characteristics, so this is the first report of how sweet potato competition affects characteristics such as flowering and seed production in *M. micrantha*. These findings are important to further elucidate the competitive interaction and mechanisms between sweet potato and *M. micrantha* and provide insights for similar ecological control methods that could be applied to other invasive alien species.

Methods
Study site
The study site was located in Longchuan County (24°08′–24°39′ N, 97°17′–97°39′ E), in the western end of Yunnan Province, Southwest China. This area is characterized by a typical tropical climate, having a rainy season featuring heavy rainfall with 90 % relative humidity alternating with a dry season [30]. Rainfall averages 15450 mm per year and the annual mean temperature is 18.9 °C. Recently, the range of *M. micrantha* has been expanding rapidly within Longchuan County, as the plant has invaded agricultural areas and forest margins [7].

Study species
Mikania micrantha is one of the most serious invasive species in Longchuan County where this study took place. This perennial weed exhibits a climbing growth form in forests, orchards and shrublands, but on roadsides, in open wasteland areas without crops, and other areas without woody vegetation, it takes on a prostrate form. It has infested sugarcane, orange, banana, coffee, pineapple, bamboo, sweet potato, maize crops, as well as artificial pasture and secondary forest in the study area [7]. *M. micrantha* can invade disturbed environments via light weight wind-dispersed seeds that are produced in great numbers, as high as 170,000 m^{-2} [32]. At a local level, vegetative reproduction is responsible for most population growth as facilitated by rooting of stem fragments [3].

Sweet potato, native to the American tropics, is one of the main food and cash crops in tropical and subtropical regions of Yunnan Province. It is also grown in many other regions of China and other subtropical or warm-temperate regions of the world as a food source. In Longchuan County, local villagers have grown it for over 100 years [30]. This herbaceous perennial vine usually exhibits a prostrate growth form in agricultural areas, so its niche is similar to that of *M. micrantha*. Because of its purple root, it is also known as purple sweet potato. The aboveground parts of the plant are used for livestock fodder and its roots are used for human eating. It is propagated by seed or by clonal means, with 20–50 cm fragments with 3–5 nodes typically planted [33].

Experiment design and data collection

The experiments were conducted during the May, 2014-October, 2015 growing season within maize and sweet potato intercropping land in the vicinity of Zhangfeng Town, Longchuan County, utilizing a de Wit replacement series method [34]. On 7 May 2014, whole *M. micrantha* plants (including roots) were collected from a *M. micrantha* population located in a nearby forest margin and whole sweet potato plants were collected from farmland, respectively. To ensure relative uniformity among the experimental stock, one-node segments (fresh weight 3.0–3.5 g, 7–8 cm pieces) were taken from central stem portions of relatively young plants of similar size from both species. The segments were placed in Hoagland's solution [35] and grown for 10 days. On 17 May 2014, the sprouts derived from cuttings of both species were transplanted in the field test plots. Six ratios of sweet potato and *M. micrantha* plants were utilized (3:1, 2:1, 1:1, 1:2, 1:3, 0:4) while maintaining a constant planting density of 20 plants m^{-2} (0.25 × 0.20 m spacing). All plots were arranged in a complete randomized design with 4 replicates utilizing 25 m^2 plots (5 × 5 m). All sweet potato and *M. micrantha* plants were distributed evenly within the plot. During the experiment, the two species exhibited prostrate growth. The plots were not weeded and no fertilizers were used.

From October 2014 to February 2015, flowering and fruiting phenology were recorded at 3 days intervals for *M. micrantha*, including the dates of initial and last budding, flowering, fruiting and seed set. During peak flowering times, we marked 20 inflorescences of *M. micrantha* in each plot and recorded the number of flower visits per insect visitor on each inflorescence between 9:00 and 18:00 (time of maximum pollinator activity) for two continuous days with all plots monitored simultaneously over the same two days. According to [36] each flower visitor was classified as belonging to Apidae, Calliphoridae, Syrphidae (generally the most frequent visitors to *M. micrantha*), or another pollinator group beyond these three taxa. At the same time, twenty plants of each species were selected randomly and harvested within the middle region of each plot. *M. micrantha* plants were carefully removed and separated, and number of inflorescences and flowers were recorded. The fresh weight of inflorescences and flowers and total biomass of each *M. micrantha* plant were measured. On 28 January 2015, seed production of *M. micrantha* was measured in the study plots after flowering had waned, but prior to seed dispersal. Another twenty plants of each species were selected randomly and harvested within the middle region of each plot. The number, size (length) and dry weight of *M. micrantha* seeds were measured. Sixty days later, germination rates of *M. micrantha* seeds from each plot were tested in the laboratory.

During the spring, summer and fall of 2015 (March–October), four small quadrats (1 × 1 m) were selected randomly and marked in each plot. Seedlings were identified as either produced from germinating seeds (sexual) or vegetative growth (asexual). The number of new sexual and asexual *M. micrantha* seedlings was monitored in each quadrat monthly. We did not remove the seedlings that were counted but rather kept track of the total that emerged through the season, month by month. Seedling mortality was also recorded for the *M. micrantha* that emerged.

Data analyses

Reproductive characteristics of *M. micrantha* [37, 38] were calculated in each plot with the following parameters: (1) Reproductive allocation (g·g^{-1}) = inflorescence biomass/total biomass of each plant, (2) Reproductive investment (g·g^{-1}) = flower biomass/total biomass of each plant, (3) Reproductive index (g·g^{-1}) = flower biomass/inflorescence biomass of each plant, (4) Reproductive ratio (flower·mg^{-1}) = flower number/inflorescence biomass of each plant and (5) Reproductive efficiency index (flower·mg^{-1}) = flower number/total biomass of each plant.

All growth variables (flowering, bud formation and fruiting duration, inflorescence number, flower number, germination, and biomass of inflorescences, and flowers and seeds) of *M. micrantha* plants were analyzed by analysis of variance (one-way ANOVA) using IBM SPSS 22.0 software. If significant differences were detected with the ANOVA, Tukey's HSD, Post Hoc Multiple Comparisons, Homogeneity of Variance tests were used to detect differences among treatments at a 5 % level of significance.

Results

Reproductive phenology of *Mikania micrantha*

Budding of *M. micrantha* occurred at the end of September, and flowering and fruiting occurred from October to February in our study area (Table 1). The peak bloom occurred between mid-November and early December. Fruiting started as early as the end of November, and almost all fruits had dropped by early February. The flowers opened throughout the day, but the majority did so in the morning. In monoculture, the duration of bud formation, flowers and fruits of *M. micrantha* was 27.25 ± 0.05 d, 79.75 ± 1.71 d, and 75.75 ± 1.71 d, respectively (Table 1). With increased sweet potato: *M. micrantha* ratios, the initial date of budding, flowering and fruiting of *M. micrantha* was significantly delayed, and the duration of flowering and fruiting of

Table 1 Flowering and fruiting phenology of *Mikania micrantha* growing as a monoculture or under mixed culture conditions

Variables	Ratios (sweet potato: *M. micrantha*)					
	3:1	2:1	1:1	1:2	1:3	0:4
Initial budding date	10 October	5 October	2 October	28 September	26 September	26 September
Initial flowering date	16 November	12 November	4 November	27 October	24 October	24 October
Initial fruiting date	20 November	16 November	11 November	4 November	2 November	1 November
Duration of bud formation (d)	37.75 ± 0.96a	34.75 ± 0.96b	31.25 ± 0.96c	30.25 ± 0.96c	28.25 ± 0.96 cd	27.25 ± 0.05d
Duration of flowering (d)	48.25 ± 1.26f	54.75 ± 1.50e	60.50 ± 1.29d	68.25 ± 2.22c	73.75 ± 2.06b	79.75 ± 1.71a
Duration of fruiting (d)	48.25 ± 1.89e	55.00 ± 2.16d	59.50 ± 1.29c	66.50 ± 1.29b	70.00 ± 1.63b	75.75 ± 1.71a

Data are expressed as mean ± standard deviation. The different letters within same row signify significantly different at P < 0.05

M. micrantha significantly declined, but duration of bud formation of *M. micrantha* was markedly increased (P < 0.05).

A total of about 47 flower visits of *M. micrantha* were observed for each inflorescence per day, with Apidae (bees), Calliphoridae (flies), Syrphidae (flies) and other pollinators accounting for 61.2, 22.0, 12.4, and 4.4 % of total visits, respectively, in monoculture (Table 2). Overall flower visiting behavior and visitation rate of Apidae, Calliphoridae and Syrphidae per inflorescence were substantially reduced with increasing proportions of sweet potato (P < 0.05).

Reproductive characteristics of *Mikania micrantha*

In monoculture, the biomass of plant, inflorescences and flowers of *M. micrantha* was 301.61 ± 5.19 g, 54.66 ± 1.14 g, and 40.99 ± 1.01 g, respectively; the inflorescence and flower numbers of *M. micrantha* were 2647.8 ± 55.3 and 10,587.7 ± 239.3, respectively. In mixed culture, the total biomass of plant, biomass of inflorescences and flowers, and numbers of inflorescences and flowers of *M. micrantha* were significantly (P < 0.05) suppressed with decreasing proportions of *M. micrantha* (Table 3). With proportional increases in sweet potato, reproductive allocation, reproductive

investment and reproductive index of *M. micrantha* were significantly lower (P < 0.05). The reproductive ratio of *M. micrantha* did not differ significantly among treatments (Table 3). With decreasing proportions of *M. micrantha*, the reproductive efficiency index of *M. micrantha* was reduced to a certain extent by sweet potato but the trend was not clear.

For a ratio of sweet potato to *M. micrantha* of 3:1, the number and biomass of *M. micrantha* seeds were reduced by a factor of more than 100 compared to *M. micrantha* in monoculture, i.e., 17,632.6 ± 479.8 vs. 171.7 ± 4.3 and 1.772 ± 0.042 g vs. 0.014 ± 0.000 g, respectively (Table 4). The number, biomass, length, set rate, germination rate, and 1000-grain dry weight of *M. micrantha* seeds were significantly suppressed (P < 0.05) with decreasing proportions of *M. micrantha*.

Seedling population dynamics of *Mikania micrantha*

Sexual seedling populations of *M. micrantha* germinated for 6 months (March–August), primarily occurring between May–June. Asexual seedling populations first arose in March, and then increased in density monthly from March to October. In monoculture, sexual population and asexual population densities were 89.25 ± 4.35 m^{-2} and 134.75 ± 4.99 m^{-2}, respectively

Table 2 Total number of visits (visits per day and inflorescence) by the four pollinator groups to *Mikania micrantha* growing as a monoculture or under mixed culture conditions

Pollinators	Ratios (sweet potato: *M. micrantha*)					
	3:1	2:1	1:1	1:2	1:3	0:4
Apidae	8.75 ± 0.95d	10.50 ± 1.11d	13.50 ± 1.04c	24.50 ± 1.14b	27.50 ± 1.43a	28.50 ± 1.40a
Calliphoridae	4.92 ± 0.70c	6.00 ± 0.63c	6.50 ± 0.50bc	8.02 ± 0.75b	9.74 ± 0.88a	10.26 ± 0.96a
Syrphidae	1.26 ± 0.46c	1.54 ± 0.45c	2.28 ± 0.40c	3.76 ± 0.39b	5.63 ± 0.54a	5.75 ± 0.85a
Other pollinator	2.53 ± 0.48a	1.63 ± 0.57a	1.76 ± 0.41a	2.23 ± 0.73a	1.77 ± 0.92a	2.04 ± 0.67a
Total	17.46 ± 1.19d	19.66 ± 2.60d	24.04 ± 1.76c	38.50 ± 1.55b	44.63 ± 2.13a	46.55 ± 0.95a

Data are expressed as mean ± standard deviation. Different letters within the same row signify significant differences at P < 0.05

Table 3 Flowering characteristics of *Mikania micrantha* growing as a monoculture or under mixed culture conditions

Variables	Ratios (sweet potato: *M. micrantha*)					
	3:1	2:1	1:1	1:2	1:3	0:4
Total biomass (g)	25.16 ± 1.10f	35.88 ± 0.96e	54.80 ± 0.80d	82.21 ± 1.36c	104.41 ± 2.62b	301.61 ± 5.19a
Flower number	429.9 ± 10.3f	1168.3 ± 82.7e	2486.2 ± 62.9d	3508.8 ± 53.8c	4705.9 ± 106.5b	10,587.7 ± 239.3a
Inflorescence number	108.3 ± 3.7f	289.8 ± 14.1e	613.9 ± 12.1d	868.8 ± 16.1c	1179.4 ± 22.5b	2647.8 ± 55.3a
Flower biomass (g)	0.63 ± 0.03e	1.55 ± 0.04e	3.86 ± 0.06d	7.32 ± 0.06c	11.39 ± 0.29b	40.99 ± 1.01a
Inflorescence biomass (g)	1.24 ± 0.04e	2.28 ± 0.07e	6.86 ± 0.09d	11.21 ± 0.13c	14.95 ± 0.27b	54.66 ± 1.14a
Reproductive allocation (g·g^{-1})	0.049 ± 0.004f	0.064 ± 0.002e	0.125 ± 0.002d	0.136 ± 0.001c	0.143 ± 0.002b	0.181 ± 0.002a
Reproductive investment (g·g^{-1})	0.025 ± 0.002f	0.043 ± 0.001e	0.071 ± 0.001d	0.089 ± 0.002c	0.109 ± 0.001b	0.136 ± 0.002a
Reproductive index (g·g^{-1})	0.512 ± 0.012d	0.678 ± 0.005b	0.563 ± 0.003c	0.653 ± 0.010b	0.762 ± 0.012a	0.750 ± 0.018a
Reproductive ratio (flower·mg^{-1})	3.970 ± 0.081a	4.028 ± 0.089a	4.049 ± 0.025a	4.039 ± 0.025a	3.990 ± 0.015a	3.999 ± 0.041a
Reproductive efficiency index (flower·mg^{-1})	0.017 ± 0.001d	0.033 ± 0.002c	0.045 ± 0.001a	0.043 ± 0.000b	0.045 ± 0.001ab	0.035 ± 0.000c

Data are expressed as mean ± standard deviation. The different letters within same row signify significant differences at P < 0.05

Table 4 Characteristics of *Mikania micrantha* seed growing as a monoculture or under mixed culture conditions

Variables	Ratios (sweet potato: *M. micrantha*)					
	3:1	2:1	1:1	1:2	1:3	0:4
Seed number	171.7 ± 4.3f	702.9 ± 8.2e	1995.1 ± 54.9d	4066.9 ± 94.8c	6797.6 ± 92.2b	17,632.6 ± 479.8a
Seed biomass (g)	0.014 ± 0.000f	0.058 ± 0.001e	0.171 ± 0.004d	0.362 ± 0.008c	0.654 ± 0.011b	1.772 ± 0.042a
Seed length (mm)	0.750 ± 0.008f	0.808 ± 0.013e	0.860 ± 0.008d	0.900 ± 0.012c	1.058 ± 0.013b	1.290 ± 0.014a
Seed set rate (%)	7.99 ± 0.26f	12.07 ± 0.76e	16.05 ± 0.13d	23.18 ± 0.27c	28.90 ± 0.53b	33.31 ± 0.64a
Germination rate (%)	16.88 ± 0.48f	20.63 ± 0.63e	25.13 ± 0.85d	32.63 ± 1.65c	53.13 ± 1.60b	68.25 ± 1.04a
1000-grain dry weight (g)	0.078 ± 0.001f	0.082 ± 0.001e	0.086 ± 0.001d	0.089 ± 0.001c	0.096 ± 0.001b	0.101 ± 0.001a

Data are expressed as mean ± standard deviation. Different letters within the same row signify significant differences at P < 0.05

(Table 5). During growth some seedlings died; the sexual seedlings mostly died from May–June, and asexual seedlings did so between July–September. In all treatments, the asexual seedling population was comprised significantly higher densities (P < 0.05) than the sexual seedling population, and the asexual mortality rate was much lower (P < 0.05) than sexual mortality rate. With proportional increases in sweet potato, the total population, sexual population and asexual population of *M. micrantha* significantly declined (P < 0.05); higher seedling mortality rates were also associated with greater proportions of sweet potato.

Discussion

Our research showed that the biomass, flowering, seed and seedling characteristics of *M. micrantha* were reduced when grown in association with sweet potato. A previous study showed that plant height, branch, leaf, stem node, adventitious root, and biomass of *M. micrantha* were suppressed significantly by sweet potato competition [18]. The present study found that the biomass of plant, inflorescences, flowers, and seeds of *M. micrantha*

were also significantly suppressed with increasing proportions of sweet potato. Moreover, with decreasing number and biomass of inflorescences and flowers, the reproductive allocation, reproductive investment and reproductive index of *M. micrantha* were also significantly reduced. The net result of the presence of sweet potato was reduced reproductive potential of M. micrantha and like other invasive species, its reproductive ability, including flowering characteristics, seed dispersal and seed germination parameters, is associated with its invasiveness [39–43].

Flowering phenology is affected by number, timing and duration of flowers [44]. These factors are not only constrained by genetics and phylogeny, but also affected by environmental conditions, such as sunlight, temperature, nutrients, and competition [42]. The present study found that flowering phenology of *M. micrantha* was significantly delayed, duration of flowering and fruiting was significantly reduced and duration of bud formation was markedly increased with increased sweet potato proportions.

Along with reduced biomass and delayed flowering, another major factor reducing reproductive output in M.

Table 5 Population densities m^{-2} from sexual and asexual reproduction of *Mikania micrantha* growing as a monoculture or under mixed culture conditions

Variables	Ratios (sweet potato: *M. micrantha*)					
	3:1	2:1	1:1	1:2	1:3	0:4
Total population	42.00 ± 2.83f	65.25 ± 2.99e	102.50 ± 3.70d	166.75 ± 6.02c	191.50 ± 4.51b	224.00 ± 6.06a
Sexual population	9.25 ± 1.71f	21.00 ± 2.16e	31.50 ± 2.08d	55.75 ± 3.30c	70.50 ± 1.29b	89.25 ± 4.35a
Asexual population	32.75 ± 1.71f	44.25 ± 2.63e	71.00 ± 2.83d	111.00 ± 3.92c	121.00 ± 3.37b	134.75 ± 4.99a
Sexual mortality rate (%)	0.629 ± 0.092a	0.596 ± 0.080a	0.438 ± 0.036b	0.340 ± 0.035bc	0.291 ± 0.022c	0.241 ± 0.019c
Asexual mortality rate (%)	0.358 ± 0.034a	0.265 ± 0.008b	0.152 ± 0.019c	0.077 ± 0.014d	0.056 ± 0.008de	0.030 ± 0.001e

Data are expressed as mean ± standard deviation. The different letters within same row signify significant differences at P < 0.05

micrantha observed in our study was the negative impact of sweet potato on pollinator visitation. *M. micrantha* depends on insects for sexual reproduction as it is self-incompatible [45]. For insect-pollinated plant species, competition for pollinators may lead to changes in visitation frequency or pollinator composition [46–48] and consequently, a lowered reproductive output [48]. In this study, species of Apidae (bee species), Calliphoridae (flies), and Syrphidae (flies) were the most abundant visitors to *M. micrantha* flowers and were observed to have the longest foraging time of all floral visitors, which is consistent with the results of other studies [36, 49]. Overall flower visits of Apidae, Calliphoridae and Syrphidae per inflorescence were reduced significantly with increasing proportions of sweet potato. From October to February in the study area, the temperature gradually became lower and the delayed flowering phenology of *M. micrantha* corresponded with reduced insect activity. Moreover, in mixed culture, 70–90 % of *M. micrantha* stems and leaves were covered by sweet potato [18], thus reducing insect visitation via diminished visibility of *M. micrantha* flowers. Because flowering in both sweet potato and *M. micrantha* occurs at virtually the same time, pollinators visited both species during the monitoring period; however, because the number of flowers per shoot was at least 15 times greater for *M. micrantha* than sweet potato [18], the main influence on pollinator visitation was *M. micrantha* flower number. *M. micrantha* can produce a large number of flowers and small, light, and wind-dispersed seeds [32]. The negative correlation we observed between seed set and pollinator visitation in *M. micrantha* is consistent with the commonly observed link between pollinator visitation rate and seed set [50].

Nutrient availability also influences reproductive output by *M. micrantha*, with fewer flowers, lower seed setting percentage, lower 1000-grain weight and shorter flowering duration observed in plants growing in nutrient-deficient soils with suboptimal fertility, but soils with an overabundance of nutrients (e.g., silt from ponds or dump sites) also resulted in fewer flowers and low seed set [51]. Meanwhile, plants growing in an open habitat had more flowers with longer flowering duration, and under shade the 1000-grain weight was shown to have a slight increase, but light that was too bright or too dim was not conducive to seed set [51, 52]. Light was found to affect fruiting; for example, a photoperiod of 12 h/day resulted in 68.4 % of flowers producing fruit [53]. The present study found that the number, biomass, length, set rate, germination rate, and 1000-grain dry weight of *M. micrantha* seeds was significantly suppressed with decreasing proportions of *M. micrantha*. This is because *M. micrantha* plants covered by dense carpets of sweet potato received fewer pollinator visits and produced fewer seeds. Furthermore, sweet potato exhibited greater absorption of soil nutrients than *M. micrantha* [18]; the resulting lack of nutrients likely lead to reduced 1000-grain weight and germination rate of *M. micrantha* in the presence of sweet potato.

The potential for high levels of sexual and/or vegetative reproduction by *M. micrantha* is formidable [22]. It has transient soil seed bank and persistent soil seed bank, and some seeds would germinate given ideal germination conditions such as season, temperature, moisture; otherwise the seeds would remain dormant [54, 55]. Large numbers of seeds of *M. micrantha* were concentrated primarily in the 0–5 cm soil layer, which contained 98 % of the total seeds present in the soil [55]. Vegetative propagation of *M. micrantha* from stem fragments that root easily at the nodes and from vegetative ramets arising from rosettes can be considered at least as important as reproduction by seeds [22, 56]. The seedling is the most vulnerable stage in the life history of *M. micrantha* and seedlings suffer a high level of mortality under natural conditions [54]. The present study found that with proportional increases in sweet potato, both sexual and asexual seedling populations of *M. micrantha* were significantly suppressed, corresponding to increased mortality with increasing levels of sweet potato competition.

Thus, the best time to control *M. micrantha* is during the seedling period and control measures should be comprehensive involving both herbicides and appropriate cultural techniques [31].

Conclusion

The competitive advantage of sweet potato over *M. micrantha* could be used to reduce *M. micrantha* growth and reproductive ability in tropical and subtropical agricultural regions suitable for cultivation of sweet potato. Both plants have similar growth forms and climatic requirements, and sweet potato is a high value crop, and thus we recommend planting sweet potato in areas infested by *M. micrantha*, perhaps as part of a rotation involving more vulnerable crops. Sweet potato could even be planted in habitats such as waste areas not currently cultivated in order to reduce *M. micrantha* populations. Our results showed that various components of reproduction for *M. micrantha* were significantly reduced by suppression of plant growth; the original data is available online [57]. Flowering phenology was impacted by sweet potato competition, and delayed flowering phenology, reduced duration of flowering and fruiting and increased duration of bud formation resulted in reduced pollinator visits and seed set for *M. micrantha*. Finally, high cover of sweet potato shading *M. micrantha* plants also reduced pollinator visits, seeds number, and seedling populations of *M. micrantha*. Thus the mechanism by which sweet potato reduced sexual reproduction in *M. micrantha* was essentially twofold: causing a delay in flowering phenology and reducing pollinator visits. In addition to utilizing sweet potato, research in this study and other recent studies revealed that control of *M. micrantha* ideally should take place during the seedling period when *M. micrantha* is most vulnerable and should be comprehensive for optimal results, employing both chemical and cultural control. Thus in the case of our study region in southern Asia, the most effective timing of control is in the peak of sexual seedling emergence in May–June. The potential for utilizing a crop like sweet potato to compete with an invasive plant may well apply to many other agronomic settings where other management techniques (e.g., chemical control, mechanical control or classical biological control) are unreliable or are associated with environmental concerns. The scenario of controlling *M. micrantha* by utilizing a crop with a similar growth form may provide a useful model for similar management strategies in other systems.

Authors' contributions

SCS and FDZ conceived and designed the experiments; SCS, GFX, GMJ, SFL, YXY, ADC, and FDZ performed the experiments; SCS and DRC analyzed the data and wrote the draft; HKN designed and commented on the manuscript. All authors read and approved the final manuscript.

Author details
[1] Agricultural Environment and Resource Research Institute, Yunnan Academy of Agricultural Sciences, Kunming 650205, Yunnan, China. [2] Biology Department, Trinity Western University, 7600 Glover Road, Langley, BC V2Y1Y1, Canada. [3] Department of Applied Biological Science, Faculty of Agriculture, Kagawa University, Miki, Kagawa 761-0795, Japan.

Acknowledgements
We wish thank Yang Jian, Dong Jianping and Gao Rui, from the Plant Protection Station of Longchuan County, Dehong Prefecture of Yunnan Province for their great field support. We would like to thank two anonymous reviewers for their helpful comments, which considerably improved the manuscript.

Competing interests
All authors declare that they have no competing interests.

Funding
This research was supported by Yunnan Provincial Key Fund Program (2010CC002), Middle-aged and Young Academic Leader Training Foundation of Yunnan Province (2014HB039) and International Science & Technology Cooperation Program of Yunnan Provincial Science and Technology Department (2014IA009).

References
1. Alpert P, Bone E, Holzapfel C. Invasiveness, invisibility and the role of environmental stress in the spread of nonnative plants. Perspect Plant Ecol Evol Syst. 2000;3:52–66. doi:10.1078/1433-8319-00004.
2. Richardson DM, Allsopp N, D'Antonio CM, Milton SJ, Rejmánek M. Plant invasions—the role of mutualisms. Biol Rev. 2000;75:65–93. doi:10.1111/j.1469-185X.1999.tb00041.x.
3. Swamy PS, Ramakrishnan PS. Effect of fire on population dynamics of *Mikania micrantha* H.B.K. during early succession after slash-and-burn agriculture (Jhum) in northeast India. Weed Res. 1987;27:397–404. doi:10.1111/j.1365-3180.1987.tb01590.x.
4. Britton AJ, Marrs RH, Carey PD, Pakeman RJ. Comparison of techniques to increase *Calluna vulgaris* cover on heath land invaded by grasses in Breckland, south east England. Biol Conserv. 2000;95:227–32. doi:10.1016/S0006-3207(00)00047-1.
5. Timmins SM. How weed lists help protect native biodiversity in New Zealand. Weed Technol. 2004;18:1292–300. doi:10.1614/0890-037X(2004)018[1292:HWLHPN]2.0.CO;2.
6. Paynter Q, Flanagan GJ. Integrating herbicide and mechanical control treatments with fire and biological control to manage an invasive wetland shrub, *Mimosa pigra*. J Appl Ecol. 2004;41:615–29. doi:10.1111/j.0021-8901.2004.00931.x.
7. Shen SC, Xu GF, Zhang FD, Jin GM, Liu SF, Liu MY, Chen AD, Zhang YH. Harmful effects and chemical control study of *Mikania micrantha* H.B.K. in Yunnan, Southwest China. Afr J Agr Res. 2013;8:5554–61. doi:10.5897/AJAR2013.7688.
8. Barreto RW, Evans HC. The mycobiota of the weed *Mikrania micrantha* in southern Brazil with particular reference to fungal pathogens for biological control. Mycol Res. 1995;99:343–52. doi:10.1016/S0953-7562(09)80911-8.

9. Mack RN, Simberloff D, Lonsdale WM, Evans H, Clout M, Bazzaz FA. Biotic invasions: causes, epidemiology, global consequences, and control. Ecol Appl. 2000;10:689–710. doi:10.1890/1051-0761(2000)010[0689:BICEGC]2.0.CO;2.

10. Messing RH, Wright MG. Biological control of invasive species: solution or pollution? Front Ecol Environ. 2006;4:132–40. doi:10.1890/1540-9295(2006)004[0132:BCOISS]2.0.CO;2.

11. Yu H, Liu J, He WM, Miao SL, Dong M. Cuscuta australis restrains three exotic invasive plants and benefits native species. Biol Invasions. 2011;13:747–56. doi:10.1007/s10530-010-9865-x.

12. Marrs RH, Frost AJ. A microcosm approach to the detection of the effects of herbicide spray drift in plant communities. J Environ Manag. 1997;50:369–88. doi:10.1006/jema.1996.9984.

13. Milligan AL, Putwain PD, Marrs RH. A field assessment of the role of selective herbicides in the restoration of British moorland dominated by Molinia. Biol Conserv. 2003;109:369–79. doi:10.1016/S0006-3207(02)00163-5.

14. Mason TJ, French K. Management regimes for a plant invader differentially impact resident communities. Biol Conserv. 2007;136:246–59. doi:10.1016/j.biocon.2006.11.023.

15. Ding H, Xu HG, Liu ZL. Impacts of invasion of Eupatorium adenophorum on vegetation diversity. J Ecol Rural Environ. 2007;23:29–32.

16. Liu J, Wisniewski M, Droby S, Tian SP, Hershkovitz V, Tworkoski T. Effect of heat shock treatment on stress tolerance and biocontrol efficacy of Metschnikowia fructicola. Microbiol Ecol. 2011;76:145–55. doi:10.1111/j.1574-6941.2010.01037.x.

17. Lugo AE. The apparent paradox of reestablishing species richness on degraded lands with tree monocultures. For Ecol Manag. 1997;99:9–19. doi:10.1016/S0378-1127(97)00191-6.

18. Shen SC, Xu GF, Clements DR, Jin GM, Chen AD, Zhang FD, Hisashi KN. Suppression of the invasive plant mile-a-minute (Mikania micrantha) by local crop sweet potato (Ipomoea batatas) by means of higher growth rate and competition for soil nutrients. BMC Ecol. 2015;15:1–10. doi:10.1186/s12898-014-0033-5.

19. Li WH, Luo JN, Tian XS, Chow WS, Sun ZY, Zhang TJ, Peng SL, Peng CL. A new strategy for controlling invasive weeds: selecting valuable native plants to defeat them. Sci Rep. 2015;5:11004. doi:10.1038/srep11004.

20. Cao YS, Wang T, Xiao YA, Zhou B. The interspecific competition between Humulus scandens and Alternanthera philoxeroides. J Plant Interact. 2014;9:194–9. doi:10.1080/17429145.2013.808767.

21. Gosper CR, Vivian-Smith G. Approaches to selecting native plant replacements for fleshy-fruited invasive species. Restor Ecol. 2009;17:196–204. doi:10.1111/j.1526-100X.2008.00374.x.

22. Zhang LY, Ye WH, Cao HL, Feng HL. Mikania micrantha H.B.K. in China—an overview. Weed Res. 2004;44:42–9. doi:10.1111/j.1365-3180.2003.00371.x.

23. Lowe S, Browne M, Boudjelas S, Poorter MD. 100 of the World's Worst Invasive Alien Species. A Selection from the Global Invasive Species Database. Auckland: IUCN/SSC Invasive Species Specialist Group (ISSG); 2001.

24. Manrique V, Diaz R, Cuda JP, Overholt WA. Suitability of a new plant invader as a target for biological control in Florida. Invas Plant Sci Mana. 2011;4:1–10. doi:10.1614/IPSM-D-10-00040.1.

25. Day MD, Kawi A, Kurika K, Dewhurst CF, Waisale S, Saul-Maora J, Fidelis J, Bokosou J, Moxon J, Orapa W, Senaratne KAD. Mikania micrantha Kunth (Asteraceae) (mile-a-minute): its distribution and physical and socioeconomic impacts in Papua New Guinea. Pac Sci. 2012;66:213–23. doi:10.2984/66.2.8.

26. Zan QJ, Wang YJ, Wang BS, Liao WB, Li MG. The distribution and harm of the exotic weed Mikania micrantha. Chin J Ecol. 2000;19:58–61.

27. Shen SC, Xu GF, Clements DR, Jin GM, Liu SF, Zhang FD, Yang YX, Chen AD, Hisashi KN. Effects of invasive plant Mikania micrantha on plant community and diversity in farming systems. Asian J Plant Sci. 2015;14:27–33. doi:10.3923/ajps.2015.27.33.

28. Yu H, Liu J, He WM, Miao SL, Dong M. Native Cuscuta campestris restrains exotic Mikania micrantha and enhances soil resources beneficial to natives in the invaded communities. Biol Invasions. 2009;11:835–44. doi:10.1007/s10530-008-9297-z.

29. Xu GF, Zhang FD, Li TL, Shen SC, Zhang YH. Effects of 5 species and planting density on Mikania micrantha H.B.K growth and competitive traits. Ecol Environ Sci. 2011;20:798–804.

30. Shen SC, Xu GF, Zhang FD, Li TL, Zhang YH. Competitive effect of Ipomoea batatas to Mikania micrantha. Chin J Ecol. 2012;31:850–5.

31. Shen SC, Xu GF, Clements DR, Jin GM, Liu SF, Chen AD, Zhang FD, Hisashi KN. Control of invasive plant mile-a-minute (Mikania micrantha) with the local crop sweet potato (Ipomoea batatas) and applications of the herbicide bentazon. Asian J Plant Sci. 2014;13:59–65. doi:10.3923/ajps.2014.59.65.

32. Kuorr YL, Chen TY, Lingg CC. Using a periodic cutting method and allelopathy to control the invasive vine, Mikania micrantha H.B.K. Taiwan J For Sci. 2002;17:171–81.

33. Sihachakr D, Haïcour R, Cavalcante JM, Umboh I, Nzoghé D, Servaes A, Servaes A, Ducreux G. Plant regeneration in sweet potato (Ipomoea batatas L., Convolvulaceae). Euphytica. 1997;96:143–52. doi:10.1023/A:1002997319342.

34. De Wit CT. On Competition. Verslagen Landbouwkundige Onderzoekigen. 1960;66:1–82.

35. Hoagland DR, Arnon DI. The water-culture method for growing plants without soil. Agricultural Experiment Station Circular. Berkley: College of Agriculture University of California; 1950.

36. Hong L, Shen H, Ye WH, Cao HL. Study of pollinating insects of Mikania micrantha H.B.K and their foraging behavior. J South China Normal Univ (Nat Sci Ed). 2011;43:98–102. doi:10.6054/j.jscnun.2011.11.016.

37. Liu ZH, Du GZ, Chen JK. Size-dependent reproductive allocation of Ligularia virgaurea in different habitats. Acta Phytoeol Sin. 2002;26:44–50.

38. Obeso JR. The cost of reproduction in plants. New Phytol. 2002;155:321–48. doi:10.1046/j.1469-8137.2002.00477.x.

39. Lodge DM. Biological invasions: lessens for ecology. Trends Ecol Evol. 1993;8:133–7. doi:10.1016/0169-5347(93)90025-K.

40. Van Kleunen M, Weber E, Fischer M. A meta-analysis of trait differences between invasive and non-invasive plant species. Ecol Lett. 2010;13:235–45. doi:10.1111/j.1461-0248.2009.01418.x.

41. Reichard SH, Hamilton CW. Predicting invasions of woody plants introduced into North America. Conserv Biol. 1997;11:193–203. doi:10.1046/j.1523-1739.1997.95473.x.

42. Wilke BJ, Irwin RE. Variation in the phenology and abundance of flowering by native and exotic plants in subalpine meadows. Biol Invasions. 2010;12:2363–72. doi:10.1007/s10530-009-9649-3.

43. Rejmánek M. A theory of seed plant invasiveness: the first sketch. Biol Conserv. 1996;78:171–81. doi:10.1016/0006-3207(96)00026-2.

44. McIntosh ME. Flowering phenology and reproductive output in two sister species of Ferocactus (Cactaceae). Plant Ecol. 2002;159:1–13. doi:10.1023/A:1015589002987.

45. Hong L, Shen H, Ye WH. Self-incompatibility in Mikania micrantha in South China. Weed Res. 2007;47:280–3. doi:10.1111/j.1365-3180.2007.00575.x.

46. Chittka L, Schürkens S. Successful invasion of a floral market. Nature. 2001;411:653. doi:10.1038/35079676.

47. Bartomeus I, Vilà M, Santamaría L. Contrasting effects of invasive plants in plant-pollinator networks. Oecologia. 2008;155:761–70. doi:10.1007/s00442-007-0946-1.

48. Morales CL, Traveset A. A meta-analysis of impacts of alien vs. native plants on pollinator visitation and reproductive success of co-flowering native plants. Ecol Lett. 2009;12:716–28. doi:10.1111/j.1461-0248.2009.01319.x.

49. Macanawai AR, Day MD, Tumaneng-Diete T, Adkins SW. The impact of rainfall upon pollination and reproduction of Mikania micrantha in Viti Levu, Fiji. Proceedings of the 23rd Asian-Pacific Weed Science Society Conference. Cairns; 2011.

50. Ashman TL, Knight TM, Steets JA, Amarasekare P, Burd M, Campbell DR, Dudash MR, Johnston MO, Mazer SJ, Mitchell RJ, Morgan MT, Wilson WG. Pollen limitation of plant reproduction: ecological and evolutionary causes and consequences. Ecology. 2004;85:2408–21. doi:10.1890/03-8024.

51. Yang QH, Feng HL, Ye WH, Cao HL, Deng X, Xu KY. An investigation of the effects of environmental factors on the flowering and seed setting of Mikania micrantha H.B.K (Compositae). J Trop Subtrop Bot. 2003;11:123–6.

52. Zhang WY, Li GM, Wang BS, Zan QJ, Wang YJ. Seed production characteristics of an exotic weed Mikania micrantha. J Wuhan Bot Res. 2003;21:143–7.

53. Hu YJ, But PPH. A study on life cycle and response to herbicides on Mikania micrantha. Acta Sci Nat Univ Sunyatseni. 1994;33:88–95.

54. Zhang WY, Wang BS, Zhang JL, Li MG, Zan QJ, Wang YJ. Study on the structure and dynamics of seedlings of *Mikania micrantha* populations. Acta Sci Nat Univ Sunyatseni. 2002;41:64–6.

55. Shen SC, Xu GF, Zhang FD, Li TL, Jin GM, Zhang YH. Characteristics of the seed banks and seedling banks of *Mikania micrantha*-invaded soils different in type of habitat. J Ecol Rural Environ. 2013;29:483–8.

56. Li X, Shen Y, Huang Q, Fan Z, Huang D. Regeneration capacity of small clonal fragments of the invasive *Mikania micrantha* H.B.K.: effects of burial depth and stolon internode length. PLoS ONE. 2013;8:e84657. doi:10.1371/journal.pone.0084657.

57. Shen Sh-C, Xu G-F, Clements DR, Jin G-M, Yang Y, Zhang F-D, Kato-Nuguchi H. Data from: suppression of reproductive characteristics of the invasive plant Mikania micrantha by sweet potato competition. Dryad Digital Repository. 2016. doi:10.5061/dryad.9522r.

11

A metabarcoding framework for facilitated survey of endolithic phototrophs with *tuf*A

Thomas Sauvage[1*], William E. Schmidt[1], Shoichiro Suda[2] and Suzanne Fredericq[1]

Abstract

Background: In spite of their ecological importance as primary producers and microbioeroders of marine calcium carbonate ($CaCO_3$) substrata, endolithic phototrophs spanning both prokaryotic (the cyanobacteria) and eukaryotic algae lack established molecular resources for their facilitated survey with high throughput sequencing. Here, the development of a metabarcoding framework for the elongation factor EF-T*tu* (*tuf*A) was tested on four Illumina-sequenced marine $CaCO_3$ microfloras for the characterization of their endolithic phototrophs, especially the abundant bioeroding *Ostreobium* spp. (Ulvophyceae). The framework consists of novel *tuf*A degenerate primers and a comprehensive database enabling Operational Taxonomic Unit (OTU) identification at multiple taxonomic ranks with percent identity thresholds determined herein.

Results: The newly established *tuf*A database comprises 4057 non-redundant sequences (from 1339 eukaryotic and prokaryotic phototrophs, and 2718 prokaryotic heterotrophs) including 27 classes in 10 phyla of phototrophic diversity summarized from data mining on GenBank®, our barcoding of >150 clones produced from coral reef microfloras, and >300 eukaryotic phototrophs (>230 Ulvophyceae including >100 '*Ostreobium*' spp., and >70 Florideophyceae, Phaeophyceae and miscellaneous taxa). Illumina metabarcoding with the newly designed primers resulted in 802 robust OTUs including 618 phototrophs and 184 heterotrophs (77 and 23 % of OTUs, respectively). Phototrophic OTUs belonged to 14 classes of phototrophs found in seven phyla, and represented ~98 % of all reads. The phylogenetic profiles of coral reef microfloras showed few OTUs in large abundance (proportion of reads) for the Chlorophyta (Ulvophyceae, i.e. *Ostreobium* and *Phaeophila*), the Rhodophyta (Florideophyceae) and Haptophyta (Coccolithophyceae), and a large diversity (richness) of OTUs in lower abundance for the Cyanophyta (Cyanophyceae) and the Ochrophyta (the diatoms, 'Bacillariophyta'). The bioerosive '*Ostreobium*' spp. represented four families in a large clade of subordinal divergence, i.e. the Ostreobidineae, and a fifth, phylogenetically remote family in the suborder Halimedineae (provisionally assigned as the 'Pseudostreobiaceae'). Together they harbor 85–95 delimited cryptic species of endolithic microsiphons.

Conclusions: The novel degenerate primers allowed for amplification of endolithic phototrophs across a wide phylogenetic breadth as well as their recovery in very large proportions of reads (overall 98 %) and diversity (overall 77 % of OTUs). The established companion *tuf*A database and determined identity thresholds allow for OTU identification at multiple taxonomic ranks to facilitate the monitoring of phototrophic assemblages via metabarcoding, especially endolithic communities rich in bioeroding Ulvophyceae, such as those harboring '*Ostreobium*' spp., *Phaeophila* spp. and associated algal diversity.

---

*Correspondence: tomsauv@gmail.com
[1] Department of Biology, University of Louisiana at Lafayette, Lafayette, LA 70503, USA
Full list of author information is available at the end of the article

Keywords: Algae, Amplicon Metagenomics, Bryopsidales, Calcium carbonate, Coral reef, Illumina MiSeq, Metabarcoding, Next generation sequencing, *Ostreobium*, *Phaeophila*, Phototrophs, Plastid, Rhodoliths, *tuf*A, Ulvophyceae

Background

Endolithic phototrophs are major primary producers [68] and microbioeroders in marine carbonate substrata [98]. They may colonize the surface and cavities of the substratum (chaesmoendoliths and cryptoendoliths, respectively), as well as actively penetrate it (euendoliths) wherever sufficient light penetrates for photosynthesis. They may be found in aragonite, calcite, or a combination of both, in e.g. live and dead corals, mollusk shells, and crustose coralline algae (CCA) [97]. In calcium carbonate $CaCO_3$-building ecosystems, such as CCA ridges, coral reefs, oyster reefs and rhodolith beds, euendolithic phototrophs play a critical role in the dynamic balance between constructive (accretion) and destructive (dissolution) processes [36]. With upcoming global changes, this balance may be negatively affected by enhancing the bioerosive power of boring phototrophs, as measured under projected ocean acidification [higher partial pressure of carbon dioxide (pCO_2)] and increased temperature regimes [81, 82, 99]. The prospect of accelerated biogenic $CaCO_3$ dissolution poses strong concerns for the future maintenance of the structural integrity and functionality of these ecosystems considering the biodiverse assemblages of micro- and macro-organisms they support [72, 73].

Commonly reported euendolithic phototrophs include the eukaryotic green algal genera *Ostreobium* Bornet and Flahault (order Bryopsidales) and *Phaeophila* Hauck ('Ulvales-Ulothrichales') (both in class Ulvophyceae, phylum Chlorophyta), and prokaryotic (eubacterial)

blue-green algal genera (all in class Cyanophyceae, phylum Cyanophyta) such as *Mastigocoleus* Lagerheim ex Bornet and Flahault, and *Plectonema* Thuret ex Gomont [98]. They also include microscopic alternate life stages of otherwise conspicuous alga, e.g. the *Conchocelis*-stage of the red alga *Porphyra* C. Agardh (Bangiophyceae, Rhodophyta) and endolithic vegetative networks underlying diminutive epilithic Bryopsidales, e.g. *Pseudochlorodesmis* Børgesen and *Caulerpa ambigua* Okamura [1, 50]. Among the above, *Ostreobium* microsiphons are omnipresent agents of bioerosion [98], although the microfilaments of the genus *Phaeophila* are also often reported (e.g. see [12, 79]). The molecular diversity of *Ostreobium* spp. remains particularly unexplored toward establishing comprehensive sequence reference databases for the profiling of endolithic communities via metabacording (e.g. [14]).

The skeleton of reef-building scleractian coral species (Cnidaria) is abundantly colonized by *Ostreobium* [54, 60, 61] (Fig. 1), in which it develops a dense green layer underlying the animal tissue and where, as part of the coral holobiont, it may play a role as a nutritional ally (i.e. metabolite translocation, [31]). Previously, Gutner-Hoch and Fine [38] investigated the molecular diversity of this 'green layer' with *rbc*L in two coral species from the Red Sea in order to gain insights into potential patterns of association of *Ostreobium* haplotypes with coral species. While these authors reported some possible haplotype-to-coral species distributional patterns, they also revealed multiple *rbc*L haplotypes of *Ostreobium*,

Fig. 1 *Ostreobium* in situ in the Ryukus. **a** Coral reef habitat where microfloras harboring *Ostreobium* spp. underlay live coral tissue (*picture background*) and limestone often covered by epilithic turf algae and crustose coralline algae (*foreground*). **b** Fragmented coral colony showing *Ostreobium*'s '*green layer*' found below live coral tissue, here *Porites* sp. **c** Calcium carbonate colonized by *Ostreobium* sp. microsiphons (*Scale bar* approximately 20 µm)

whose taxonomic breadth remains unknown (with regard to species diversity and the taxonomic rank attributed to this novel diversity) from a lack of phylogenetic context in their analysis. Earlier, Verbruggen et al. [102] suggested that the genus *Ostreobium* might actually represent an entire suborder of microsiphonous species that they informally proposed as the 'Ostreobidineae' (next to two other informally accepted suborders in the Bryopsidales, the Bryopsidineae and Halimedineae, see [41]) based on the early branching of a single *Ostreobium* specimen in a comprehensive multi-marker phylogeny of the Bryopsidales [Chloroplast 16S ribosomal DNA (rDNA), RuBisCO Large subunit (*rbc*L), and elongation EF-*Tu* (*tuf*A)]; however, this remains to be substantiated. Overall, considering the high density of *Ostreobium* microsiphons in the so-called coral 'green layer', this microhabitat is particularly convenient to target in order to rapidly build reference barcode data sets.

Metabarcoding represents a novel terminology [96] for amplicon-based metagenomics (e.g. usually targeting 16S, [18]) in contrast to whole-genome metagenomics [65]. Important limitations to metabarcoding include the availability of universal primers amplifying the targeted diversity with minimal taxon bias [51, 59], the clustering of next-generation reads into biologically relevant Operational Taxonomic Unit (OTU) through the elimination of sequencing error/noise (e.g. [25, 26, 88]), and building up taxonomy-curated reference sequence database for OTU identification/annotation (e.g. Greengenes, [21]; Ribosomal Database Project [RDP], [13]; silva, [77]). Currently, a metabarcoding framework specifically developed to target lower phototrophs (i.e. the algae sensu lato: prokaryotic blue-green algae and eukaryotic algae) and their recovery in large proportions of reads is inexistent and is sorely needed to facilitate the monitoring of endolithic phototrophs and other microbial/algal assemblages. A recent progress toward metabarcoding phototrophs was made with the establishment of a curated chloroplast 16S rDNA database (primarily for phytoplankton taxa, PHYTO-REF, [17]); however, primers used to amplify 16S are generally highly conserved in prokaryotes and tend to recover low proportions of phototrophic organisms from environmental mixtures where microbial DNA from heterotrophic phyla is inherently overdominant. For instance, in a study of freshwater phytoplankton communities with 16S (V3–V4 region), only 9 % of reads represented phototrophic organisms [27]. Likewise, 16S libraries sequenced from coral tissue and their underlying endolithic communities enumerated very few phototrophs in comparison to heterotrophs (e.g. see [59, 85, 94]).

A candidate DNA marker for metabarcoding phototrophs is the gene encoding the protein chain elongation factor EF-*Tu*, or *tuf*A, whose role in the RNA translation machinery is deeply conserved among eubacteria and their eukaryotic endosymbiotic offshoot, the organelles [52], especially in the chloroplast (translation mechanisms are modified in the mitochondrion, see [95]). Iwabe et al. [42] first used the elongation factors EF-*Tu* (and its homolog in archaea and eukaryotic nucleus, the elongation factor 1 alpha, EF-1a, [40]) to examine deep (domain) phylogenetic relationships (EF-*Tu* and EF-1a show some amino acid conservation but their DNA sequences are highly divergent). Later, Delwiche et al. [19] demonstrated the cyanobacterial (Cyanophyta) origin of all plastids in a single-gene phylogeny of *tuf*A rooted with heterotrophs. Subsequently, *tuf*A has gained much popularity in phylogenetic and systematic studies of diverse phototrophs (e.g. [5, 29, 66, 84, 106]) and was also recommended as a standard marker for the routine barcoding of the Chlorophyta [86] for its high amplification rate (95 %, except in the Cladophorales, Ulvophyceae), and faster evolving rate relative to other commonly used markers in this phylum [e.g. *rbc*L or rDNA markers such as the large subunit (LSU), 23S universal plastid amplicon (UPA) and the internal transcribed spacer (ITS)]. In early branching members of the Streptophyta (the Charophytes *Mesostigma* and *Chlorokybus*), *tuf*A is chloroplast-encoded while in the remainder of this phylum (i.e. the higher phototrophs), it is nuclear-encoded [3]. In some eubacteria (Gram-negative), including some Cyanophyta, e.g. in the Oscillatoriales (e.g. *Oscillatoria nigro-viridis* PCC7112), *tuf*A paralogs known as elongation factor EF-*Tu* 2 or *tuf*B may exist [45, 53, 90]; however, these usually undergo concerted evolution and thus exhibit very low divergence [100]. Considering the universality of *tuf*A and the large amount of data available on Genbank® for the order Bryopsidales (>1300 accessions), this marker is thus well-suited for the barcoding of *Ostreobium* spp. and metabarcoding of microfloras dominated by its microsiphons and associated algal diversity.

Here, we established a metabacording framework consisting of newly developed primers, a curated database of phototrophic diversity (GenBank® data, a clone library of endolithic phototrophs, and new reference barcodes for *Ostreobium* spp. and related taxa), a provisional classification scheme for cryptic endoliths in the order Bryopsidales, and recommended identity thresholds for the taxonomic annotation of phototrophs at high levels (domain to class) and at lower levels in the Ulvophyceae (order to family, i.e. in the Bryopsidales and Ulvales). This framework, geared toward molecular ecology studies of endolithic phototroph assemblages rich in bioeroding Ulvophycean taxa (e.g. *Ostreobium* spp.), is tested on four Illumina-metabarcoded $CaCO_3$ microfloras.

Fig. 2 Free-living *Ostreobium* in culture. **a** Specimen belonging in provisional family 'Hamidaceae' (TS1385, note that some chloroplast-depleted siphons may falsely appear septated), **b** 'Maedaceae' (TS1410B), and **c** 'Odoaceae' (TS1408). No picture is available for the 'Unarizakiaceae'. *Scale bar* 50 μm

Methods

Building up *Ostreobium* and reference *tuf*A diversity

Ostreobium specimens were sequenced primarily from the CaCO$_3$ of densely colonized coral skeletons collected throughout the Ryukyu archipelago and culture starters established from them (Fig. 2, Additional file 1: Table S1). A few additional specimens originated from endolithic siphons underlying rhodolith-forming encrusting red algae from the Gulf of Mexico (GM) and also from starters established from oyster shells collected in Florida and corals from miscellaneous localities (Additional file 1: Table S1). For culturing, colonized blocks of CaCO$_3$ were subsetted to ~0.04–0.25 cm^2 and incubated for 10–20 days in 60 mL polypropylene cups (Diamond™) with 20–30 mL half strength Provasoli enriched seawater medium [75] supplemented with 1.25 mg/L germanium dioxide (GeO$_2$) under light–dark 14:10 cycles (50 μmol m^{-2} s^{-1}) and at room temperature (22 ± 1 °C). Specimens were extracted with a DNeasy Plant Mini Kit (Qiagen, Valencia, CA, USA) and *tuf*A amplified by polymerase chain reaction (PCR) as previously published [29, 39, 87] and/or with novel primers in various combination (Table 1). Successful PCR products were Sanger-sequenced commercially and chromatograms assembled in Sequencher v.5.1 (Gene Codes, Ann Arbor, Michigan, USA). To further increase reference sequence context, *tuf*A barcodes were also generated for macroscopic (>5–20 cm, e.g. *Codium, Halimeda, Rhipilia*) and diminutive members of the Bryopsidales (<2 cm) (e.g. the polyphyletic '*Pseudochlorodesmis*' species complex, [103]; and the monophyletic *Caulerpa* '*ambigua*' species complex, [23]). Likewise, miscellaneous 'Ulvales-Ulothrichales' that occasionally emerged in cultures (Additional file 1: Table S1) as well as several Florideophyceae (Rhodophyta) and few Phaeophyceae (Ochrophyta)

Table 1 Newly designed PCR primers for *tuf*A barcoding and metabarcoding

Name	Sequence (5′–3′)	bp	GC (%)	Tm (°C)
*tu*470F	TTTTAATGGCTGTCGAAAATGTTG	24	33.3	52.8
*tu*bryoF	GCAGATGGTCCAATGCCWCAAAC	23	52.2	59
*tu*bryoR	CCWGGTTTAGCTAAAACCATNCC	23	45.7	54.9
*env_tuf*AF[a]	TGGGTDGAHAADATTTWYNMNY-TRATGR	28	33.3	46.5–62.8
*env_tuf*AR[a]	TNACATCHGTWGTWCKNACATA-RAAYTG	28	35.1	49.8–60.6

[a] Metabarcoding primers

maintained in the algal collections at the University of Louisiana at Lafayette (LAF) were also barcoded (Additional file 2: Table S2). Corals and reef substratum collections in the Ryukyu archipelago were conducted under Permit 24-3 delivered by the Okinawa Prefecture Fishery Control 33-2-40. Collections in Florida's coastal waters were permitted by a 'Saltwater fishing' license from the Fish and Wildlife (#1000427446) and those from Garden Key, Dry Tortugas by the U.S. National Park Service (#DRTO-2013-SCI-0015).

Microflora samples

Several microflora specimens were selected for environmental sequencing to build a clone library and for metabarcoding (Table 2, Fig. 3). These consisted of limestone fragments from reef rubble or substratum adjacent to coral colonies originating from the Ryukyu archipelago (JP01, JP03, JP04, JP06, JP07 and JP25), the Florida Keys (FL01 and FL02) and a northwestern Gulf of Mexico rhodolith (GM14). These specimens were lightly drilled within CaCO$_3$ patches devoid of crustose

Table 2 Limestone samples processed via cloning and metabarcoding

Sample	Clon./Metab.[a]	Site	Depth (m)	Date	Substratum
FL01	−/+	Big pine key, Florida	4	09/2013	Reef rubble
FL02	−/+	Big pine key, Florida	4	09/2013	Reef rubble
JP01	+/−	Ryukyus[c]	<5	07/2012	Reef rubble
JP03	+/−	Ryukyus	<5	07/2012	Reef rubble
JP04	+/−	Ryukyus	<5	07/2012	Reef rubble
JP06	+/−	Ryukyus	<5	07/2012	Reef matrix
JP07	+/+	Ryukyus	<5	07/2012	Reef matrix/rubble
JP25	+/−	Ryukyus	<5	07/2012	Reef rubble
GM14	−/+	Ewing Bank, NWGM[b]	57	08/2008	Rhodolith
E09	+/−	–	–	03/2013	Aquarium window
S15	+/−	NWGM	65	08/2008	*Lithophyllum* sp.

[a] Cloning and/or metabarcoding: performed (+), not performed (−)

[b] Northwestern Gulf of Mexico

[c] Mixed locations and sampling dates within the archipelago

Fig. 3 Drilled surfaces of microfloras samples FL02 and GM14. Note the dense (FL02) vs. light (GM14) phototroph colonization. *Scale bar* approximately 1.5 cm

epiliths with a sterile 1.6 mm (1/16″) bit mounted on a Flex-Shaft Attachment (Model 225) powered by a Dremel 3000 rotary tool (Dremel®, Racine, WI, USA). For each microflora sample, DNA from a total of 40-80 mg of pulverized $CaCO_3$ obtained from multiple drills was extracted with a PowerSoil DNA Isolation kit (MO BIO Laboratories, Carlsbad, CA, USA). Samples JP01 to JP07 were each extracted from limestone fragments pooled from multiple locations in the Ryukyus to maximize *tuf*A sequence diversity. Four of the above microflora samples were metabarcoded (JP07, FL01, FL02 and GM14); samples JP07 and FL02 were densely colonized with endolithic taxa whereas samples FL01 and GM14 were more lightly colonized (Fig. 3) and were thus metabarcoded at different sequencing depths (see "*tuf*A microflora assays" section). In preliminary cloning assays, two miscellaneous environmental samples comprising an aquarium window scrap exhibiting encrusting Ulvophyceae (E09), and endoliths underlaying a crustose coralline *Lithophyllum* sp. specimen (S15) were also processed (Table 2).

tufA microflora assays

Degenerate primers (*env_tuf*AF/*env_tuf*AR, see Table 1) were designed with HYDEN [58] on a phylogenetically diverse alignment of phototroph sequences. These primers target a 462 base pair (bp) amplicon (407 bp without incorporated primers) that is nested in the 3′ half of *tuf*A (see Fig. 4, created with WebLogo 3.4, [15]). The produced amplicon does not overlap with the 5′ intron found in the Euglenophyta [67], and is devoid of codon insertion/deletions (some exist within the amplicon in heterotrophic bacteria). PCR products for cloning or Illumina-metabarcoding were amplified from DNA extracts on a low temperature/long annealing and long extension cycle to maximize diversity recovery (3 min at 95 °C, 1 min steps at 94, 42, and 72 °C for 40 cycles and a 5 min final 72 °C extension). Samples generally required 1:10th to 1:100th dilution for successful amplification due to DNA extract concentration variation and/or the potential presence of PCR inhibitors commonly found in algal samples (such as polysaccharides and natural products, [105]). For cloning, PCR products were separated with a TOPO® TA Cloning® Kit (One Shot® Top 10 chemically competent *E. coli*, Invitrogen™, Life Technologies, Grand Island, NY) following the manufacturer's protocol. Clones were grown on LB agar plates containing 50 µg/mL Ampicillin and 40 µg/mL X-Gal. PCR was performed on white colonies with primers *env_tuf*AF and M13R (M13 Reverse priming site on TOPO® vector) and conditions as above. Colony PCR products of correct size were sequenced commercially and assembled in Sequencher v5.1. For metabarcoding, DNA extracts were shipped to MRDNA (http://www.mrdnalab.com, Shallowater, TX, USA), where PCR products were amplified using a HotStarTaq Plus Master Mix Kit (Qiagen, Valencia, CA,

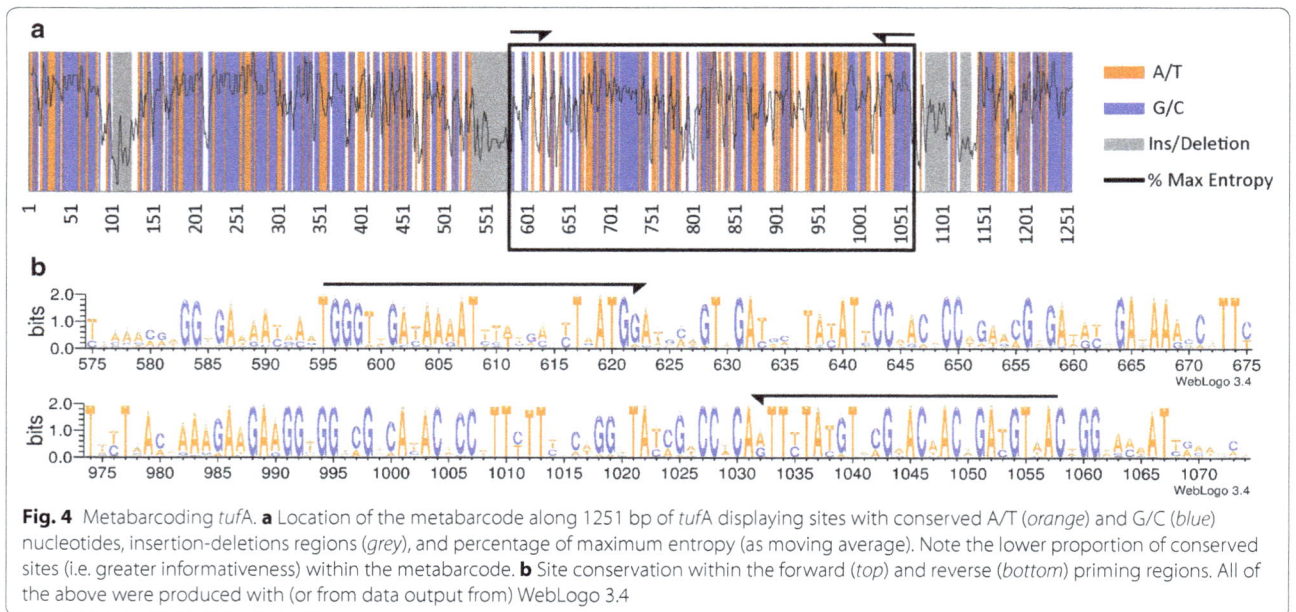

Fig. 4 Metabarcoding *tuf*A. **a** Location of the metabarcode along 1251 bp of *tuf*A displaying sites with conserved A/T (*orange*) and G/C (*blue*) nucleotides, insertion-deletions regions (*grey*), and percentage of maximum entropy (as moving average). Note the lower proportion of conserved sites (i.e. greater informativeness) within the metabarcode. **b** Site conservation within the forward (*top*) and reverse (*bottom*) priming regions. All of the above were produced with (or from data output from) WebLogo 3.4

USA) with indexed *env_tuf*AF primers (Additional file 3: Table S3). The amplification of the densely colonized samples JP07 and FL02 at MRDNA produced strong PCR products, while the more lightly colonized samples GM14 and FL01 resulted in weaker PCR products (in congruence with PCR testing conducted at LAF prior to the shipping of these samples). Thus, JP07 and FL02 products were pooled in higher proportion for deep sequencing while products from FL01 and GM14 were normalized and pooled with microbiota assays from other customers (i.e. 16S rDNA) to produce a nominal 20,000 reads per assay (as routinely performed at MRDNA). The prepared library (TruSeq DNA Sample Prep Kit for 2 × 250 bp paired-ends) was sequenced on the lane of an Illumina MiSeq Platform (Illumina Inc, San Diego, CA, USA).

De novo clustering of *tuf*A OTUs

Raw Miseq reads were processed with the USEARCH pipeline (http://www.drive5.com/uparse/) [26] to overlap paired-ends (*-fastq_mergepairs*, Q = 2), demultiplexing (including stripping of indexes and forward primers, *fastq_strip_barcode_relabel.py* script), filter for high quality reads (*-fastq_filter*) of appropriate size (expected error E >0.5, min 400 bp, max 440 bp), global trimming (5′ cropping of 20 bp, 3′ cropping of 50 bp), and removal of noisy reads after dereplication (singletons, doubletons and tripletons found across the entire data set), as well as chimeras and contaminants (non-*tuf*A reads) with UCHIME (command *-uchime_ref*) and UBLAST (*-ublast*) (Table 3). The dereplication file (with remaining

Table 3 Remaining reads throughout the UPARSE pipeline as counts and percentage of raw reads

Raw	4,917,888	100 %
Paired-ends merging	4,720,138	96
Demultiplexing	1,928,898	39
Read quality filter	1,465,438	30
Read size filter[a]	1,331,438	27
Denoising[b]	837,069	17
Decontamination[c]	824,355	17

[a] Reads used for OTU mapping

[b] Singletons, doubletons and tripletons

[c] Non-*tuf*A reads and chimeric reads

reads, average read length of 375 bp) was then clustered with UPARSE (*-cluster_otus*) at the recommended 97 % global threshold and with SWARM [62] at multiple local thresholds (d = 1 to 16) to explore its clustering optima (none published). UPARSE (97 %) and SWARM (d = 10) outputs were then parsed to identify robust core OTUs generated by both algorithms to follow recent recommendations for reproducibility with the use of multiple algorithms [88]. Core OTUs were then mapped at the 97 % level (with *-usearch_global*) against the quality-filtered (merged) reads (including singletons, doubletons and tripletons) to produce an OTU abundance table (*uc2otutab.py* script). Cumulative and non-cumulative rank-abundance curves were built from this table to explore the microflora's assemblage structure.

tufA reference database

GenBank®'s *tuf*A sequences were summarized into a non-redundant local database augmented with newly generated Sanger data (including clones). We retrieved with BLASTn the 20,000 closest matches to multiple *tuf*A sequences representing major branches of photosynthetic algal diversity (eukaryotic and prokaryotic). Search results were pooled, filtered for sequences >300 bp, and dereplicated (100 % identity). Following preliminary tree building, sequences of prokaryotic heterotrophs (eubacteria other than cyanobacteria) were segregated and used for separate searches as above. The numerous resulting heterotroph sequences were dereplicated at 99 %. We excluded the non-photosynthetic apicoplast-encoded *tuf*A of the Apicomplexa [11] and the nuclear-encoded *tuf*A of the Streptophyta [3] because of conflicting phylogenetic signal (see [19, 43, 49]) and irrelevance to the target group/habitat investigated herein (i.e. lower phototrophs inhabiting $CaCO_3$). Nonetheless, the deepest branch of the Streptophyta, i.e. the Charophyte algae *Mesostigma* and *Chlorokybus* [56, 101] were maintained in the database because their *tuf*A is chloroplast-encoded. The only two sequences available for the terrestrial order Trentepohliales (*Cephaleuros* and *Trentepohlia*, Ulvophyceae) were also excluded for their large divergence with other eukaryotic *tuf*A and ambiguous phylogenetic placement (until genome sequencing clarifies the organellar localization of their *tuf*A, i.e. nuclear or chloroplastic). The final sorted *tuf*A database comprised 4057 sequences (1339 phototrophs and 2718 heterotrophs). We followed PHYTO-REF for the classification of Diatoms (namely a single phylum-class noted as 'Bacillariophyta' within the Ochrophyta) and otherwise followed the Algaebase class-scheme [37]. Within the Ulvophyceae, we adopted the combined notation 'Ulvales-Ulothrichales' [55]. A summary of the database sequence content is provided in Additional file 4: Table S4.

tufA phylogenies

Phylogenetic reconstruction of the *tuf*A database was performed to visualize its diversity and for clade-based (topological) identification of clones and OTU sequences (n = 802). In order to produce a diverse heterotroph outgroup while greatly reducing sequence number for efficient computation time, we 'framed' bacterial reference diversity by selecting the most distant haplotypes in large clades observed in preliminary trees (n = 556 heterotroph sequences kept). The final dataset comprising 2697 sequences was translated into amino acids and aligned with MUSCLE [24]. Regions with poor homology (codon insertion/deletions) were cropped between sites of conserved amino acids (see Fig. 4) and the sequences translated back to nucleic acids. The final alignment

(891 bp) was then ran with RAxML-HPC2 on the CIP-RES computer cluster (http://www.phylo.org) with a GTR+I+G model of evolution partitioned per codon position, 200 topological searches from random restarts and 1000 bootstrap replicates for node support estimation. To further detail diversity in euendolithic Ulvophyceae, and for species delimitation analyses (see below), the above phototroph alignment was subsetted for the orders Bryopsidales and 'Ulvales-Ulothrichales' and few outgroup taxa. Previously dereplicated barcode data were reintroduced to show sampling effort in these orders. The final alignment comprising 906 sequences was used for phylogenetic analysis as listed above (with 1000 restarts and 1000 bootstrap replicates) and species delimitation analyses (see below). All trees were edited in iTOL [57].

Ulvophycean species delimitation

Branch lengths were extracted from the RAxML Ulvophyceae tree with function *cophenetic.phylo* of the package APE in R [69, 78] to produce a distance matrix as input for the standalone version of the Automatic Barcode Gap Discovery software (ABGD, [76]). The latter was run with minimum (pmin) and maximum (pmax) intraspecific distance priors comprised between 0.001 and 1 in 100 steps, and with a 0.5 relative gap width. Alternative species boundaries hypotheses were produced with the general mixed yule coalescence (GMYC) model with the package SPLITS in R [35], with the single threshold method based on an ultrametric tree generated in BEAST v2.0 [8] using a relaxed log-normal clock with a constant population coalescent as prior, and a GTR+I+G model of evolution partitioned per codon position. Markov Chain Monte Carlo (MCMC) chains were run for 30 million generations (sampled every 1000th generations) and the quality of the run assessed in Tracer v1.6 [80] to ensure that effective sample size (ESS) values were >200 with the default burnin (3000 trees).

Divergence and identity thresholds

The *tuf*A database alignment was cropped to the metabarcode length (i.e. clustered OTUs of 375 bp) and a pairwise percent identity matrix created with function *dist.dna* in package APE (R) (a few sequences with >20 bp of missing data on their 5' or 3' side were excluded in order to avoid inflation of computed percent identity values). From this distance matrix, boxplots depicting the amount of divergence within individual families, suborders and orders of the Ulvophyceae ('Ulvales-Ulothrichales' and Bryopsidales) were drawn in order to examine their validity. Next, to define clade-based (conservative) identity thresholds for the rapid annotation of OTU at multiple taxonomic ranks, the function *sppDist* of the package SPIDER in R [10] was used to compute the overall intra- and inter- clade divergence values for all phototrophs at high-levels ranks (domain, phylum, and class clades) and at lower levels for the Ulvophyceae (order, suborder, and family clades, as well as molecular species clades delimited by ABGD and GMYC). *In silico* (relaxed) thresholds for the above ranks were also assessed by scoring taxonomic discrepancies between an OTU's best hit (obtained with function *-usearch_global*) and its topological reference taxonomy. From this procedure, the distribution of correctly and incorrectly classified hits was reported (for all phototrophs and the Ulvophyceae). Finally, to assess the overall performance of the database in light of annotation thresholds defined above (i.e. "How distant are OTUs from database-based sequences?" and "Do abundant OTUs score highly against the database?"), the distributions of OTUs' best hit (percent identity) against the database was plotted for all phototrophic and Ulvophycean OTUs, as well as for the most abundant OTUs (i.e. those representing ≥1 % of reads) in each of the metabarcoded samples.

Availability of supporting data

The annotated database and annotated core OTUs are available in the Dryad repository, doi:10.5061/dryad.6cj8h. Newly barcoded specimens and clones are deposited in Genbank® (KU361834-KU362236) and the raw Illumina dataset on the National center for Biotechnology and Information (NCBI) Sequence Read Archive (#SRP067712).

Results

DNA Sequencing

A total of 316 specimens were barcoded with *tuf*A (including 238 Ulvophyceae, 72 Florideophyceae, and six Phaeophyceae) among which 104 were new representatives of 'Ostreobium' spp. (47 from cultures) (Additional file 1: Table S1, Additional file 2: Table S2). Cloning resulted in the sequencing of 153 bacterial colonies, which after dereplication represented a library of 86 unique *tuf*A sequences, including 37 Chlorophyta, 25 Cyanophyta, 15 Rhodophyta, one Haptophyta and eight Ochrophyta (see Additional file 5: Table S5 for class details). The shared Miseq run output was a total of 4,917,888 raw paired-end reads for the four microfloras tested. Upon data processing throughout the USEARCH pipeline, read numbers gradually decreased, leaving 27 % of the raw data for OTU abundance mapping and ~17 % for clustering (Table 4). Clustering at multiple levels with SWARM (d = 1–16) output from 2483 to 740 OTUs. SWARM indicated a clear clustering optima between local thresholds of d = 10–13 (i.e. up to 13 steps from a given centroid), where the number of common OTUs with UPARSE did not vary (Table 5). Clustering at the 97 % recommended

Table 4 OTU counts and corresponding read counts per microflora sample

	OTUs	Reads
Mapped	802	1,261,195
Corrected[a]	802	1,260,811
FL01	28	23,785
FL02	317	558,503
JP07	609	672,119
GM14	38	6404

[a] Table cells with read counts <3 were removed

global threshold with UPARSE generated 822 OTUs, the majority but 20 were also produced by SWARM. The 802 core OTUs common to SWARM and UPARSE mapped to 1,260,811 (~94 %) of the 1,331,438 reads available (70,243 reads unmapped) (Table 4). The number of OTUs and their final read counts for each of the microflora samples clearly reflected the differential sequencing depth carried at MRDNA. Non-cumulative ranked abundance curves demonstrated 3–5 OTUs strongly dominating each of the microfloras and the presence of numerous low abundance OTUs (i.e. long tails, Fig. 5). Cumulative curves (ranked OTUs) demonstrated rapid plateauing of mapped read abundance (Fig. 5 and Additional file 6: Figure S1). Phototrophs represented 618 out of the 802 robust OTUs, i.e. 77 % of OTUs (53.3–84.6 % depending on the sample) and overall 98 % of mapped reads (83.2–99.4 % depending on the microflora sample) (Fig. 13, Additional file 7: Table S6).

Phototroph phylogeny

Phylogenetic reconstruction of phototrophic *tuf*A diversity recovered topological features congruent with those presented in Delwiche et al. [19], albeit with much denser taxon sampling. The resulting tree was characterized by a poorly resolved backbone with higher resolution at lower taxonomic levels. Phylum para/polyphyly was caused by the nesting of secondary endosymbiotic phyla (Ochrophyta, Haptophyta, Cryptophyta; and Euglenophyta and 'Chlorarachniophyta') among extent members of early (single-celled) lineages of the Rhodophyta and Chlorophyta [46] from which they diversified (respectively). A few classes are also para/polyphyletic in congruence with known taxonomic discrepancies (e.g. see [34]). Subclasses of the Cyanophyta (not shown) are also mostly polyphyletic (e.g. Synechococcophycideae) and distributed between two major clades (as in e.g. [63, 83]). Overall, the phototrophic diversity comprised in the database includes 27 classes in 10 phyla (Fig. 6, Additional file 4: Table S4). Several phylogenetic landmarks in the evolution of photosynthetic organisms are highlighted on the tree (Fig. 6).

Table 5 Number of OTUs common to UPARSE and SWARM output at multiple clustering thresholds

	UPARSE	99 %	98 %	97 %	96 %	95 %
SWARM		1070	918	*822*	761	698
d01	2483	1044	913	*818*	752	696
d04	1416	963	902	*813*	750	693
d08	1208	851	845	*806*	748	690
d09	1135	835	829	*804*	747	689
d10	*1093*	*833*	*827*	802	*747*	689
d11	*1066*	*833*	*827*	802	*747*	689
d12	*1047*	*833*	*827*	802	*747*	689
d13	*1032*	*833*	*827*	802	*747*	689
d14	788	735	735	*735*	719	676
d15	763	722	722	*723*	708	673
d16	740	706	706	*707*	696	668

Recommended global threshold for UPARSE (97 %) and determined local threshold optima for SWARM (d10–d13) are in bold italics. The 802 OTUs kept for subsequent analyses are italicized

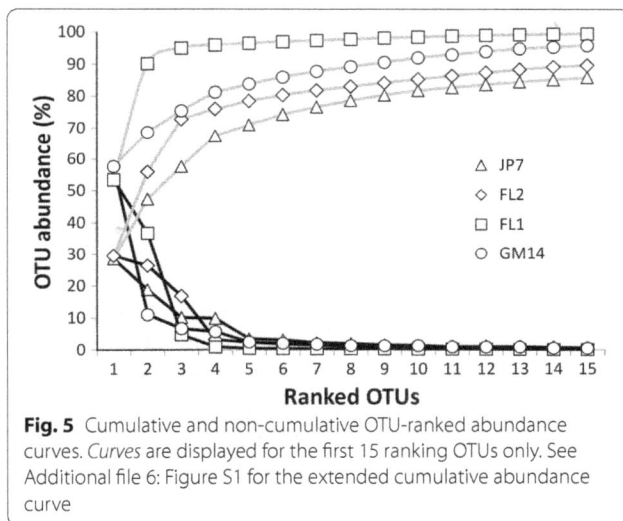

Fig. 5 Cumulative and non-cumulative OTU-ranked abundance curves. *Curves* are displayed for the first 15 ranking OTUs only. See Additional file 6: Figure S1 for the extended cumulative abundance curve

file 9: Figure S2, Additional file 8: Table S7). Their abundance profiles were also similar except at the family level (Fig. 13). By comparison, the less dense microfloras FL01 and GM14 (i.e. those with low sequencing depth) showed skewed profiles toward particular groups and much fewer OTUs (33 phototrophic OTUs, 13 unique to these two samples, 20 shared with others). In the coral reef samples (i.e. FL01, FL02, and JP07), the Cyanophyceae (Cyanophyta) and 'Bacillariophyta' (Ochrophyta) were found in low read abundance but revealed tremendous OTU diversity (267 and 162 OTUs, respectively), while the Coccolithophyceae (Haptophyta), the Pedinophyceae and Ulvophyceae (Chlorophyta) and the Florideophyceae (Rhodophyta) were found in large read abundance but were much less diverse (i.e. few OTUs, Fig. 6). In the rhodolith sample (GM14), the Florideophyceae were particularly abundant.

OTU richness and abundance
The phylogenetic distribution of OTUs (richness and abundance) is clearly visible on the phototroph tree (see dashed lines and bars' height, Fig. 6) and allowed their classification (on a topological basis) to 14 classes in seven phyla. All but 22 eukaryotic OTUs could not be assigned with confidence to a specific class within the phylum they branched in (Additional file 8: Table S7); however, these OTUs represented very few reads (<0.25 % of all mapped reads). The densely colonized microfloras FL02 and JP07 (i.e. those sequenced deeply) exhibited particularly similar richness profiles for phylum, class, and Ulvophycean family (Fig. 13). Together, these two microfloras harbored 605 phototrophic OTUs, 141 of which were assigned to both samples following demultiplexing (Additional

Ulvophyceae phylogeny
The phylogenetic estimation made on the Ulvophyceae dataset (Fig. 7) resulted in nearly identical branching features than within the larger tree except for a few differences in the Bryopsidales (shown for the Ostreobidineae, see Fig. 8). The Bryopsidales and the 'Ulvales-Ulothrichales' radiations each received strong bootstrap support (95 and 82 %, respectively). Overall, the tree details 23 families, some accepted taxonomically, others provisionally delimited here based on topological features and clade divergence. 'Ostreobium' is revealed as a polyphyletic form genus representing a complex of family-level clades (Figs. 7, 8, 9). Most of its diversity is monophyletic within a large, confirmed subordinal-level clade, the Ostreobidineae (previously proposed by [102]), whose divergence is particularly consistent with other

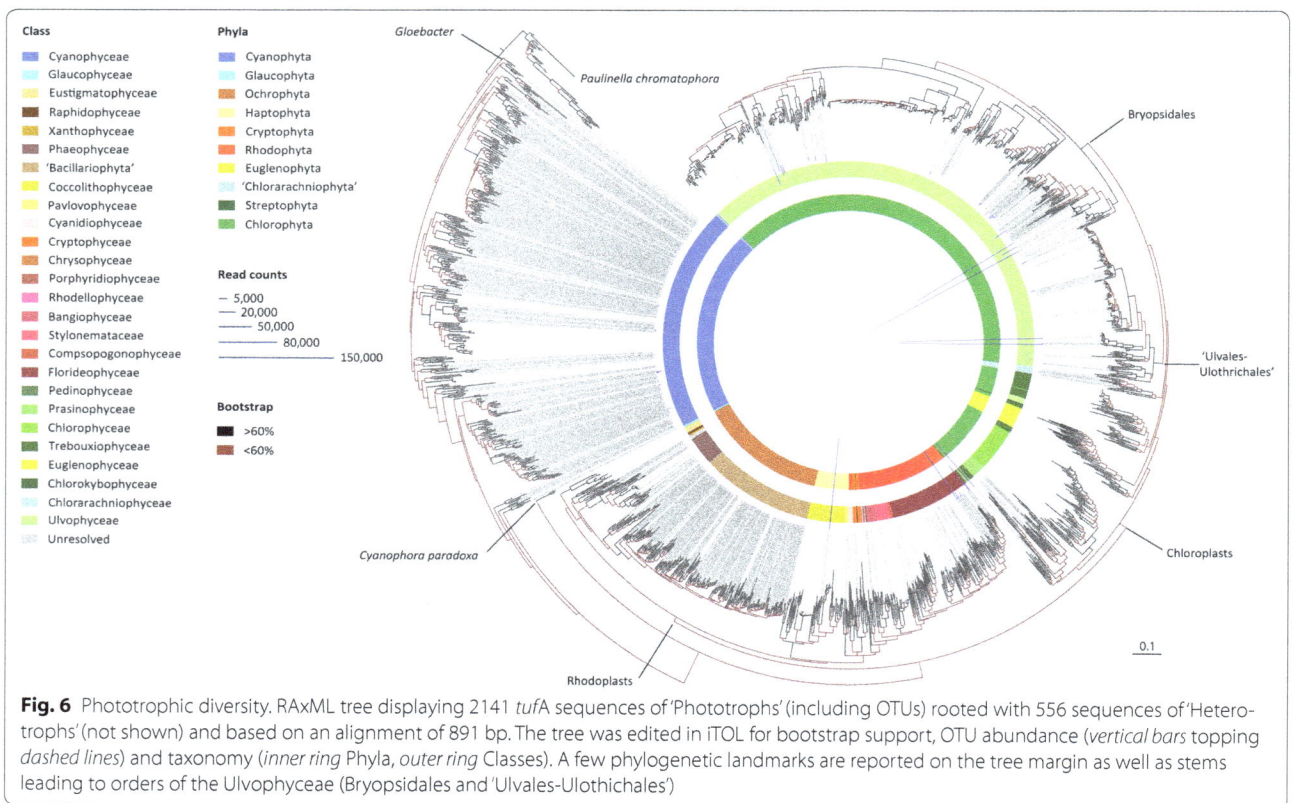

Fig. 6 Phototrophic diversity. RAxML tree displaying 2141 *tuf*A sequences of 'Phototrophs' (including OTUs) rooted with 556 sequences of 'Hetero-trophs' (not shown) and based on an alignment of 891 bp. The tree was edited in iTOL for bootstrap support, OTU abundance (*vertical bars* topping *dashed lines*) and taxonomy (*inner ring* Phyla, *outer ring* Classes). A few phylogenetic landmarks are reported on the tree margin as well as stems leading to orders of the Ulvophyceae (Bryopsidales and 'Ulvales-Ulothrichales')

suborders, the Bryopsidineae and the Halimedineae (Fig. 9). In this suborder, four 'Ostreobium' families are provisionally delimited and named from their location of collection in the Ryukyus for the purpose of molecular classification (*nomina nuda* 'Hamidaceae', 'Maedaceae', 'Odoaceae' and 'Unarizakiaceae', Fig. 9). Few OTUs are found in the phylogenetic vicinity of the 'Maedaceae' and presently cannot be assigned to a family clade (Figs. 7, 8). Additional 'Ostreobium' diversity is found in a remote family-level clade of the suborder Halimedineae provisionally assigned for taxonomic simplicity to the 'Pseudostreobiaceae' (Figs. 7, 8, 9). Our barcoding efforts also further documented members of the diminutive epilithic, polyphyletic form genus 'Pseudochlorodesmis' delimited here in two provisionally family-level clades named based on their taxonomic history [103], i.e. the 'Pseudochlorodesmidaceae' and Siphonogramenaceae' (Figs. 7, 8, 9). Other 'Pseudochlorodesmis' spp. (see Additional file 1: Table S1 for details) represented primordia of otherwise conspicuous taxa in families of the Halimedineae (e.g. in *Halimeda*, see [64]) and neotenic thalli within the Rhipiliaceae [103] and the Caulerpaceae (a sister taxon to the C. 'ambigua' complex, [23]). Barcoding of diminutive, simple Bryopsidineae also revealed numerous unresolved/unknown taxa in the phylogenetic vicinity of the

Derbesiaceae and Bryopsidaceae (Figs. 7, 8). Several epilithic families are found endolithically as seen by the presence of OTUs in their clade (Fig. 7, e.g. the Bryopsidaceae, Derbesiaceae and Halimedaceae) supposedly as reproductive cells, germlings and/or endolithic siphonous networks).

Delimited species

The GMYC analysis delimited 504 hypothetical species in the Ulvophyceae (LGMYC = 7663.072> L0 = 7 577.771.182, P = 0) while with ABGD the output was 666 spp. Species groups delimited by the two methods (GMYC-ABGD) represented 349–485 spp. in the Bryopsidales, and 147–172 spp. in the 'Ulvales-Ulothrichales' (Fig. 7). The majority of discrepancies arose in the Bryopsidales in well-sampled families of the suborder Halimedineae (overall 195–308 spp.), namely in the Halimedaceae (56–88 spp.) and especially in the rapidly diversifying Caulerpaceae (67–132 spp.), where GMYC seemed to overlump and ABGD oversplit terminal clades. By contrast, species delimitation showed greater congruence between the two methods in the remainder of the Ulvophyceae, in the 'Ulvales-Ulothrichales' (Ulvellaceae: 56–65 spp., Ulvaceae: 43–52 spp., Phaeophilaceae: 13–18 spp.) and in the Bryopsidales within the suborders

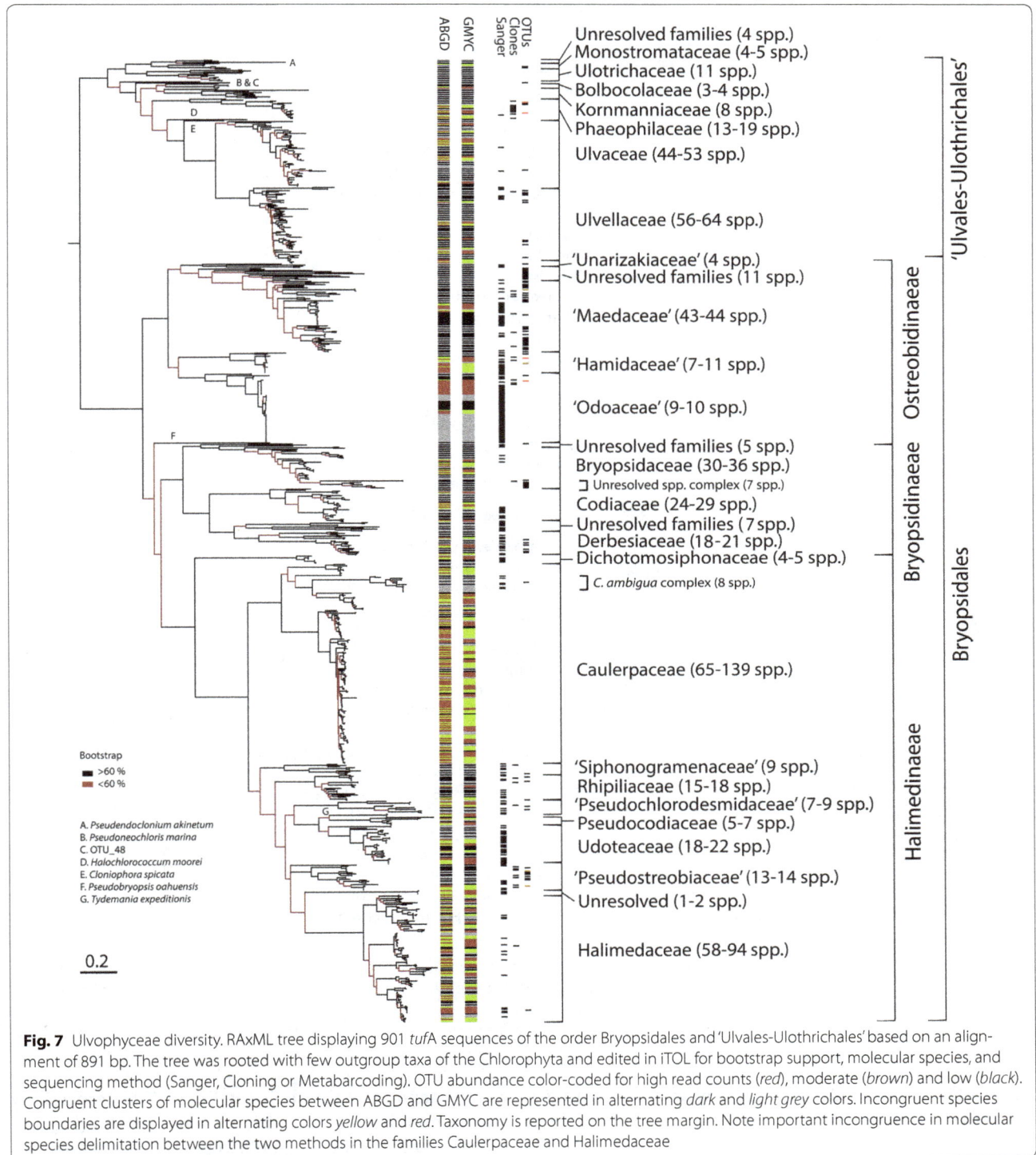

Fig. 7 Ulvophyceae diversity. RAxML tree displaying 901 *tuf*A sequences of the order Bryopsidales and 'Ulvales-Ulothrichales' based on an alignment of 891 bp. The tree was rooted with few outgroup taxa of the Chlorophyta and edited in iTOL for bootstrap support, molecular species, and sequencing method (Sanger, Cloning or Metabarcoding). OTU abundance color-coded for high read counts (*red*), moderate (*brown*) and low (*black*). Congruent clusters of molecular species between ABGD and GMYC are represented in alternating *dark* and *light grey* colors. Incongruent species boundaries are displayed in alternating colors *yellow* and *red*. Taxonomy is reported on the tree margin. Note important incongruence in molecular species delimitation between the two methods in the families Caulerpaceae and Halimedaceae

Bryopsidineae (overall 85–91 spp.) and Ostreobidineae (overall 72–81 spp.; 53–58 spp. in the Ryukyus alone). The remote *Ostreobium* clade found in the Halimedineae, i.e. the 'Pseudostreobiaceae', harbors 13–14 spp.

Molecular species divergence computed over the metabarcode length (375 bp) indicated species boundaries lying between 99.73 and 97.87 % with the GMYC method (i.e. 1–8 bp substitutions), and between 99.73 and 99.47 %

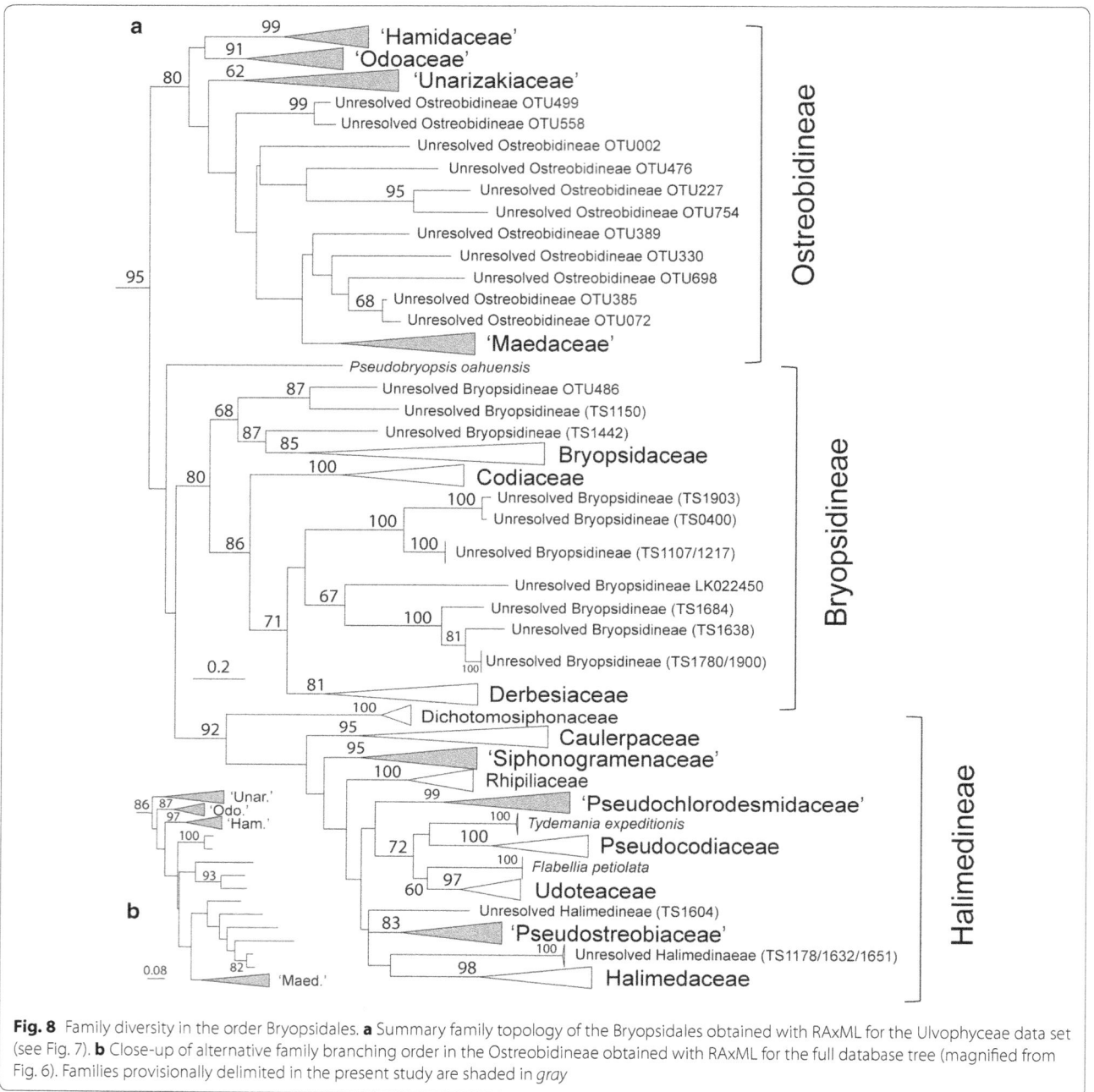

Fig. 8 Family diversity in the order Bryopsidales. **a** Summary family topology of the Bryopsidales obtained with RAxML for the Ulvophyceae data set (see Fig. 7). **b** Close-up of alternative family branching order in the Ostreobidineae obtained with RAxML for the full database tree (magnified from Fig. 6). Families provisionally delimited in the present study are shaded in *gray*

with ABGD (1–2 bp substitutions) (not shown, see Additional file 10: Figure S3).

Identity Threshold for annotation

Examination of taxonomic rank divergence obtained with *sppDist* within the Ulvophyceae (Fig. 10) indicated conservative annotation thresholds for family, suborder and order boundaries at 92, 84 and 79 %, respectively (values rounded up from 91.2, 83.2 and 78.9 %, respectively) and

relaxed annotation thresholds at 85, 79, 77 %, respectively (values rounded up from 84.5, 78.4, and 76.3 %) (Fig. 11). For the full breadth of phototrophic OTUs, conservative annotation thresholds for class, phylum and domain boundaries laid at 86, 83 and 82 %, respectively (values rounded up from 85.1, 82.2 and 81.9 %), while relaxed annotation thresholds laid at 84, 83 and 77 %, respectively (values rounded up from 84.0, 82.7 and 76.3 %) (Figs. 10, 11, respectively). Overall, phototrophic OTUs matched

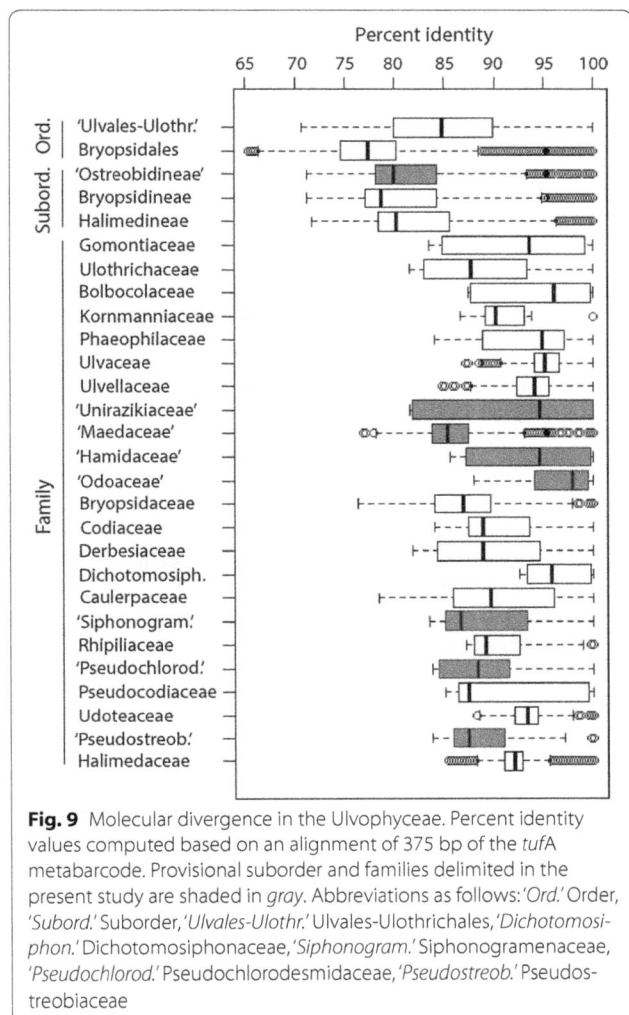

Fig. 9 Molecular divergence in the Ulvophyceae. Percent identity values computed based on an alignment of 375 bp of the *tuf*A metabarcode. Provisional suborder and families delimited in the present study are shaded in *gray*. Abbreviations as follows: '*Ord.*' Order, '*Subord.*' Suborder, '*Ulvales-Ulothr.*' Ulvales-Ulothrichales, '*Dichotomosiphon.*' Dichotomosiphonaceae, '*Siphonogram.*' Siphonogramenaceae, '*Pseudochlorod.*' Pseudochlorodesmidaceae, '*Pseudostreob.*' Pseudostreobiaceae

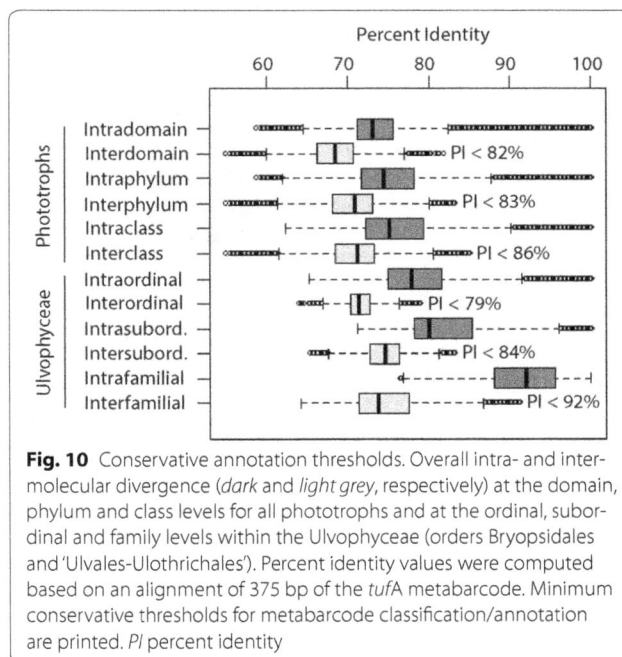

Fig. 10 Conservative annotation thresholds. Overall intra- and inter-molecular divergence (*dark* and *light grey*, respectively) at the domain, phylum and class levels for all phototrophs and at the ordinal, subordinal and family levels within the Ulvophyceae (orders Bryopsidales and 'Ulvales-Ulothrichales'). Percent identity values were computed based on an alignment of 375 bp of the *tuf*A metabarcode. Minimum conservative thresholds for metabarcode classification/annotation are printed. *PI* percent identity

the database at high identity levels with the majority of hits ≥84 % (see lower boxplot quartiles, Fig. 12) in the range of class annotation (i.e. relaxed = 84 %, conservative = 86 %). For the Ulvophyceae, the majority of hits were ≥87 % (see lower boxplot quartiles, Fig. 12) in the range of family-level thresholds (i.e. relaxed = 85 %, conservative = 92 %). The most abundant OTUs, those driving abundance profiles (i.e. >1 % abundance, Fig. 13 and Additional file 11: Figure S4), hit the database at even higher identity values (lower quartile >91 %, Fig. 12) and displayed the same distribution for the phototrophs (n = 32 OTUs) or the Ulvophyceae-only (n = 19 OTUs).

Discussion

Bioinformatics

Following demultiplexing, we noted important read losses (>50 %) (Table 3) probably caused by the systematic exclusion of reads containing base pair error(s) within their index in the early steps of the UPARSE pipeline

(no mismatch allowed). Here, demultiplexing losses may have also been exarcerbated by the length of our amplicon (467 bp) for the chemistry used (2 × 250 bp) since paired-ends could not overlap over the index regions. An additional consequence of potential increased substitution errors in the index region along with tag jumping in Illumina-based metabarcoding [89], is the misassignment of sequence-to-sample (i.e. the false assignment of an OTU to a given sample), which may inflate OTU diversity across samples and thus bias richness profiles. However, since index error/jumping generally concerns only a small proportion of reads, false assignment is less likely to bias abundance profiles (Fig. 13), especially when those profiles are underlayed by very abundant OTUs such as shown here (Fig. 5 and Additional file 11: Figure S4). Although we noted some possible tag error/jumping in OTUs retrieved in the deeply sequenced microfloras FL02 and JP07 (Additional file 12: Figure S5), numerous unique OTUs were also found in each of these samples (Additional file 9: Figure S2), (Fig. 13). If there is any bias, we suspect that the richness profile reported for the Ulvophyceae (Fig. 13, bottom left histogram) was the most affected since this class included OTUs found in very large proportion of reads that were more likely to propagate ('bleed') to other samples via tag error/jump. Hence, in this regard, abundance profiles are in our opinion a much more robust depiction of OTUs actually present in a given sample.

Currently, tag-jumps represent a non-negligible issue of the metabarcoding approach. Indeed, Esling et al. [28]

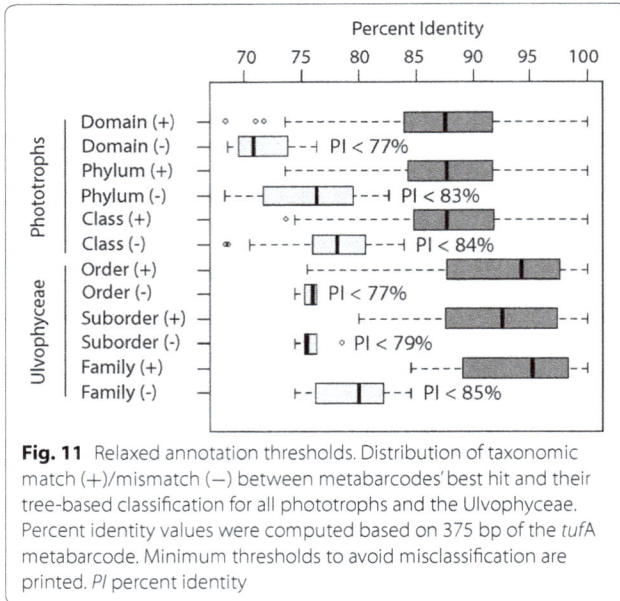

Fig. 11 Relaxed annotation thresholds. Distribution of taxonomic match (+)/mismatch (−) between metabarcodes' best hit and their tree-based classification for all phototrophs and the Ulvophyceae. Percent identity values were computed based on 375 bp of the *tuf*A metabarcode. Minimum thresholds to avoid misclassification are printed. *PI* percent identity

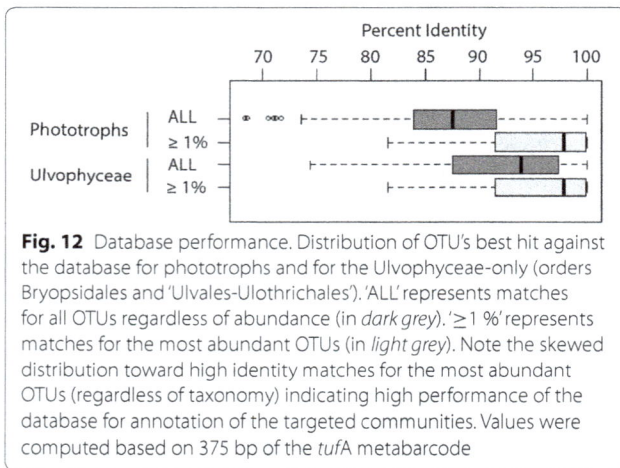

Fig. 12 Database performance. Distribution of OTU's best hit against the database for phototrophs and for the Ulvophyceae-only (orders Bryopsidales and 'Ulvales-Ulothrichales'). 'ALL' represents matches for all OTUs regardless of abundance (in *dark grey*). '≥1 %' represents matches for the most abundant OTUs (in *light grey*). Note the skewed distribution toward high identity matches for the most abundant OTUs (regardless of taxonomy) indicating high performance of the database for annotation of the targeted communities. Values were computed based on 375 bp of the *tuf*A metabarcode

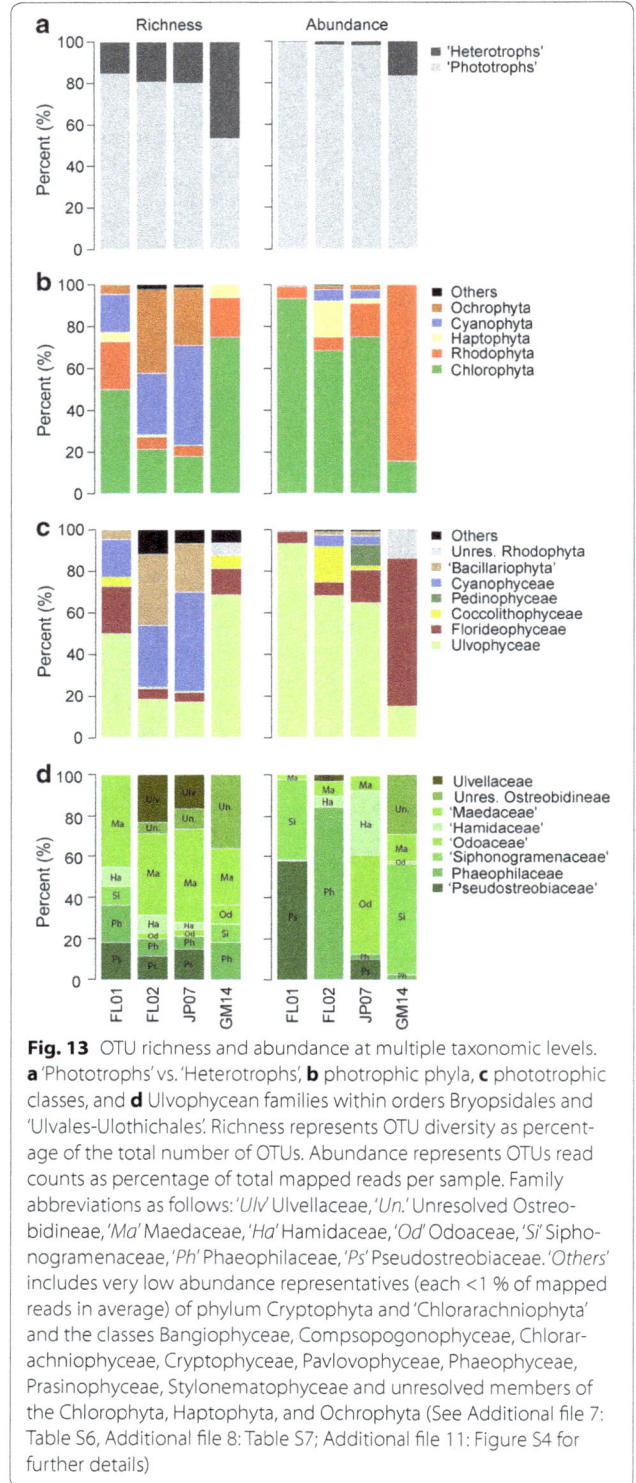

Fig. 13 OTU richness and abundance at multiple taxonomic levels. **a** 'Phototrophs' vs. 'Heterotrophs', **b** photrophic phyla, **c** phototrophic classes, and **d** Ulvophycean families within orders Bryopsidales and 'Ulvales-Ulothichales'. Richness represents OTU diversity as percentage of the total number of OTUs. Abundance represents OTUs read counts as percentage of total mapped reads per sample. Family abbreviations as follows: '*Ulv*' Ulvellaceae, '*Un.*' Unresolved Ostreobidineae, '*Ma*' Maedaceae, '*Ha*' Hamidaceae, '*Od*' Odoaceae, '*Si*' Siphonogramenaceae, '*Ph*' Phaeophilaceae, '*Ps*' Pseudostreobiaceae. '*Others*' includes very low abundance representatives (each <1 % of mapped reads in average) of phylum Cryptophyta and 'Chlorarachniophyta' and the classes Bangiophyceae, Compsopogonophyceae, Chlorarachniophyceae, Cryptophyceae, Pavlovophyceae, Phaeophyceae, Prasinophyceae, Stylonematophyceae and unresolved members of the Chlorophyta, Haptophyta, and Ochrophyta (See Additional file 7: Table S6, Additional file 8: Table S7; Additional file 11: Figure S4 for further details)

demonstrated that libraries prepared with single index primers (as done here) and saturated double index primers result in undetectable read cross-contamination. This is particularly problematic when presence-absence of OTUs (i.e. richness) is critical for the outcome of a particular study (e.g. keystone OTU, geographic variation). To mitigate cross-contamination, these authors showed that multiplexing samples following Latin Square designs can optimize mistag detection. Recently, Kitson et al. [48] also reported on the development of nested metabarcode tagging (2 × 2 indexes) that apparently lead to accurate sample/index tracking. Sequencing depth and evenness of the community found within a sample are also nontrivial issues affecting community profiles (richness and abundance) obtained via metabarcoding. On the one

hand, samples sequenced evenly (i.e. normalized across samples for similar sequencing depth) may not show all diversity present for samples that are strongly dominated by a few taxa 'masking' low abundance OTUs [2]. On the other hand, samples sequenced unevenly (as done here) may falsely exhibit higher diversity simply because they were sequenced at greater depth, which promotes biodiversity recovery [91]. In spite of these concerns, we are confident that our deeply sequenced samples JP07 and FL02 truly harbor more diversity than FL01 and GM14, especially considering the dense endolithic colonization observed for these samples, and that the preparation of JP07 consisted in pooling subsamples of microfloras from multiple locations across the Ryukyus archipelago to specifically maximize biodiversity recovery (see "Methods" section). To further support the above, we subsampled JP07 and FL02 to a shallow sequencing depth of 20,000 reads as reported for FL01 and GM14 (Table 4). Following OTU clustering (not shown), we observed a ~2–3 fold higher diversity for JP07 and FL02 than for FL01 and GM14, namely 87 and 66 OTUs vs. 28 and 38 OTUs, respectively. We nonetheless recognize that sequencing FL01 and GM14 deeper could have retrieved some additional OTU diversity.

Overall, sequencing future projects with the now available 2×300 bp chemistries should greatly improve upon demultiplexing issues caused by index errors. In spite of the noted important read losses, our data set contained very high read redundancy and still comprised important sequencing noise (as seen from overly abundant singletons, doubletons and tripletons resulting from the deep sequencing of FL02 and JP07, Table 3) and thus included much sufficient data to assess our microfloras' diversity. Using a combination of superior algorithms with different *de novo* clustering strategies (as recommended in [88]) such as UPARSE's global and SWARM's local threshold algorithms, allowed us to filter for high quality core OTUs produced by both pipelines. As a result of this dual-clustering approach, our OTUs are very close to biological sequences (i.e. contain 0, 1 or 2 incorrect bases as shown for clones and Illumina core OTUs obtained for JP07, our only sample sequenced with both methods, (see Additional file 13: Figure S6), in congruence with the ≤ 1 % error rate reported on a mock community for UPARSE alone [26]. Overall, the phylogenetic breadth documented by our *de novo* OTUs (prokaryotic and eukaryotic diversity), added to the negligible proportion of identified contaminants (i.e. non-*tuf*A, <1 % of filtered reads, Table 3), together demonstrate the specificity of our newly designed degenerate primers to target *tuf*A from phototrophs that represented the majority of retrieved OTUs (77 % overall) and mapped reads (>98 % overall) (Fig. 13 and Additional file 7: Table S6). The priming region of *env_tuf*AF (Fig. 4) actually overlaps with an extended stretch of codon insertion/deletions in numerous heterotrophs, which may further favor annealing, and therefore amplification, toward phototrophic

diversity. For samples with very low phototroph content and/or containing inhibitors (i.e. secondary metabolites [105]), target capture via hybridization with *tuf*A probes may provide a valuable alternative to amplification, as well as enhance rare OTUs detection limits. Some studies have performed target capture metabarcoding, such as for instance Patel et al. [71], Denonfoux et al. [20] and more recently Dowle et al. [22].

Endolithic Ulvophyceae

The molecular diversity and phylogenetic breadth of the boring microsiphons referred to as *Ostreobium* spp. had clearly been overlooked in the literature and lacked *tuf*A referencing (a single sequence—FJ535859 was available on GenBank® prior to our study). We document for this form genus an impressive polyphyletic cryptic diversity encompassing an estimated 85–95 species-level entities (Fig. 7), for which we delimited five provisional families to facilitate the profiling of endolithic communities (Figs. 9, 13). These families exhibit molecular divergence comparable to conspicuous Bryopsidalean families that are well circumscribed morphologically (e.g. the Halimedaceae, the Caulerpaceae, etc., Fig. 9). Our phylogeny also reveals unresolved branches at the base of the 'Maedaceae' (Ostreobidineae) (Figs. 7, 8), which pending further sampling, may lead to the delimitation of additional families. In this endeavor, clarifying the molecular identity of the generitype *O. quecketti* [7] from its type locality (Le Croisic, Brittany, France) should allow recircumscription of the family Ostreobiaceae P.C. Silva (proposed in [70]; validated in [9]) toward establishing a more stable classification. We expect that sequencing of endolithic communities worldwide will further increase species diversity of euendolithic microsiphons. Finally, although field collections were focused on 'Ostreobium' spp., our assessment also recovered common euendolithic microfilamentous genera of the 'Ulvales-Ulothrichales', namely the Phaeophilaceae and Ulvellaceae, in culture and from cloning and metabarcoding (Fig. 7, Additional file 1: Table S1, Additional file 5: Table S5 and Additional file 8: Table S7).

Database performance

Thorough analyses of our newly established database resources for *tuf*A permitted referencing the taxonomy of OTUs based on their phylogenetic position, as well as defining identity thresholds for their rapid annotation in the context of environmental biomonitoring studies via metabarcoding. Although the backbone of our phototroph tree was overall poorly resolved (Fig. 6, as in [19]), topological relationships obtained among phyla/classes with *tuf*A were congruent with currently accepted hypotheses of chloroplasts/rhodoplasts and secondary endosymbiotic lineages evolution [74], allowing visualization of the database and OTUs in a phylogenetically sound frame-work (Fig. 6 and Additional file 14: Figure S7). Relying on our divergence analyses, we were able to establish in silico (relaxed) and clade-based (conservative) thresholds for OTU annotation at high taxonomy levels (Class 84–86 %, Phylum 83–83 %, and Domain 77–82 %, Figs. 10, 11) and lower taxonomy levels in the Ulvophyceae (species 97–99 %, family 85–92 %, suborder 79–84 % (for the Bry-opsidales) and order 77–79 %, Figs. 10, 11). The fact that high-level taxonomy thresholds exceeded lower-level ones (i.e. within the Ulvophyceae) reflects the combina-tion of group-specific differences in *tuf*A's evolutionary rate (e.g. due to genome architecture, or lineage diversifi-cation age), low biodiversity sampling in some clades, tax-onomic discrepancies, and poorly resolved relationships (i.e. paraphyletic/polyphyletic classes and phyla). Overall, our database achieved matching of OTUs at high identity values (the majority matching at >85 %, Fig. 12), which as a single threshold holds the potential to accurately anno-tate OTUs to both family (for Ulvophyceae representa-tives) and class-levels (all phototrophs). Here, the majority of hits below 85 % belonged to the class Cyanophyceae (not shown), which as the sole representative in the phy-lum Cyanophyta (and 'Phototrophs' in the prokaryotic domain) may thus be annotated with identity thresholds of 82 % (Fig. 10), and eventually further relaxed down to 77 % (Fig. 11). Overall, we recommend applying a range of thresholds (based on the above relaxed and conservative thresholds) and investigate with tree methods the phylo-genetic position of OTUs that may cause profile dispari-ties in order to correct their taxonomy. As demonstrated in the present study, our 375 bp metabarcode packs suffi-cient phylogenetic signal for model-based tree estimation when included in well-sampled sequence alignments (e.g. as implemented here within RAxML). Finally, matches in the 97–99 % range could be used for molecular species annotation in the Ulvophyceae (as estimated by GMYC or ABGD analyses, see Fig. 7 and Additional file 10: Figure S3), providing that profiling at this level becomes neces-sary and that a molecular-based taxonomy becomes avail-able for this cryptic diversity (e.g. [44]).

Community profile

In spite of a possible, minimal bias in richness pro-file introduced via tag error/jump, the profiling of the microfloras revealed a core assemblage of phyla, classes and some Ulvophycean families inhabiting endolithic communities, whose abundance varies, but appears uni-versal (Fig. 13). Similarities in the phyla/classes rich-ness and abundance profiles of the OTU-rich coral reef samples FL02 and JP07 are interesting considering their distant geographic origin (Pacific vs. Atlantic) and the fact that for the preparation of JP07, multiple $CaCO_3$

samples collected from across the Ryukyu archipelago had been pooled together to maximize the recovery of endolithic biodiversity (see "Methods" section). In this regard, JP07 assemblage may thus be viewed as an aver-age phyla/classes profile for shallow coral reefs of the Ryukyus (e.g. abundance of <5 % 'Bacillariophyta', <5 % Coccolithophyceae, <5 % Cyanophyceae, <20 % Florideo-ophyceae and >65 % Ulvophyceae, and miscellaneous groups), whose applicability to other tropical areas such as in the Atlantic (FL02) would appear relevant. By con-trast with these two samples, the skewed profiles of the less densely colonized microfloras FL01 and GM14 may reflect immature or developing communities undergo-ing succession following for example, recent niche open-ing or disturbance. The rhodolith microflora of GM14 was particularly Rhodophyta-rich from an abundant crust-forming *Rhizophyllis* sp. OTU (Florideophyceae) and an early branching OTU that we could not presently resolve within the phylum (Fig. 13, Additional file 11: Figure S4). Although we were careful to sample $CaCO_3$ patches devoid of conspicuous encrusting red algae for the preparation of the microfloras' DNA extracts, we cannot exclude that accidental drilling of *Rhizophyl-lis* (Rhizophyllidaceae) (present on the surface of GM14 upon this sample reexamination) could account for these reads rather than propagules of this taxon present endo-lithically. Likewise, the most abundant Rhodophyta in other microflora samples were also encrusting taxa, the families 'Corallinaceae-Hapalidiaceae' and Peysson-neliaceae (Florideophyceae), but other non-encrusting taxa were also present (in class Bangiophyceae, Comp-sopogonophyceae, Stylonematophyceae, and within the Florideophyceae, e.g. the order Ceramiales, and family Dumontiaceae, Additional file 7: Table S6, Additional file 8: Table S7). Lastly, GM14 also exhibited a lower phototroph/heterotroph ratio than other microfloras (Fig. 13), which may be explained by its depth of collec-tion in the mesophotic zone (65 m) where light availabil-ity may limit photosynthesis and algal $CaCO_3$ penetration more importantly than in shallow waters.

Evolutionary perspectives

Our barcoding approach coupled with metabarcoding brought to light the widespread use of the endolithic niche as a habitat by the Bryopsidales (including otherwise epi-lithic taxa, see OTU and clones' distribution in Fig. 7). For instance, several macroscopic taxa of the Bryopsi-dineae (e.g. Derbesiaceae, Bryopsidaceae) and the Hali-medineae (e.g. Caulerpaceae, Halimedaceae, Rhipiliaceae, 'Siphonogramenaceae and Pseudochlorodesmidaceae') are present and may occupy the substratum in the form of germlings, reproductive cells, and endolithic sipho-nous/microsiphonous networks, the latter possibly used for lateral vegetative dispersal, thallus regeneration and access to nutrient-rich sediment trapped in cavities. Overall, the Bryopsidales seem to present a tight evolutionary link with the endolithic niche as an ancestral habitat, as evidenced by the early branching of the Ostreobidineae in multi-marker studies [102], also supported here with *tuf*A (bootstrap >95 %, Fig. 8). We hypothesize that environmental constraints associated with the epilithic niche such as high light/UV, direct water flow, and macroscopic predators (which themselves diversified and specialized over time, [16], such as kleptoplastic sacoglossan sea slugs, [39]) may have sparked the diversification of morphologically complex and conspicuous families from ancestral microsiphons living at the interface between the water column and the endolithic niche. In epilithic families able to exploit the endolithic niche with networks of siphons, this feature may thus represent a symplesiomorphy for some (e.g. the 'Siphonogramenaceae and 'Pseudochlorodesmidaceae') and a possibly re-acquired trait for others (i.e. the *Caulerpa* 'ambigua' complex in the Caulerpaceae) as suggested from the long branches leading to such lineage (i.e. accelerated evolution following key innovation into a new adaptive zone, Simpson [92, 93]) (Figs. 7, 8). Aside from the Bryopsidales, the phylogenetic extent of phototrophic OTUs also found in the endolithic niche seemingly in the form of chaesmo- and cryptoendolithic reproductive cells or germlings, perennial boring euendoliths (endolithic Ulvophyceae) as well as transient ones stages (e.g. some alternate life history stages of the Florideophyceae, Rhodophyta), highlights the critical importance of the $CaCO_3$ substratum in algal evolution as a seedbank for life cycle completion and survival of diverse algal taxa [30, 32].

Conclusions

In summary, we provide a flexible and comprehensive metabarcoding framework including primers, reference data and annotation thresholds for the facilitated recovery and rapid profiling of phototrophs found in endolithic communities. As a new resource for environmental biomonitoring, the framework is timely to enable the use of high throughput sequencing to accelerate biodiversity characterization of microbial/algal assemblages from endolithic communities found in coral reef and rhodolith ecosystems in the context of global change studies (e.g. bioerosion, distributional shifts), holobiont functioning and anthropogenic degradation (e.g. eutrophication, overfishing). Our framework could also find useful applications for water quality studies related to public health, such as the monitoring of river eutrophication based on the Diatom Index (i.e. composition of the 'Bacillariophyta' [47, 104]) and the detection of toxic Cyanobacteria (i.e. the Cyanophyceae) in freshwater and coastal systems [4, 6] pending further curation and increase in reference

data for these particular groups, which our primers retrieved very efficiently. Overall, *tuf*A metabarcoding assays could be performed on numerous types of samples harboring algal phototrophs including for instance water column, biofilms, periphyton, soil and ice samples and herbivorous organisms (e.g. vertebrate/invertebrate stomach contents, kleptoplastidic slugs tissue). Further studies of endolithic communities in coral reef and rhodolith ecosystems may reveal potential bioindicators of ecosystem degradation (taxonomic group/OTU ratio), whose efficient monitoring and detection may allow better management and conservation practices.

Additional files

Additional file 1: Table S1. Barcoded Ulvophyceae. Collection information for Ulvophyceae specimens newly sequenced for *tuf*A. To ease public database searches, Genbank® identifiers include both morphological and molecular identification (in brackets []) at the genus or family level (when abbreviated). Polyphyletic genera, species complexes or combined orders indicated in between quotes. All specimens from shallow waters (<0–5 m) otherwise indicated in footnotes. Endolithic habitat abbreviated as follows: $CaCO_3$ open reef substratum, *Corall.* Florideophycean crusts of the order Corallinales, *Peyss.* Florideophycean crusts of the order Peyssonneliales, *LPS* large polyp coral species (e.g. *Galaxea* sp., *Favia* sp.), *Shell* Oyster shell (primarily) or other bivalve, *SPS* small polyp coral species (e.g. *Porites* sp., *Montipora* sp.), *na* epilithic, epiphytic, or not recorded. Collector (Coll.) initials as follows: DP = D. Pence, HS = H. Spalding, JR = J. Richards, KI = K. Ikemoto, MDR = M. Diaz-Ruiz, MS = M. Star, NP = N. Pyron, TS = T. Sauvage, WES = W.E. Schmidt, ZM = Z. McCorkhill. Other abbreviations: Cult. = Cultured, IUI = Interuniversity Institute for Marine Science, INVEMAR = Instituto de investigaciones marinas y costeras.

Additional file 2: Table S2. Barcoded Florideophyceae, Phaeophyceae and miscellaneous taxa. Collection information for specimens of the Florideophyceae and Phaeophyceae sequenced for *tuf*A in the present study (as well as a miscellaneous Prasinophyceae). Polyphyletic orders or family and combined orders are indicated in between quotes. All specimens from shallow waters (0–10 m) otherwise indicated in footnotes. Collector (Coll.) initials as follows: CP = C. Pueschel, CS = C. Stoude, DG = D. Gabriel, DK = D. Krayesky, DWF = D.W. Freshwater, EC = E. Coppejans, ED = E. Deslandes, JC = J. Cabioch, JH = J. Hughey, JR = J. Richards, JRu = J. Rueness, JZ = J. Zertuche, MG = M. Guiry, MHH = M. H. Hommersand, MJW = M. J. Wynne, MY = M. Yoshizaki, OC = O. Camacho, SF = S. Fredericq.

Additional file 3: Table S3. Indexed forward primers used for metabarcoding. Indexes in bold.

Additional file 4: Table S4. Summary taxonomy and content of the non-redundant *tuf*A database. Diversity and sequence counts for (A) Heterotrophs and Phototrophs, (B) Phyla and Classes for all phototrophs and (C) Orders/suborders and families for the Ulvophyceae (Genus taxonomy not shown for conciseness). Taxonomic groupings with unsettled naming such as combined orders, provisional families, and polyphyletic families (i.e. 'Kornamanniaceae' and 'Ulothrichaceae') are noted between quotes.

Additional file 5: Table S5. Clone taxonomy. Detailed reference taxonomy for 86 phototrophic clones from microfloras sampled in the Ryukyu archipelago (JP01, JP03, JP04, JP06, JP07, JP25), an aquarium window scrap (E09) and a Gulf of Mexico crustose coralline *Lithophyllum* sp. rhodolith. Taxonomic groupings with unsettled naming such as combined orders, provisional families, and polyphyletic or monophyletic species or genera complexes, are noted between quotes.

Additional file 6: Figure S1. Cumulative OTU-ranked abundance curves. Note the much larger OTU diversity in densely colonized JP07 and FL02 vs. lightly colonized microfloras FL01 and GM14. Cumulative abundances values below 80 % not displayed for figure succinctness.

Additional file 7: Table S6. Summary of microfloras richness and abundance at multiple taxonomic levels. (A–B) OTU counts and read abundance for 'Phototrophs' vs. 'Heterotrophs' group. (C–D) OTU counts and read abundance for phototrophic phyla and classes. (E–F) OTU counts and read abundance for Ulvophyceaen families. Note that OTUs may be found more than once across samples.

Additional file 8: Table S7. OTUs taxonomy and their microflora assignment. Detailed reference taxonomy for 802 core OTUs as determined from their phylogenetic position (see Fig. 6). Taxonomic groupings with unsettled naming such as combined orders, provisional families, and polyphyletic or monophyletic species or genera complexes, are noted between quotes.

Additional file 9: Figure S2. Overlapping OTU diversity. Venn diagram depicting common OTU diversity (i.e. at the circles' intersection) and unique OTUs in each of FL02 and JP07 for all phototrophs and several OTU-rich classes. Overlapping OTUs may represent diversity that is both truly found in both samples (i.e. pantropical) and resulting from false assignment caused by index tag jumping or tag error (see Additional file 12: Figure S5), which may potentially inflate richness estimates. Note, however, the still important OTU richness unique to both FL02 and JP07, especially the OTU-rich classes Cyanophyceae and 'Bacillariophyta'.

Additional file 10: Figure S3. Molecular species divergence in the Ulvophyceae. Distribution of intra- and interspecific divergence for molecular species of the orders Bryopsidales and 'Ulvales-Ulothrichales' delimited with ABGD and GMYC. Percent identity values were computed based on an alignment of 375 bp of the *tuf*A metabarcode. Maximum thresholds for molecular species classification/annotation are printed. *PI* percent identity.

Additional file 11: Figure S4. OTU abundance profiles. OTUs underlying abundance patterns in Fig. 13. OTUs found in large read abundance across multiple microflora samples are color-coded in blue, green, orange and red. Other abundant OTUs are color-coded in white and those below 2 % abundance in shades of grey. The OTUs low-level taxonomy and their percent match to the database are listed in the legend.

Additional file 12: Figure S5. Example of putative index tag jumping or tag error. Comparative read abundance for seven OTUs demultiplexed (i.e. assigned) to both FL02 and JP07. For illustration purpose, only OTUs with >5 % of mapped reads in either FL02 or JP07 are shown (these correspond to seven out of the 141 common OTUs, see Additional file 9: Figure S2). Note the drastic abundance disparity illustrative of or potential of false assignment via index tag jumping or tag error (e.g. OTU 511 representing ≪0.1 % reads in FL02 and 10.4 % of reads in JP07). By contrast, note the more balanced abundance of OTU 355, 356, and 506, which may reflect the true presence of common OTUs in both samples (i.e. pantropical OTUs).

Additional file 13: Figure S6. OTUs vs. clone identity and coverage. Comparison of all phototrophic clones obtained from JP07 with their matching OTUs. Note high identity of clones and OTUs based on 375 bp, with 0, 1 or 2 base pair variation (100, 99.7 and 99.5 % respectively). Also note that clone H and its corresponding OTU_364 share the highest coverage, although further cloning would be necessary to accurately assess congruence in sequence coverage between the two methods.

Additional file 14: Figure S7. *tuf*A database and higher taxonomy overview. Unrooted RAxML tree comprising 556 'Heterotrophs' and 2141 'Phototrophs' *tuf*A sequences aligned on 891 bp. 'Heterotrophs' include multiple phyla of the Eubacteria. 'Phototrophs' include the prokaryotic eubacterial phylum Cyanophyta and several eukaryotic phyla, including the Archaeplastida (Chlorophyta, Glaucophyta and Rhodophyta) and secondary endosymbiotic phyla. 'Chloroplasts' include the Chlorophyta and the secondary endosymbiotic phyla Euglenophyta and Chlorarachniophyta. 'Rhodoplasts' include the Rhodophyta and the secondary endosymbiotic phyla of the Cryptophyta, Haptophyta and Ochrophyta.

Abbreviations

ABGD: automatic barcode gap discovery; bp: base pairs; $CaCO_3$: calcium carbonate; CCA: crustose coralline algae; EF-1a: elongation factor 1 alpha; ESS: effective sample size; GeO_2: germanium dioxide; GMYC: general mixed yule coalescence; ITS: internal transcribed spacer; LAF: University of Louisiana at Lafayette; LSU: large subunit; MCMC: Markov Chain Monte Carlo; NCBI: National Center for Biotechnology Information; pCO_2: partial pressure of carbon dioxide; PCR: polymerase chain reaction; PI: percent identity; OTU: Operational Taxonomic Unit; *rbc*L: RuBisCO large subunit; RDP: ribosomal database project; sp., spp.: species; *tuf*A: elongation factor EF-Tu; *tuf*B: elongation factor EF-*Tu* 2; UPA: universal plastid amplicon.

Authors' contributions

TS conducted field collections and culturing, summarized Genbank® content for *tuf*A, and designed and tested novel degenerate primers. TS and WES generated reference Sanger data (including cloning data), performed bioinformatics and phylogenetics. TS conceptualized and drafted the manuscript. WES, SF and SS contributed to data interpretation, taxonomic expertise, collections and revision of the manuscript. SS hosted TS in Okinawa, Japan. All authors read and approved the final manuscript.

Author details

[1] Department of Biology, University of Louisiana at Lafayette, Lafayette, LA 70503, USA. [2] Department of Marine Science, Biology and Chemistry, University of the Ryukyus, Nishihara, Okinawa 903-0213, Japan.

Acknowledgements

Field Sampling, Sanger sequencing, cloning, and next generation sequencing were primarily supported by the JSPS/CNRS Summer 2012 Fellowship #SP12209 awarded to TS and from the "International Hub Project for Climate Change and Coral Reef/Island Dynamics of the University of the Ryukyus, Okinawa, Japan" Grant to SS, with additional contribution from the University of Louisiana at Lafayette Graduate Student Organization, Student Government Association and Center for Ecology and Environmental Technology (Grants-In-Aid of Research) and a Link Foundation/Smithsonian Marine Station 2013 Fellowship awarded to TS. Sequencing of Florideophyceae and Phaeophyceae specimens were partly funded by the National Science Foundation (DEB-0936855, DEB-1027107) and a Gulf of Mexico Research Initiative Coastal Water Consortium (GoMRI-I) Grant to SF. TS thanks Susan Brawley for the invitation to attend a bioinformatics training workshop offered during the *Porphyra* Algal Genomics' Research Collaboration Network (NSF 0741907), and EuroMarine/EMBRC-France for further training provided during the 2015 Marine Ecological and Evolutionary Genomics in Roscoff, France. We are also thankful for help in the field from Yuhi Seto and Ooshi Hiraishi in the Ryukyus, and Olga Camacho, Natalia Arakaki and Joe Richards for miscellaneous specimens. We thank Alison Sherwood and Heather Spalding for specimens databased in Hawaii. We are grateful for constructive criticism provided by two anonymous reviewers on an early version of the manuscript.

Competing interests

The authors declare that they have no competing interests.

References

1. Abbott IA, Huisman JM. Marine green and brown algae of the Hawaiian Islands. Honolulu: Bishop Museum Press; 2004. p. 260.
2. Adams RI, Amend AS, Taylor JW, Bruns T. A unique signal distorts the perception of species richness and composition in high-throughput sequencing surveys of microbial communities: a case study of fungi in indoor dust. Microb Ecol. 2013;66:735–41.
3. Baldauf S, Palmer JD. Evolutionary transfer of the chloroplast *tuf*A gene to the nucleus. Nature. 1990;344:262–5.
4. Bartram J, Carmichael WW, Chorus I, Jones G, Skulberg OM. Introduction. In: Chorus I, Bartram J, editors. Toxic cyanobacteria in water: a guide to their public health consequences, monitoring and management. London: WHO and E&FN Spon; 1999. p. 1–14.

5. Bendif EM, Probert I, Carmichael M, Romac S, Hagino K, de Vargas C. Genetic delineation between and within the widespread coccolithophore morpho-species *Emiliania huxleyi* and *Gephyrocapsa oceanica* (Haptophyta). J Phycol. 2014;50:140–8. doi:10.1111/jpy.12147.

6. Bláha L, Babica P, Maršálek B. Toxins produced in cyanobacterial water blooms—toxicity and risks. Interdiscip Toxicol. 2009;2:36–41. doi:10.2478/v10102-009-0006-2.

7. Bornet É, Flahault C. Sur quelques plantes vivantes dans le test calcaire des mollusques. Bull Soc Bot France. 1889;36:167–76.

8. Bouckaert R, Heled J, Kühnert D, Vaughan T, Wu C-H, et al. BEAST 2: a software platform for Bayesian evolutionary analysis. PLoS Comput Biol. 2014;10:e1003537. doi:10.1371/journal.pcbi.1003537.

9. Brodie J, Maggs CA, John DM. Green seaweeds of Britain and Ireland. London: British Phycological Society; 2007. p. 1–242.

10. Brown SDJ, Collins RA, Boyer S, Lefort MC, Malumbres-olarte J, Vink CJ, Cruickshank RH. Spider: an R package for the analysis of species identity and evolution, with particular reference to DNA barcoding. Mol Ecol Res. 2012;12:562–5.

11. Chaubey S, Kumar A, Singh D, Habib S. The apicoplast of *Plasmodium falciparum* is translationally active. Mol Microbiol. 2005;56:81–9.

12. Chazottes V, Le Campion-Alsumard T, Peyrot-Clausade M. Bioerosion rates on coral reef: interactions between macroborers, microborers and grazers (Moorea, French Polynesia). Palaeogeogr Palaeoclimat Palaeoecol. 1995;113:189–98.

13. Cole JR, Wang Q, Fish JA, Chai B, McGarrell DM, Sun Y, Brown CT, Porras-Alfaro A, Kuske CR, Tiedje JM. Ribosomal database project: data and tools for high throughput rRNA analysis. Nucl Acids Res. 2014;42(Database issue):D633–42. doi:10.1093/nar/gkt1244.

14. Cowart DA, Pinheiro M, Mouchel O, Maguer M, Grall J, Miné J, Arnaud-Haond S. Metabarcoding Is powerful yet still blind: a comparative analysis of morphological and molecular surveys of seagrass communities. PLoS One. 2015. doi:10.1371/journal.pone.0117562.

15. Crooks GE, Hon G, Chandonia JM, Brenner SE. WebLogo: a sequence logo generator. Genome Res. 2004;14:1188–90.

16. Dawkins R, Krebs JR. Arms races between and within species. Proc R Soc Lond. 1979;205:489–511. doi:10.1098/rspb.1979.0081.

17. Decelle J, Romac S, Stern RF, Bendif EM, Zingone A, Audic S, Guiry MD, Guillou L, Tessier D, Le Gall L, Gourvil P, Dos Santos AL, Probert I, Vaulot D, de Vargas, Christen R. PhytoREF: a reference database of the plastidial 16S rRNA gene of photosynthetic eukaryotes with curated taxonomy. Mol Ecol Resour. 2015;15(6):1435–45.

18. Degnan PH, Ochman H. Illumina-based analysis of microbial community diversity. ISME J. 2012;6:183–94.

19. Delwiche CF, Kuhsel M, Palmer JD. Phylogenetic analysis of *tufA* sequences indicates a cyanobacterial origin of all plastids. Mol Phylog Evol. 1995;4:110–28.

20. Denonfoux J, Parisot N, DugatBony E, Biderre-Petit C, Boucher D, Morgavi DP, Paslier D, Peyretaillade E, Peyret P. Gene capture coupled to high-throughput sequencing as a strategy for targeted metagenome exploration. DNA Res. 2013;20(2):185–6.

21. DeSantis TZ, Hugenholtz P, Larsen N, Rojas M, Brodie EL, Keller K, Huber T, Dalevi D, Hu P, Andersen GL. Greengenes, a chimera-checked 16S rRNA gene database and workbench compatible with ARB. Appl Env Microbiol. 2006;72:5069–72.

22. Dowle EJ, Pochon X, Banks JC, Shearer K, Wood SA. Targeted gene enrichment and high-throughput sequencing for environmental biomonitoring: a case study using freshwater macroinvertebrates. Mol Ecol Res. 2015. doi:10.1111/1755-0998.12488.

23. Draisma SGA, van Reine WFP, Sauvage T, Belton GS, Gurgel CFD, Lim PE, Phang SM. A re-assessment of the infra-generic classification of the gneus *Caulerpa* (Caulerpaceae, Chlorophyta) inferred from a time-calibrated molecular phylogeny. J Phycol. 2014;50:1020–34.

24. Edgar RC. MUSCLE: multiple sequence alignment with high accuracy and high throughput. Nucl Acids Res. 2004;32:1792–7.

25. Edgar RC. Search and clustering orders of magnitude faster than BLAST. Bioinformatics. 2010;26:2460–1.

26. Edgar RC. UPARSE: highly accurate OTU sequences from microbial amplicon reads. Nat Meth. 2013;10:996–8.

27. Eiler A, Drakare S, Bertilsson S, Pernthaler J, Peura S, Rofner C, Simek K, Yang Y, Znachor P, Lindström ES. Unveiling distribution patterns of freshwater phytoplankton by a next generation sequencing based approach. PLoS One. 2013;8:e53516. doi:10.1371/journal.pone.0053516.

28. Esling P, Lejzerowicz F, Pawlowski J. Accurate multiplexing and filtering for high-throughput amplicon-sequencing. Nucleic Acids Res. 2015;43(5):2513–24. doi:10.1093/nar/gkv107

29. Famà P, Wysor B, Kooistra WHCF, Zuccarello GC. Molecular phylogeny of the genus *Caulerpa* (*Caulerpales*, Chlorophyta) inferred from chloroplast *tufA* gene. J Phycol. 2002;38:1040–50.

30. Felder DL, Thoma BP, Schmidt WE, Sauvage T, Self-Krayesky S, Chistoserdov A, Bracken-Grissom H, Fredericq S. Seaweeds and decapod crustaceans on Gulf deep banks after the Macondo Oil Spill. Bioscience. 2014;64:808–19.

31. Fine M, Loya Y. Endolithic algae: an alternative source of photoassimilates during coral bleaching. Proc R Soc Lond B Biol Sci. 2002;269:1205–10.

32. Fredericq S, Arakaki N, Camacho O, Gabriel D, Krayesky D, Self-Krayesky S, Rees G, Richards J, Sauvage T, Venera-Ponton T, Schmidt WE. A dynamic approach to the study of rhodoliths: a case study for the Northwestern Gulf of Mexico. Crypt Algol. 2014;35:77–98.

33. Fresnel J, Probert I. The ultrastructure and life cycle of the coastal coccolithophorid *Ochrosphaera neapolitana* (Prymnesiophyceae). Europ J Phycol. 2005;40:105–22.

34. Fučíková K, Leliaert F, Cooper ED, Škaloud P, D'hondt S, De Clerck O, et al. New phylogenetic hypothesis for the core Chlorophyta based on chloroplast sequence data. Front Ecol Evol. 2014;2:63.

35. Fujisawa T, Barraclough TC. Delimiting species using single-locus data and the generalized mixed Yule coalescent approach: a revised method and evaluation on simulated data sets. Syst Biol. 2013;62:707–24.

36. Glynn PW. Bioerosion and coral reef growth: a dynamic balance. In: Birkeland C, editor. Life and death of coral reefs. New York: Chapman and Hall; 1997. p. 68–95.

37. Guiry MD, Guiry GM. AlgaeBase. World-wide electronic publication. 2016; National University of Ireland, Galway. http://www.algaebase.org.

38. Gutner-Hoch E, Fine M. Genotypic diversity and distribution of *Ostreobium quekettii* within scleractinian corals. Coral Reefs. 2011;30:643–50.

39. Händeler K, Wägele H, Wahrmund U, Rüdinger M, Knoop V. Slugs' last meals: kleptoplastidsin Sacoglossa (Opisthobranchia, Gastropoda): molecular identification of sequestered chloroplasts from different algal origins. Mol Ecol Res. 2010. doi:10.1111/j.1755-0998.2010.02853.x.

40. Hartman H, Favaretto P, Smith TF. The archaeal origins of the eukaryotic translational system. Archaea. 2006;2:1–9.

41. Hillis-Colinvaux L. Systematics of the siphonales. In: Irvine DEG, John DM, editors. Systematics of the green algae. London: Academic Press; 1984. p. 271–91.

42. Iwabe N, Kuma KI, Hasegawa M, Osawa S, Miyata T. Evolutionary relationship of archaebacteria, eubacteria, and eukaryotes inferred from phylogenetic trees of duplicated genes. Proc Natl Acad Sci USA. 1989;86:9355–9.

43. Janouskovec J, Horák A, Oborník M, Lukes J, Keeling PJ. A common red algal origin of the apicomplexan, dinoflagellate, and heterokont plastids. Proc Natl Acad Sci USA. 2010;107:10949–54.

44. Jörger KM, Schrödl M. How to describe a cryptic species? Practical challenges of molecular taxonomy. Front Zool. 2013;10:59. doi:10.1186/1742-9994-10-59.

45. Ke D, Boissinot M, Huletsky A, Picard FJ, Frenette J, Ouellette M, Roy PH, Bergeron MG. Evidence for horizontal gene transfer in evolution of elongation factor *Tu* in enterococci. J Bacteriol. 2000;182:6913–20.

46. Keeling PJ. Chromalveolates and the evolution of plastids by secondary endosymbiosis. J Eukaryot Microbiol. 2009;56:1–8.

47. Kelly MG, Whitton BA. The Trophic Diatom Index: a new index for monitoring eutrophication in rivers. J App Phycol. 1995;7:433–44.

48. Kitson JJN, Hahn C, Sands RJ, Straw NA, Evans DM, Lunt DH. Nested metabarcode tagging: a robust tool for studying species interactions in ecology and evolution. BioRxiv. 2016; doi:10.1101/035071.

49. Köhler S, Delwiche CF, Denny PW, Tilney LG, Webster P, Wilson RJM, Palmer JD, Roos DS. A plastid of probable green algal origin in apicomplexan parasites. Science. 1997;275:1485–9.

50. Kraft GT. Algae of Australia: marine benthic algae of Lord Howe Island and the southern great barrier reef 1: green algae, 1. Collingwood: Australian Biological Resources Study and CSIRO Publishing; 2007. p. 365.

51. Kumar PS, Brooker MR, Dowd SE, Camerlengo T. Target region selection is a critical determinant of community fingerprints generated by 16S Pyrosequencing. PLoS One. 2011;6:e20956. doi:10.1371/journal.pone.0020956.

52. Lapointe J, Braker-Gigras L. Translation mechanisms. Molecular biology intelligence unit. New York: Springer; 2014. p. 468.

53. Lathe WC III, Bork P. Evolution of *tuf* genes: ancient duplication, differential loss and gene conversion. FEBS Lett. 2001;502:113–6.

54. Le Campion-Alsumard T, Golubic S, Hutchings P. Microbial endoliths in the skeletons of live and dead corals: *Porites lobata* (Moorea, French Polynesia). Mar Ecol Progr Ser. 1995;117:149–57.

55. Leliaert F, Lopez-Bautista JM. The chloroplast genomes of *Bryopsis plumosa* and *Tydemania expeditiones* (Bryopsidales, Chlorophyta): compact genomes and genes of bacterial origin. BMC Genom. 2015;16:204. doi:10.1186/s12864-015-1418-3.

56. Lemieux C, Otis C, Turmel M. A clade uniting the green algae *Mesostigma viride* and *Chlorokybus atmophyticus* represents the deepest branch of the Streptophyta in chloroplast genome-based phylogenies. BMC Biol. 2007;5:2. doi:10.1186/1741-7007-5-2.

57. Letunic I, Bork P. Interactive tree of life (itol): an online tool for phylogenetic tree display and annotation. Bioinformatics. 2007;21:127–8.

58. Linhart C, Shamir R. Degenerate primer design: theoretical analysis and the HYDEN program. Methods Mol Biol. 2007;402:221–44.

59. Logares R, Sunagawa S, Salazar G, Cornejo-Castillo FM, Ferrera I, Sarmento H, Hingamp P, Ogata H, de Vargas C, Lima-Mendez G, Raes J, Poulain J, Jaillon O, Wincker P, Kandels-Lewis S, Karsenti E, Bork P, Acinas SG. Metagenomic 16S rDNA illumina tags are a powerful alternative to amplicon sequencing to explore diversity and structure of microbial communities. Env Microbiol. 2014;16:2659–71.

60. Lukas KJ. Taxonomy and ecology of the endolithic microflora of reef corals, with a review of the literature on endolithic microphytes. PhD Dissertation Univ Rhode Island, 1973; p. 159.

61. Lukas KJ. Two species of the Chlorophyte genus *Ostreobium* from skeletons of Atlantic and Caribbean reef corals. J Phycol. 1974;10:331–5.

62. Mahé M, Rognes T, Quince C, de Vargas C, Dunthorn M. Swarm: robust and fast clustering method for amplicon-based studies. PeerJ. 2014;2:e593. doi:10.7717/peerj.593.

63. Marin B, Nowack EC, Melkonian M. A plastid in the making: evidence for a second primary endosymbiosis. Protist. 2005;156:425–32.

64. Meinesz A. Sur le cycle de l'*Halimeda tuna* (Ellis et Solander) Lamouroux (Udoteaceae, Caulerpales). C R Acad Sci Paris. 1972;275:1363–5.

65. Mendoza MLZ, Sicheritz-Pontén T, Gilbert MT. Environmental genes and genomes: understanding the differences and challenges in the approaches and software for their analyses. Brief Bioinform. 2015. doi:10.1093/bib/bbv001.

66. Moniz MBJ, Guiry MD, Rindi F. *TufA* phylogeny and species boundaries in the green algal order Prasiolales (Trebouxiophyceae, Chlorophyta). Phycologia. 2014;53:396–406.

67. Montandon PE, Knuchel-Aegerter C, Stutz E. *Euglena gracilis* chloroplast DNA: the untranslated leader of tufA-ORF206 gene contains an intron. Nucl Acids Res. 1987;15:7809–22.

68. Odum HT, Odum EP. Trophic structure and productivity of a windward coral reef community on Eniwetok Atoll. Ecol Monographs. 1955;25:291–320.

69. Paradis E, Claude J, Strimmer K. APE: analyses of phylogenetics and evolution in R language. Bioinformatics. 2004;20:289–90.

70. Parker SP. Synopsis and classification of living organisms, New York: McGraw-Hill; 1982; 1:154.

71. Patel M, Gonzalez R, Halford C, Lewinski MA, Landaw EM, Churchill BM, Haake DA. Target-specific capture enhances sensitivity of electrochemical detection of bacterial pathogens. J Clin Microbiol. 2011;49:4293–6.

72. Perry CT, Murphy GN, Kench PS, Smithers SG, Edinger EB, Steneck RS, Mumby PJ. Caribbean-wide decline in carbonate production threatens coral reef growth. Nature Comm. 2013;4:1402. doi:10.1038/ncomms2409.

73. Perry CT, Murphy GN, Kench PS, Edinger EN, Smithers SG, Steneck RS, Mumby PJ. Changing dynamics of Caribbean reef carbonate budgets: emergence of reef bioeroders as critical controls on present and future reef growth potential. Proc R Soc B. 2014;281:2014–8. doi:10.1098/rspb.2014.2018.

74. Price DC, Chan CX, Yoon HS, Yang EC, Qiu H, Weber AP, Schwacke R, Gross J, Blouin NA, Lane C, Reyes-Prieto A, Durnford DG, Neilson JA, Lang BF, Burger G, Steiner JM, Löffelhardt W, Meuser JE, Posewitz MC, Ball S, Arias MC, Henrissat B, Coutinho PM, Rensing SA, Symeonidi A, Doddapaneni H, Green BR, Rajah VD, Boore J, Bhattacharya D. *Cyanophora paradoxa* genome elucidates origin of photosynthesis in algae and plants. Science. 2012;335:843–7.

75. Provasoli L. Media and prospects for the cultivation of marine algae. In: Watanabe H, Hattori A, editors. Culture and collection of Algae. Proc USA—Japan Conference. Jap Soc Plant Physiol Hakone. 1968; p. 63–75.

76. Puillandre N, Lambert A, Brouillet S, Achaz G. ABGD, automatic barcode gap discovery for primary species delimitation. Mol Ecol. 2012;21:1864–77. doi:10.1111/j.1365-294X.2011.05239.x.

77. Quast C, Pruesse E, Yilmaz P, Gerken J, Schweer T, Yarza P, Peplies J, Glöckner FO. The SILVA ribosomal RNA gene database project: improved data processing and web-based tools. Nucleic Acids Res. 2014;41:D590–6. doi:10.1093/nar/gks1219.

78. R Core Team. R: a language and environment for statistical computing. R foundation for statistical computing, Vienna: 2014. www.R-project.org.

79. Radtke G, Golubic S. Microborings in mollusc shells, Bay of Safaga, Egypt: morphometry and ichnology. Facies. 2005;51:118–34.

80. Rambaut A, Drummond A. Tracer v1.6. 2007. Available from: http://tree.bio.ed.ac.uk/software/tracer.

81. Reyes-Nivia C, Diaz-Pulido G, Kline D, Guldberg OH, Dove S. Ocean acidification and warming scenarios increase microbioerosion of coral skeletons. Glob Chang Biol. 2013;19:1919–29.

82. Reyes-Nivia C, Diaz-Pulido G, Dove S. Relative roles of endolithic algae and carbonate chemistry variability in the skeletal dissolution of crustose coralline algae. Biogeosciences. 2014;11:4615–26.

83. Rodríguez-Ezpeleta N, Brinkmann H, Burey SC, Roure B, Burger G, Löffelhardt W, Bohnert HJ, Philippe H, Lang BF. Monophyly of primary photosynthetic eukaryotes: green plants, red algae, and glaucophytes. Curr Biol. 2005;15:1325–30.

84. Rohfritsch A, Payri C, Stiger V, Bonhomme F. Molecular and morphological relationships between two closely related species, *Turbinaria ornata* and *T. conoides* (Sargassaceae, Phaeophyceae). Biochem System Ecol. 2007;35:91–8.

85. Rohwer F, Seguritan V, Azam F, Knowlton N. Diversity and distribution of coral-associated bacteria. Mar Ecol Prog Ser. 2002;243:1–10.

86. Saunders GW, Kucera H. An evaluation of *rbcL*, *tufA*, UPA, LSU and ITS as DNA barcode markers from the marine green macroalgae. Cryptog Algol. 2010;31:487–528.

87. Sauvage T, Payri P, Draisma SGA, van Prud'homme Reine WF, Verbruggen H, Sherwood A, Fredericq S. Molecular diversity of the *Caulerpa racemosa-Caulerpa peltata* complex (Caulerpaceae, Bryopsidales) in New Caledonia, with new Australasian records for *C. racemosa* var. *cylindracea*. Phycologia. 2013;52:6–13.

88. Schmidt TSB, Matias Rodrigues JF, von Mering C. Limits to robustness and reproducibility in the demarcation of operational taxonomic units. Environ Microbiol. 2015;17:1689–706.

89. Schnell IB, Bohmann K, Gilbert MT. Tag jumps illuminated—reducing sequence-to-sample misidentifications in metabarcoding studies. Mol Ecol Res. 2015. doi:10.1111/1755-0998.12402.

90. Sela S, Yogev D, Razin S, Bercovier H. Duplication of the *tuf* gene: a new insight into the phylogeny of eubacteria. J Bacteriol. 1989;171:581–4.

91. Sickel W, Ankenbrand MJ, Grimmer G, Holzschuh A, Härtel S, Lanzen J, Steffan-Dewenter I, Keller A. Increased efficiency in identifying mixed pollen samples by meta-barcoding with a dual-indexing approach. BMC Ecol. 2015;15:20. doi:10.1186/s12898-015-0051-y.

92. Simpson GG. Tempo and mode in evolution. New York: Columbia University Press; 1994.

93. Simpson GG. The major features of evolution. New York: Columbia University Press; 1953.

94. Sunagawa S, Wilson EC, Thaler M, Smith ML, Caruso C, Pringle JR, Weis VM, Medina M, Schwarz JA. Generation and analysis of transcriptomic resources for a model system on the rise: the sea anemone *Aiptasia pallida* and its dinoflagellate endosymbiont. BMC Genom. 2009;10:258. doi:10.1186/1471-2164-10-258.

95. Taanman JW. The mitochondrial genome: structure, transcription, translation and replication. Biochim Biophys Acta. 1999;1410:103–23.

96. Taberlet P, Coissac E, Pompanon F, Brochmann C, Willerslev E. Towards next-generation biodiversity assessment using DNA metabarcoding. Mol Ecol. 2012;21:2045–50.

97. Tribollet A. Dissolution of dead corals by euendolithic microorganisms across the northern Great Barrier Reef (Australia). Microbial Ecol. 2008;5:569–80.

98. Tribollet A, Langdon C, Golubic S, Atkinson M. Endolithic microflora are major primary producers in dead carbonate substrates of Hawaiian coral reefs. J Phycol. 2006;42:292–303.

99. Tribollet A, Godinot C, Atkinson M, Langdon S. Effects of elevated pCO2 on dissolution of coral carbonates by microbial euendoliths. Global Biogeochem Cycl. 2009;23:GB3008. doi:10.1029/2008GB003286.

100. Tuohy TMF, Thompson S, Gesteland RF, Hughes D, Atkins JF. The role of EF-Tu and other translation components in determining translocation step size. Biochim Biophys Acta. 1990;1050:274–8.

101. Turmel M, Brouard JS, Gagnon C, Otis C, Lemieux C. Deep division in the Chlorophyceae (Chlorophyta) revealed by chloroplast phylogenomic analyses. J Phycol. 2008;44:739–50.

102. Verbruggen H, Ashworth M, LoDuca ST, Vlaeminck C, Cocquyt E, Sauvage T, Zechman FW, Littler DS, Littler MM, Leliaert F, De Clerck O. A multi-locus time-calibrated phylogeny of the siphonous green algae. Mol Phylogen Evol. 2009;50:642–53.

103. Verbruggen H, Vlaeminck C, Sauvage T, Sherwood AR, Leliaert F, De Clerck O. Phylogenetic analysis of Pseudochlorodesmis strains reveals cryptic diversity above the family level in the siphonous green algae (Bryopsidales, Chlorophyta). J Phycol. 2009;45:726–31.

104. Visco JA, Apothéloz-Perret-Gentil L, Cordonier A, Esling P, Pillet P, Pawlowski J. Environmental monitoring: inferring the diatom index from next-generation sequencing data. Environ Sci Technol. 2015;49:7597–605. doi:10.1021/es506158m.

105. Wilson LJ, Weber XA, King TM, Fraser CI. Extraction techniques for genomic analyses of macroalgae. In Seaweed Phylogeography 2016. The Netherlands: Springer; p. 363–386.

106. Zuccarello GC, West JA, Kikuchi N. Phylogenetic relationships within the Stylonematales (Stylenomatophyceae, Rhodophyta): biogeographic patterns do not apply to Stylonema alsidii. J Phycol. 2008;44:384–93. doi:10.1111/j.1529-8817.2008.00467.x.

12

Evidence of indirect symbiont conferred protection against the predatory lady beetle *Harmonia axyridis* in the pea aphid

Jennifer L. Kovacs[1]*[iD], Candice Wolf[2], Dené Voisin[3] and Seth Wolf[2]

Abstract

Background: Defensive symbionts can provide significant fitness advantages to their hosts. Facultative symbionts can protect several species of aphid from fungal pathogens, heat shock, and parasitism by parasitoid wasps. Previous work found that two of these facultative symbionts can also indirectly protect pea aphids from predation by the lady beetle *Hippocampus convergens*. When aphids reproduce asexually, there is extremely high relatedness among aphid clone-mates and often very limited dispersal. Under these conditions, symbionts may indirectly protect aphid clone-mates from predation by negatively affecting the survival of a predator after the consumption of aphids harboring the same vertically transmitted facultative symbionts. In this study, we wanted to determine whether this indirect protection extended to another lady beetle species, *Harmonia axyridis*.

Results: We fed *Ha. axyridis* larvae aphids from one of four aphid sub-clonal symbiont lines which all originated from the same naturally symbiont free clonal aphid lineage. Three of the sub-clonal lines harbor different facultative symbionts that were introduced to the lines via microinjection. Therefore these sub-clonal lineages vary primarily in their symbiont composition, not their genetic background. We found that aphid facultative symbionts affected larval survival as well as pupal survival in their predator *Ha. axyridis*. Additionally, *Ha. axyridis* larvae fed aphids with the *Regiella* symbiont had significantly longer larval developmental times than beetle larvae fed other aphids, and females fed aphids with the *Regiella* symbiont as larvae weighed less as adults. These fitness effects were different from those previously found in another aphid predator *Hi. convergens* suggesting that the fitness effects may not be the same in different aphid predators.

Conclusions: Overall, our findings suggest that some aphid symbionts may indirectly benefit their clonal aphid hosts by negatively impacting the development and survival of a lady beetle aphid predator *Ha. axyridis*. By directly affecting the survival of predatory lady beetles, aphid facultative symbionts may increase the survival of their clone-mates that are clustered nearby and have significant impacts across multiple trophic levels. We have now found evidence for multiple aphid facultative symbionts negatively impacting the survival of a second species of aphid predatory lady beetle. These same symbionts also protect their hosts from parasitism and fungal infections, though these fitness effects seem to depend on the aphid species, predator or parasitoid species, and symbiont type. This work further demonstrates that beneficial mutualisms depend upon complex interactions between a variety of players and should be studied in multiple ecologically relevant contexts.

Keywords: Protective symbiosis, Facultative symbionts, Indirect fitness effects, Predation

*Correspondence: jkovacs@spelman.edu
[1] Spelman College, 350 Spelman Lane, S.W., Atlanta, GA 30314, USA
Full list of author information is available at the end of the article

Background

Symbionts, both obligate and facultative, can affect the fitness of their hosts in a variety of ways. Symbionts can have little or no effect on their host (often called commensals). While, others can negatively affect the fitness of their hosts, and still other symbionts are beneficial and can provide diverse fitness advantages to their hosts [1]. For example, in *Drosophila neotestacea*, a vertically transmitted symbiotic *Spiroplasma* bacteria protects its female fly hosts from being sterilized by a parasitic *Howardula* nematode [2]. While in herbivorous *Megacopta* stinkbugs, symbiotic bacteria seem to determine what host plants their hosts can utilize. Normally *Megacopta cribaria* suffers high mortality when reared on legumes. However when they are provided with symbiont capsules from the legume pest *Megacopta punctatissima* and obtain its symbiotic gut bacteria, *M. cribaria* are then able to utilize legumes as host plants [3]. In both of these examples, as well as in other cases, symbionts can have impacts beyond just their host species.

Symbionts can have effects across trophic levels, food webs, and ecological communities [2, 4–14]. They can mediate inter-specific interactions, such as competition, parasitism, and predation. For example, rove beetles in the genus *Paederus* harbor *Pseudomonas* symbionts which produce the toxic amide pederin. Pederin is highly cytotoxic and blocks protein synthesis inhibiting mitosis [15, 16]. In humans, pederin can result in itching lesions and dermatitis when it comes in contact with skin, often through the crushing of beetles [17]. *Paederus* females use their eggshells to vertically transfer *Pseudomonas* to their offspring. While pederin-provisioned larvae are just as likely to be attacked and eaten by predatory insects as *Paederus* larvae without pederin, the pederin does provide protection from predation by spiders [18]. Wolf spiders avoid *Paederus* larvae and eggs with pederin as well as *Drosophila* flies artificially provisioned with the pederin toxin [18]. In this example, the *Pseudomonas* symbiont provides a very direct defensive benefit to its host against predation by wolf spiders.

In aphids, facultative symbionts are not essential to aphid growth or survival and are not present in all aphid populations, but these facultative symbiotic bacteria can provide protection against pathogens, heat stress, and parasitoids [1, 11, 13, 19–25]. For example, when parasitoid wasps lay their eggs in aphids harboring the *Hamiltonella defensa* symbiont, those eggs fail to develop or the larvae experience very high mortality rates [26]. Additionally, Rothacher et al. [13] observed shifts in the composition of parasitoid wasp species within a natural community when black bean aphid *Aphis fabae* harbored the protective symbiont *H. defensa*. This demonstrates the potential that symbionts, particularly defensive

symbionts, can have on the structure and the diversity of food webs.

A recent study found that the survival of predatory lady beetle *Hippodamia convergens* larvae was decreased when they were fed pea aphids with facultative symbionts *H. defensa* and *Serratia symbiotica* [14]. Rather than directly protecting aphids from immediate predation, as in the case of the rove beetles, aphid symbionts may be providing indirect protection to their aphid hosts. Aphids live in tightly clustered, extremely highly related clonal groups during the parthogenetic portion of their annual life cycle [27]. By lowering predator survival, aphid symbionts could also lower the overall risk of predation for a group of clonal aphids, thereby providing indirect protection against predation by lady beetles [28–30]. We were interested in determining whether three aphid symbionts (*H. defensa*, *S. symbiotica*, and *Regiella insecticola*) negatively affected the survival of a second species of predatory lady beetle, *Harmonia axyridis*. Based on previous findings in *Hi. convergens*, we predicted that aphid secondary symbionts would lower predator survival and negatively impact larval weight and development time [14]. Our current findings suggest that multiple aphid symbionts also provide indirect protection from a second species of predatory lady beetles.

Methods
Survival experiments

Adult lady beetles (*Ha. axyridis*) were collected at Spelman College in Atlanta, GA, USA. Lady beetles were kept in mixed sex groups and maintained at 25 °C with a light regime of 16:8 Light:Dark. Adult beetles were fed aphids from genetically identical asexual aphid lineages harboring either no facultative symbionts (aphid line 5AO), the facultative symbiont *S. symbiotica* (5AR), *H. defensa* (5AT), or *R. insecticola* (5AU). All four aphid symbiont lineages (5A0, 5AR, 5AU, and 5AT) were established from the same naturally uninfected 5A clone (collected in Madison, WI, USA, June 1999). Facultative symbionts were introduced to the 5A clone through microinjection of body fluids containing symbionts (5AR & 5AU [31], 5AT, [10]). Prior to starting the experiment, lines were screened for the respective facultative symbionts using qPCR. In addition, *H. defensa* is associated with a phage, APSE, that is known to have important effect on some *Hamiltonella*-conferred phenotype. We used PCR to confirm the presence of APSE in our *Hamiltonella*-infected line ([32] unpublished data). Aphids were reared on fava seedlings (*Vicia faba* L.) at 20 °C with a light regime of 18:6 Light:Dark. New aphid bearing plants were supplied to the adult lady beetles daily.

After a week of captivity, we began removing lady beetle egg clutches from the adult cage daily and placing egg clutches in Petri dishes. Once eggs hatched, larvae were separated into individual Petri dishes. Larvae were fed aphids from the same aphid line containing the same facultative symbiont that their parents ate. We note that due to this experimental design we are unable to determine whether our observed results are due to the feeding of the symbiont to the mother (maternal effects) or to the larvae. In either case, any observed survival effect would be due to the symbiont type which was the same in both the maternal and the larval diet. Larvae were raised individually to prevent cannibalism and competition between individuals. Larvae were fed fresh aphids ad libitum, and moist cotton balls were supplied and replaced as needed. All larvae were provided approximately the same number of third and fourth instar aphids every day (~10–15 aphids per feeding during instars 1 and 2 and ~15–20 aphids during instars 3 and 4), and no major differences in feeding rates were observed between the four experimental groups, though precise daily feeding rates were not recorded in this study. Preliminary work determined that under our lab conditions *Ha. axyridis* larval development lasted 19 days from hatching to pupation. Instars one and two took 7 days, the third instar 5 days, and the fourth instar 8 days. However, exact dates for each instar molt for each individual were not recorded in this study. Time to pupation, time to emergence from pupation, and time to death were recorded daily. Larvae were weighed at day 8, and adult lady beetles were weighed upon emergence prior to additional feeding. After adult emergence individuals were sexed using dimorphic features of the distal margin of the final abdominal sternite. In a single trial lasting 40 days, 305 lady beetle larvae were observed from hatching to either adult emergence or death. Seventy-five lady beetle larvae were fed 5AT aphids, 76 were fed 5AR aphids, 75 were fed 5AU and 79 were fed 5AO aphids. All data generated and analyzed in this study are available in supplementary material files associated with this publication (Additional file 1: Data S1).

Statistical analyses

To determine whether the type of aphid symbiont in the larval diet affected overall lady beetle survival from hatching to adult emergence, we used a generalized linear model with a logit linked binomial distribution (reached adult stage = 1, died prior to reaching adult stage = 0) with "symbiont", "larval weight at day 8", and the interaction between the two as factors. The interaction term was included in the final model based on the resulting delta AICc (Akaike information criterion adjusted for sample size) scores. The model with the lowest AICc score was considered to be the "best" model. Non-significant interaction terms were retained when the delta AICc between the full model (run with the interaction term and all

other factors) and the "best" model was less than 2 and removed when it was more than 2. Post-hoc pairwise contrast analyses were performed to determine whether observed differences in pupal survival between aphid sub-clonal lines were significant.

We then further broke down survival, looking first at survival to pupation and then survival from pupation to adult emergence. The effects of aphid symbionts in the diet of *Ha. axyridis* larvae on larval survival from hatching to pupation were analyzed using a right censored Cox's proportional hazard model with "symbiont", "larval weight at day 8", and their interaction term as factors after testing the assumption of proportional hazards using cox.zph of the survival package in R [33]. All three factors were retained in the final model based on the delta AICc score between the two models (Δ AICc <2). In the case of ties, Breslow likelihood was used. In addition to the survival analysis, several correlations were run to determine the biological relevance of larval weight at day 8. Specifically, we tested for correlations between (1) larval weight at day 8 and the day an individual died as a larvae and (2) larval weight at day 8 and pupation day for those that successfully pupated. We ran an ANOVA and Tukey–Kramer HSD tests to determine whether aphid symbiont significantly affected larval weight at day 8.

Survival from pupation to adult emergence (eclosion) was analyzed using a generalized linear model with a logit linked binomial distribution run on a subset of the 232 individuals that successfully pupated. Once individuals reach pupation, it is difficult to determine whether they are alive or not. Therefore pupa were determined to have died when they did not emerge as adults after 2 weeks; pupa were recorded as either successfully emerging as adults (=1) or as being dead (=0). Based on the delta AICc scores, "symbiont" and "larval weight at day 8" were included as factors in this analysis (Δ AICc >2). Post-hoc pairwise contrast analyses were performed to determine whether observed differences in pupal survival between aphid sub-clonal lines were significant.

For those individuals surviving to pupation and to adulthood, general linear models (standardized least square) were used to determine the effect of "symbiont" and "larval weight at day 8" on lady beetle development time, both from hatching to pupation and from pupation to adult emergence separately. The interaction term between the two factors was removed from both final models based on the Δ AICc score between the two models being greater than 2. We used ANOVA's and post hoc Tukey–Kramer HSD tests to determine which groups were significantly different from one another. Correlations were also done to assess whether there were significant correlations between larval weight and development times.

For analyses of adult weight, male and female beetles were analyzed separately. Adult weights at emergence were standardized to have a mean of zero and a standard deviation of one for males and females separately. To determine the effect of aphid facultative symbionts on male adult weight, general linear models (standard least square) were run with "symbiont", "larval weight at day 8", and the interaction term as factors. For female adult weight the same models were used, but the interaction term was removed from the final model based on the Δ AICc score between the two models being greater than 2. We used ANOVA and post hoc Tukey–Kramer HSD tests to determine which groups were significantly different from one another. Correlations were also done to assess whether there were significant relationships between larval weight and adult weight in both sexes. Additionally, we tested whether the sex ratio of emerging adults was significantly different from the expected 0.50 probability for each sex using a two-sided Chi square test. All statistical analyses were performed in JMP® Pro 13.0.

Results

We found a significant effect of aphid symbiont in the diet of *Ha. axyridis* larvae on survival from hatching to adult emergence. Both aphid symbionts present in the larval diet of these beetles and their weight at day 8 significantly affected overall survival to adult emergence (Fig. 1; Table 1). In the post hoc pairwise contrast analyses, we found that individuals that were fed aphids harboring the *Serratia* symbiont and the *Regiella* symbiont were more likely to die prior to reaching adult emergence than those fed aphids with the *Hamiltonella* symbiont or those fed aphids without symbionts (*Serratia*/*Hamiltonella*: $\chi^2 = 9.21$, $p = 0.002$; *Serratia*/symbiont free: $\chi^2 = 9.85$, $p = 0.002$; *Regiella*/*Hamiltonella*: $\chi^2 = 8.14$, $p = 0.004$; *Regiella*/Symbiont free: $\chi^2 = 11.03$, $p < 0.0001$). There was no significant difference in survival between larvae fed aphids with *Hamiltonella* and those without symbionts ($\chi^2 = 0.007$, $p = 0.93$), nor was there a difference in survival rates between those fed aphids with *Regiella* and *Serratia* ($\chi^2 = 0.005$, $p = 0.95$).

Feeding *Ha. axyridis* larvae a diet of aphids with the facultative symbionts *Serratia* (5AR) and *Regiella* (5AU) significantly lowered larval survival to pupation when compared to those larvae fed symbiont-free aphids (Fig. 2; Table 1). Larvae fed aphids with *Serratia* were 9.32 times more likely to die before reaching pupation than those fed symbiont free aphids ($p = 0.013$), and larvae fed aphids with *Regiella* were 7.47 times more likely to die as larvae than larvae fed aphids without facultative symbionts ($p = 0.022$). While not statistically significant, larvae fed *Serratia* were 4.46 times more likely to die as larvae as those fed aphids with *Hamiltonella*

($p = 0.06$), and larvae fed *Regiella* were 3.57 times more likely to die as larvae as those fed aphids with *Hamiltonella* ($p = 0.10$). Larval weight at day 8 was included as a factor in the Cox's proportional hazard model due to the results of our correlations, as well as the difference in the AICc scores between the two models (Δ AICc >2). While larval weight at day 8 was not correlated with the day an individual died as a larvae ($r^2 = 0.03$, $p = 0.33$), it was significantly negatively correlated with pupation day for those individuals surviving to pupation ($r^2 = 0.22$, $p < 0.0001$). Additionally, larvae fed aphids with the *Serratia* symbiont weighed significantly more than those fed other aphids ($F_{3,256} = 9.54$, $p < 0.0001$, Tukey–Kramer HSD, $p < 0.0002$ for all comparisons to *Serratia*). However, when "symbiont" and larval weight were both added to the model, larval weight at developmental day 8 did not appear to have a significant effect on larval survival, nor was there a significant interaction between the type of aphid eaten and larval weight, suggesting that aphid symbiont is largely responsible for the differences in larval survival measured in this experiment.

For those individuals surviving to pupation, both the type of symbiont present in the larval diet ("symbiont") and larval weight significantly affected time spent as a larvae prior to pupation (Fig. 3; Table 1). Overall, there was a negative correlation between larval weight and the number of days spent as a larva prior to pupation, meaning individuals that weighed less at larval day 8 spent more time as larvae prior to pupation than those that were heavier at larval day 8 ($r^2 = 0.216$, $p < 0.0001$). Additionally, larvae fed aphids harboring *Regiella* (5AU) spent significantly longer as larvae than the other three experimental groups ($F_{3,223} = 11.26$, $p < 0.0001$, Tukey–Kramer

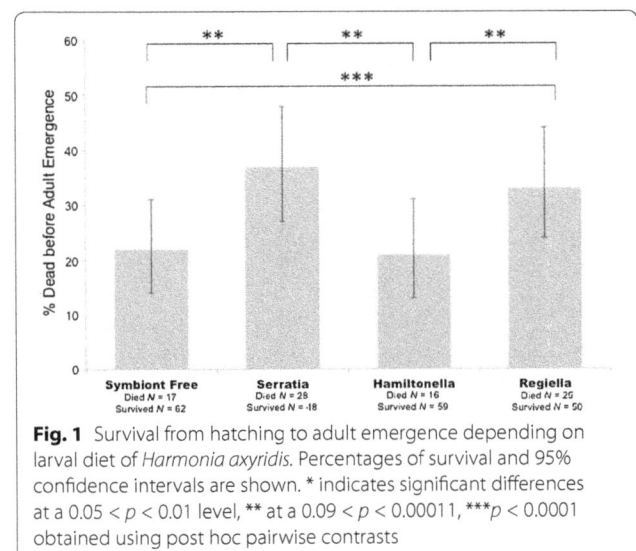

Fig. 1 Survival from hatching to adult emergence depending on larval diet of *Harmonia axyridis*. Percentages of survival and 95% confidence intervals are shown. * indicates significant differences at a $0.05 < p < 0.01$ level, ** at a $0.09 < p < 0.00011$, ***$p < 0.0001$ obtained using post hoc pairwise contrasts

Table 1 Summary of results of statistical models

Dependent variable	Name of statistical test	Statistic	Whole model	Symbiont	Weight at larval day 8	Symbiont × weight interaction
Overall survival from hatching to adult emergence	Generalized linear model (binomial distribution)	Likelihood ratio χ^2	28.10**	24.39***	5.50*	7.62
Larval survival	Cox's proportional hazards model	Likelihood ratio χ^2	13.36	8.80*	1.16	5.55
Pupal survival	Generalized linear model (binomial distribution)	Likelihood ratio χ^2	7.68	7.67*	0.11	0.50[a]
Larval development time	General linear model (least square)	$F_{4,221}$	26.85***	12.32***	67.60***	0.99[a]
Pupal development time	General linear model (least square)	$F_{4,212}$	1.70	1.20	2.81	1.18[a]
Female adult weight at emergence	General linear model (least square)	$F_{4,77}$	3.81**	4.95**	0.12	3.76[a]
Male adult weight at emergence	General linear model (least square)	$F_{7,87}$	4.35**	2.15	0.20	7.57***

Significant results for statistical analyses are indicated by asterisks

* $0.05 < p < 0.01$, ** $0.009 < p < 0.00011$, *** $p < 0.0001$

[a] Based on resulting AICc scores, non-significant interaction terms were removed from the final model when the Δ AICc >2. The results of the final model are reported for the other terms

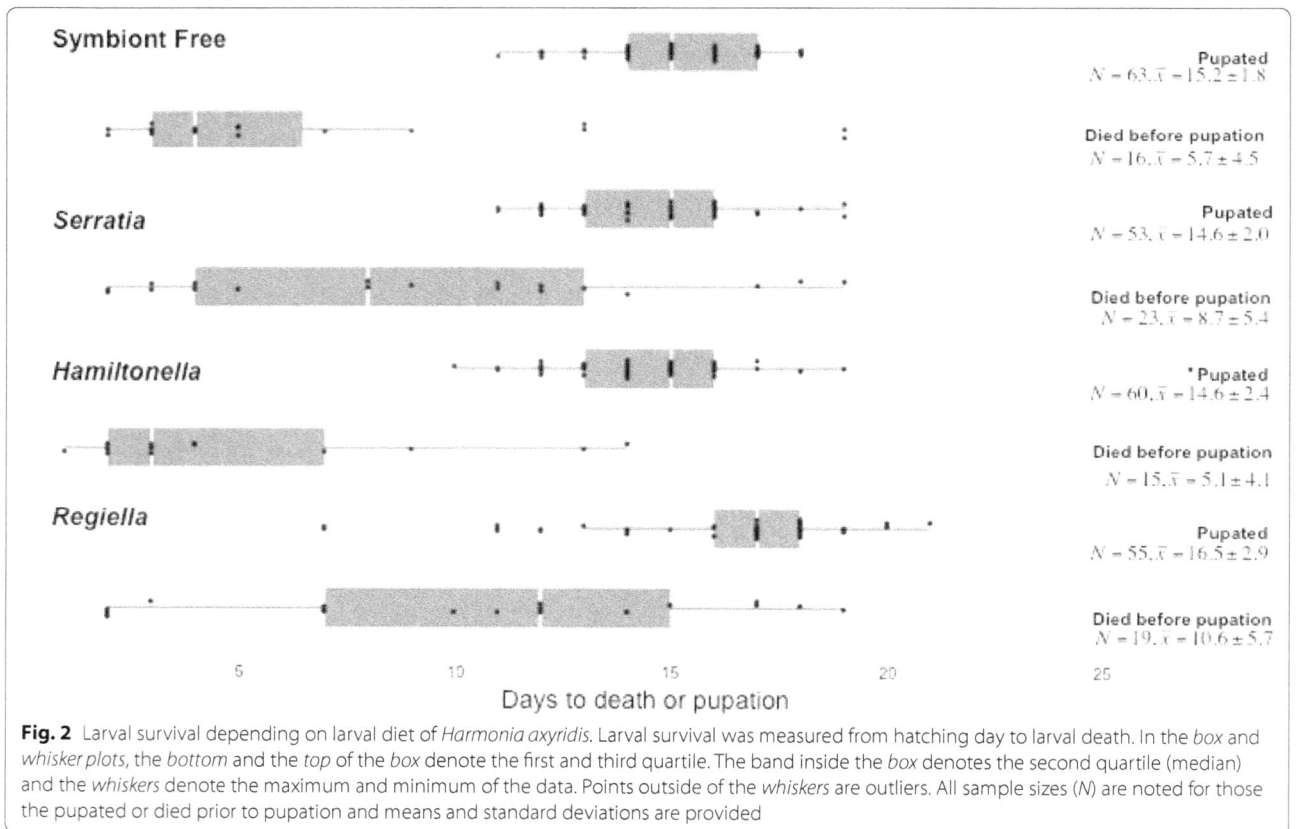

Fig. 2 Larval survival depending on larval diet of *Harmonia axyridis*. Larval survival was measured from hatching day to larval death. In the *box and whisker plots*, the *bottom* and the *top* of the *box* denote the first and third quartile. The band inside the *box* denotes the second quartile (median) and the *whiskers* denote the maximum and minimum of the data. Points outside of the *whiskers* are outliers. All sample sizes (*N*) are noted for those the pupated or died prior to pupation and means and standard deviations are provided

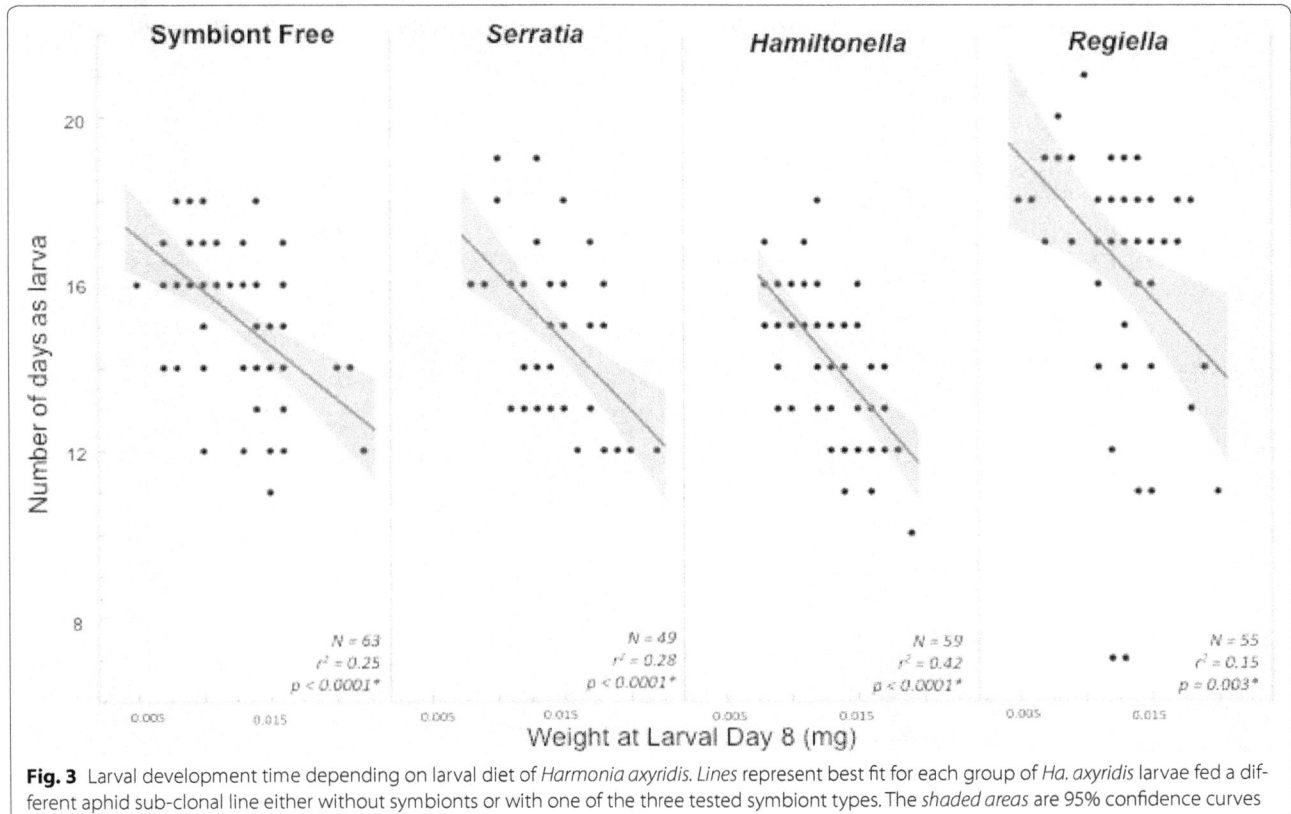

Fig. 3 Larval development time depending on larval diet of *Harmonia axyridis*. *Lines* represent best fit for each group of *Ha. axyridis* larvae fed a different aphid sub-clonal line either without symbionts or with one of the three tested symbiont types. The *shaded areas* are 95% confidence curves

HSD, *Regiella/Hamiltonella* $p < 0.0001$, *Regiella/Serratia* $p < 0.0001$, *Regiella*/symbiont free $p = 0.004$).

Individual survival from pupation to the adult stage was also affected by larval diet, though we note that overall pupal mortality was low ranging from 2 to 11% (Fig. 4). Larval weight did not significantly effect pupal survival, nor was there a significant interaction between the two effects (Fig. 4; Table 1). Post-hoc pairwise contrast analyses revealed that pupae that as larvae were fed aphids with *Regiella* were significantly more likely to survive to adult emergence than those fed aphids without symbionts (*Regiella*/symbiont free: $\chi^2 = 4.82$, $p = 0.03$). All other comparisons were not significantly different from one another ($p > 0.05$).

None of the measured variables affected the number of days an individual spent as a pupa prior to emergence as an adult, though the length of pupal development and larval weight at day 8 were significantly correlated for two of the symbiont groups (Fig. 5; Table 1). We saw no correlation between larval weight and the number of days spent as a pupa ($r^2 = 0.006$, $p = 0.24$). We also saw no effect of aphid symbiont type eaten on time to adult emergence ($F_{3,213} = 0.843$, $p = 0.47$). However, for individuals fed aphids without symbionts and those fed

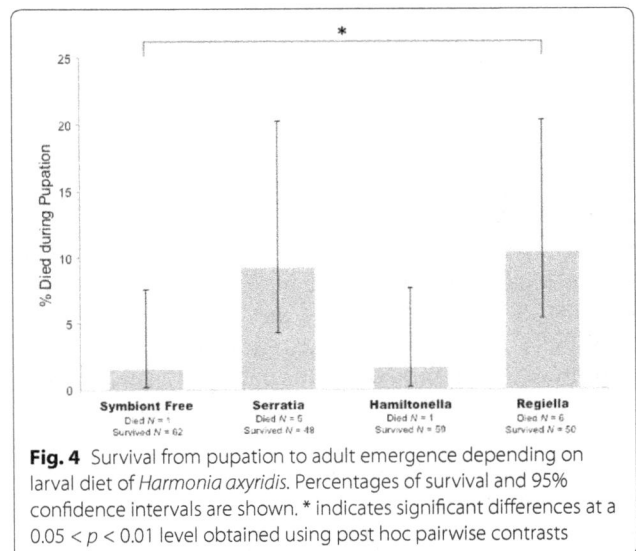

Fig. 4 Survival from pupation to adult emergence depending on larval diet of *Harmonia axyridis*. Percentages of survival and 95% confidence intervals are shown. * indicates significant differences at a $0.05 < p < 0.01$ level obtained using post hoc pairwise contrasts

aphids with the *Hamiltonella* symbiont there were significant correlations between larval weight at day 8 and the length of time spent as a pupa prior to adult emergence (Fig. 5; symbiont free: $r^2 = 0.10$, $p = 0.01$; *Hamiltonella*:

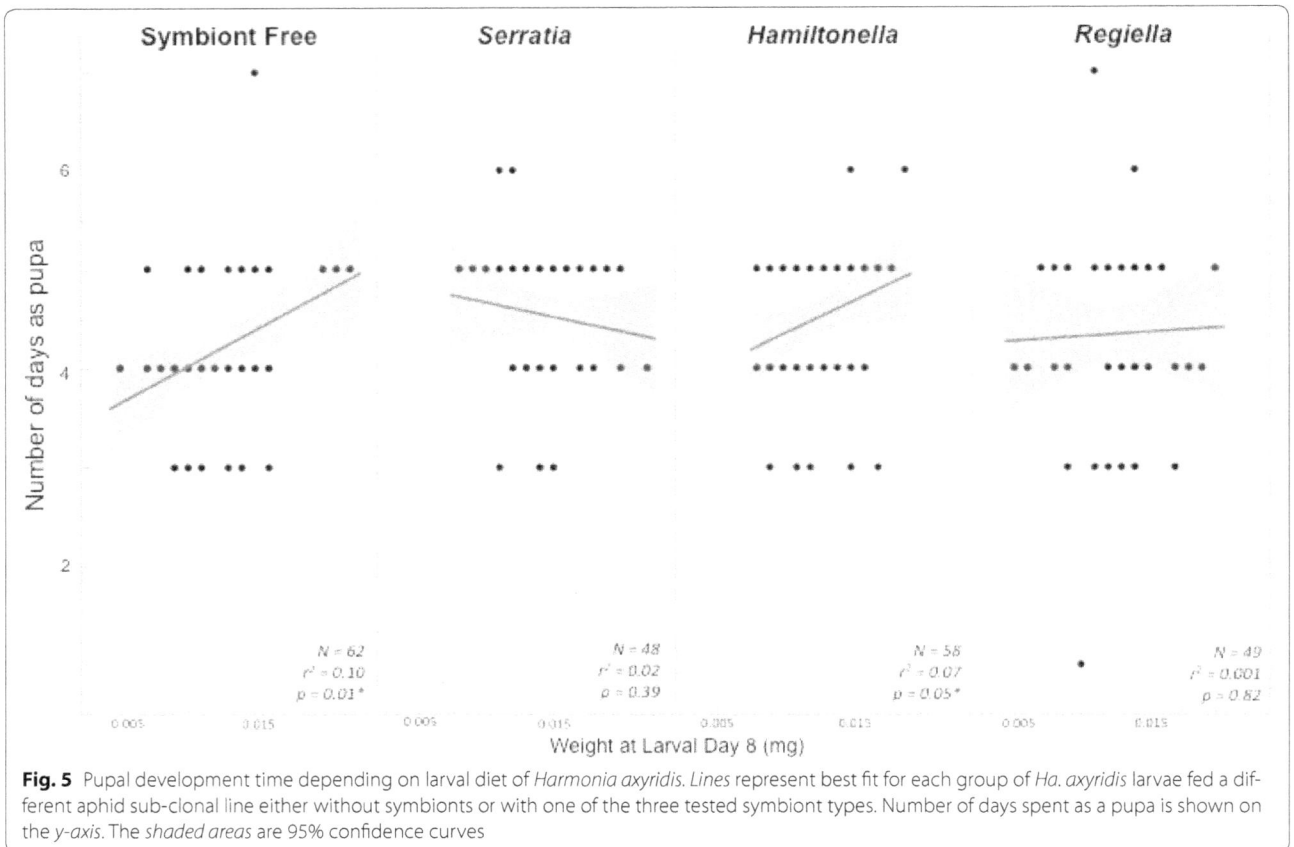

Fig. 5 Pupal development time depending on larval diet of *Harmonia axyridis*. *Lines* represent best fit for each group of *Ha. axyridis* larvae fed a different aphid sub-clonal line either without symbionts or with one of the three tested symbiont types. Number of days spent as a pupa is shown on the *y-axis*. The *shaded areas* are 95% confidence curves

$r^2 = 0.07$, $p = 0.05$). There were no correlations between larval weight and pupal development length for individuals fed aphids with the *Serratia* or *Regiella* symbionts (Fig. 5; *Serratia*: $r^2 = 0.02$, $p = 0.39$, *Regiella*: $r^2 = 0.001$, $p = 0.82$).

Aphid symbionts in the lady beetle larval diet ("symbiont") also affected weight at adult emergence for female lady beetles, but not males. Overall, larval weight was not correlated with adult weight at emergence in either sex (Fig. 6; Table 1). Females fed aphids with *Regiella* symbionts (5AU) weighed significantly less as adults than those fed aphids without symbionts or with *Hamiltonella* ($F_{3,78} = 5,14$, $p = 0.003$, Tukey–Kramer HSD, *Regiella*/*Hamiltonella*: $p = 0.003$, *Regiella*/symbiont free: $p = 0.036$). Additionally, in males, there was a significant interaction between larval weight and aphid symbiont on adult weight. This appears to be due to a negative correlation between larval weight and adult weight for male lady beetle larvae fed aphids harboring the *Hamiltonella* symbiont (5AT), while for the three other male lady beetle experimental groups there was no correlation, negative or positive, between larval weight and adult weight (Fig. 6; *Hamiltonella*:

$r^2 = 0.288$, $p = 0.002$, *Regiella*: $r^2 = 0.031$, $p = 0.532$, *Serratia*: $r^2 = 0.135$, $p = 0.071$, no symbiont: $r^2 = 0.087$, $p = 0.153$).

There was no evidence that larval aphid diet affected the survival of males and females differently. The resulting sex ratio of surviving adults was not significantly different from the expected 50/50 sex ratio for any of the experimental groups ([34], two-sided χ^2, $p > 0.05$ for all tests).

Discussion

Two of the three aphid symbionts used in this study significantly decreased both the larval and pupal survival of the aphid predator *Ha. axyridis*. Specifically, when larvae were fed aphids harboring either the *Serratia* or *Regiella* symbiont, they were significantly less likely to survive as larvae or to emerge from pupation as adults than those larvae that were fed aphids without symbionts or with the *Hamiltonella* symbiont. Additionally, *Ha. axyridis* larvae fed aphids with the *Regiella* symbiont had significantly longer larval development times than those fed aphids with no symbionts or the *Hamiltonella* or *Serratia* symbionts. Finally, adult females that had been fed

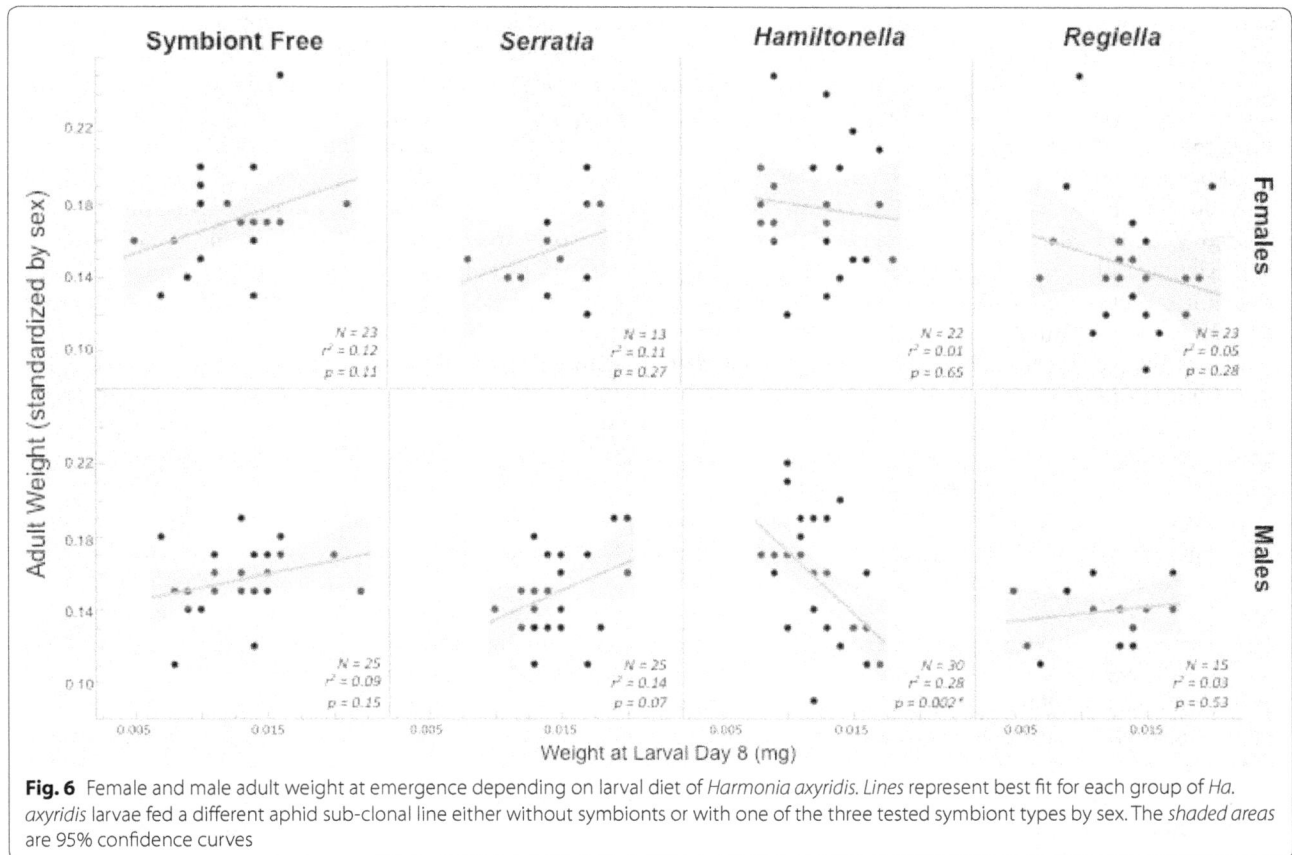

Fig. 6 Female and male adult weight at emergence depending on larval diet of *Harmonia axyridis*. *Lines* represent best fit for each group of *Ha. axyridis* larvae fed a different aphid sub-clonal line either without symbionts or with one of the three tested symbiont types by sex. The *shaded areas* are 95% confidence curves

Regiella-harboring aphids as larvae weighed significantly less than other adult female lady beetles.

These results are particularly striking due to the clonal nature of these asexual aphid lineages. All four symbiont types were established from the same clonal line (5A) and have been maintained under environmental lab conditions that ensure parthenogenetic reproduction since collection. While some mutations may have accumulated in the four sub-clonal lines since the injection of symbionts [35], the major difference between these sub-clonal aphid lines is the presence and type of symbiont. This strongly suggests that the presence of the *Regiella, Serratia*, and *Hamiltonella* aphid symbiont in the larval diet of *Ha. axyridis* are affecting lady beetle larval and pupal survival, rather than another factor associated with the aphid, like genotype.

Aphid facultative symbionts have been found to benefit their aphid hosts in a variety of ways, including protecting them from heat shock and from parasitism [1]. In nearly all of these cases, the protection directly benefits the individual aphid harboring the symbiont. In this experiment, aphid symbionts affect the fitness of the aphid predator, but only after the individual harboring

the symbiont has been eaten. In this scenario, the aphid symbionts are not providing direct protection to their hosts, but rather may be providing indirect protection to their clonal siblings harboring the same vertically transmitted facultative symbionts that are clustered nearby. Aphids reproduce parthenogenetically for the majority of their annual life cycle and in the summer months live in patches of genetically identical (or nearly identical) individuals [36]. Female lady beetles lay their eggs near aphid patches, and there is evidence to suggest that lady beetles do not disperse far during their larval and pupal stages [37]. With low dispersal of both the predator and the prey and high relatedness among clumps of aphids, we suggest that while the aphids that are eaten by lady beetles do not themselves benefit from their symbionts, by reducing the survival of local predators other aphids in their patch, clone-mates may receive indirect fitness benefits from their aphid symbionts [28, 38].

Other behaviors that may have indirect fitness effects have been observed among clusters of aphid clone-mates, and particular interest has been paid to the role of alarm pheromone in altruistic aphid defense behaviors [28, 39]. Droplets of aphid alarm pheromone that aphids smear on the predator only after being physically attacked, while not directly benefiting the individual emitting the alarm signal, may increase the likelihood that the highly related clone-mates around them escape predation [39]. Grain aphids that secrete alarm pheromones when being parasitized by wasps do not themselves seem to benefit from this action; it doesn't reduce parasitism. However, once smeared with the alarm pheromone, parasitoid wasps spend significantly more time grooming and less time ovipositing. The foraging efficiency of the parasitoid is also greatly reduced due to the defensive behaviors other aphids exhibit when exposed to the previously released alarm pheromone, such as dropping from the plant or kicking their legs [28]. All together, this can reduce parasitism within patches of clone-mates thereby providing indirect fitness benefits to the parasitized aphid. In all of these cases, lowering the predator/parasite foraging efficiency and efficacy results in a lower overall risk of predation/parasitism for a population of genetically identical aphids.

In a previous study, we found both the *Serratia* and *Hamiltonella* aphid symbionts negatively impacted the larval survival of another predatory lady beetle species, *Hi. convergens* [14]. Additionally, female *Hi. convergens* larvae fed aphids with *Serratia* or *Hamiltonella* symbionts weighed more as adults. While these experiments were not run simultaneously making direct comparisons difficult, our current findings do suggest a pattern in which the presence of aphid symbionts can significantly decrease the survival of aphid predators. It also appears that each symbiont may affect each predator differently. For example, female adult weight was significantly higher in *Hi. convergens* fed aphids with *Serratia* as larvae, but there was no difference in *Ha. axyridis* female adult weight between those fed aphids with *Serratia* and those fed symbiont free aphids. Additionally, while *Hamiltonella* negatively impacted the larval survival of *Hi. convergens*, there was no negative effect of the *Hamiltonella* symbiont on the survival of *Ha. axyridis*. Though these differences in the effects of several aphid symbionts in these two predatory lady beetle species may be due to other unmeasured differences in these two separate sets of experiments, they may also suggest variation in the mechanisms by which symbionts affect different aphid predators.

Other studies have found differences in aphid symbiont-conferred resistance to depend on a number of factors, including aphid host species, parasitoid species, other symbionts which may be present in the host, and bacteriophages associated with some symbionts [1, 10, 11, 19–23, 25, 26, 40–50]. For example, when the aphid symbiont *H. defensa* harbors a lysogenic lamdoid bacteriophage, the *Acyrthosiphon pisum* secondary

endosymbiont (APSE; [10]), it provides varying degrees of protection from some species of parasitoid wasps, but not others [10, 19, 24]. *Hamilitonella* with the APSE bacteriophage protects the pea aphid against parasitism by *Aphelinus ervi* and *Aphelinus abdominalis* wasps, but it doesn't appear to affect the survival of another pea aphid parasitoid wasp *Praon pequodorum*. This suggests that the protective toxin produced by the APSE bacteriophage does not provide general protection against all of *A. pisum*'s parasitoids [22]. Furthermore, several recent studies have demonstrated that *Hamiltonella*'s protection against parasitoids is not limited only to its mututalism with *A. pisum*. A second species of aphid, the cowpea aphid, *Aphis craccivora* is also protected from parasitism by two parasitoid wasps, *Binodoxys communis* and *Binodoxys koreanus* when it harbors *Hamiltonella* but were just as likely to be parasitized by *Lysiphelbus orientalis* and *Aphidius colemani* as symbiont free cowpea aphids. This suggests a genus-level specificity of protection in this system [44]. Interestingly, in a third aphid species, the grain aphid *Sitobion avenae*, individuals harboring *Hamiltonella* are equally susceptible to parasitism by *A. ervi* and *Ephedrus plagiator* as those without the symbiont, however, female wasps from both parasitoid species prefer to lay their eggs in symbiont-free aphids when given the choice. In this case, while *Hamiltonella* may not increase resistance against parasitism, it is still protecting its aphid host by making it less attractive to multiple parasitoid species [20]. Taken together, these studies suggest that even a single symbiont, in this case *Hamiltonella*, does not provide aphids with general protection against all parasitoid wasps. Our current study is the second to find negative fitness effects of aphid symbionts on an aphidophagous lady beetle. We have now documented this in two beetles in the family Coccinellidae. Future studies should be done to determine how general these fitness effects are across other aphidophagous predators such as lacewings or predatory midges.

In conclusion, our current study finds that two aphid secondary symbionts can significantly impact the survival of the predatory lady beetle *Ha. axyridis* during the larval and pupal development periods. They can also increase the length of development time and influence adult weight. All of these effects can have significant fitness consequences for aphid predators, which may in turn significantly affect the fitness of their aphid prey. During the summer months, parthenogenetically reproducing aphids live in clusters of highly related clonemates with low rates of dispersal. Under these conditions we could expect lower predator survival to result in a lower overall risk of predation for a population of genetically identical aphids [28]. And while the fitness effects are similar to those found in a previous study in another species of lady beetle, there does appear to be some variation in how each symbiont affects each predator species. Finally, these findings demonstrate the far-reaching effects of symbionts beyond just host fitness and survival, and suggest that symbionts have important ecological impacts along multiple food chains and across food webs.

Abbreviations

5AU: aphids from the 5A clonal line that harbor the *Regiella insecticola* facultative symbiont; 5AT: aphids from the 5A clonal line that harbor the *Hamiltonella defensa* facultative symbiont; 5AO: aphids from the 5A clonal line that do not harbor any facultative symbionts; 5AR: aphids from the 5A clonal line that harbor the *Serratia symbiotica* facultative symbiont.

Authors' contributions

JK analyzed the data, prepared the figures, wrote the manuscript, oversaw the experimental design and animal rearing, and collected the data. CW designed the experiment, headed animal rearing and the data collection effort, and completed data entry. SW helped with data collection and with animal rearing. DV helped with data entry and follow-up work. All authors read and approved the final manuscript.

Author details

[1] Spelman College, 350 Spelman Lane, S.W., Atlanta, GA 30314, USA. [2] University of Vermont, Larner College of Medicine, 89 Beaumont Ave, Burlington, VT 05405, USA. [3] Neuroscience Institute, Georgia State University, Petit Science Center, Atlanta, GA 30303, USA.

Acknowledgements

The authors would like to thank N. Gerardo, S. Birnbaum, and T. Acevedo for their extremely generous assistance with this project.

Competing interests

The authors declare that they have no competing interests.

Funding

JK was supported by start-up funds by Spelman College. CW was supported by NSF HBCU-UP ASPIRE Project Award # 0714553 through Spelman College. DV was supported by HHMI's SMART Scholars program at Spelman College.

References

1. Oliver KM, Degnan PH, Burke GR, Moran NA. Facultative symbionts in aphids and the horizontal transfer of ecologically important traits. Annu Rev Entomol. 2010;55:247–66.

2. Jaenike J, Unckless R, Cockburn SN, Boelio LM, Perlman SJ. Adaptation via symbiosis: recent spread of a *Drosophila* defensive symbiont. Science. 2010;329:212–5.

3. Hosokawa T, Kikuchi Y, Shimada M, Fukatsu T. Obligate symbiont involved in pest status of host insect. Proc Biol Sci. 2007;274:1979–84.

4. Hedges LM, Brownlie JC, O'Neill SL, Johnson KN. *Wolbachia* and virus protection in insects. Science. 2008;322:702.

5. Kellner R. Stadium specific transmission of endosymbionts needed for pederin biosynthesis in three species of *Paederus* rove beetles. Entomol Exp Appl. 2003;107:115–24.

6. Teixeira L, Ferreira A, Ashburner M. The bacterial symbiont *Wolbachia* induces resistance to RNA viral infections in *Drosophila melanogaster*. PLoS Biol. 2008;6:e2.

7. Kambris Z, Blagborough AM, Pinto SB, Blagrove MSC, Godfray HCJ, Sinden RE, et al. *Wolbachia* stimulates immune gene expression and inhibits plasmodium development in *Anopheles gambiae*. PLoS Pathog. 2010;6:e1001143.

8. Scarborough CL, Ferrari J, Godfray HCJ. Aphid protected from pathogen by endosymbiont. Science. 2005;310:1781.

9. Łukasik P, Guo H, van Asch M, Ferrari J, Godfray HCJ. Protection against a fungal pathogen conferred by the aphid facultative endosymbionts *Rickettsia* and *Spiroplasma* is expressed in multiple host genotypes and species and is not influenced by co-infection with another symbiont. J Evol Biol. 2013;26:2654–61.

10. Oliver KM, Degnan PH, Hunter MS, Moran NA. Bacteriophages encode factors required for protection in a symbiotic mutualism. Science. 2009;325:992–4.

11. Vorburger C, Gehrer L, Rodriguez P. A strain of the bacterial symbiont *Regiella insecticola* protects aphids against parasitoids. Biol Lett. 2010;6:109–11.

12. Xie J, Tiner B, Vilchez I, Mateos M. Effect of the *Drosophila* endosymbiont *Spiroplasma* on parasitoid wasp development and on the reproductive fitness of wasp-attacked fly survivors. Evol Ecol. 2011;53:1065–79.

13. Rothacher L, Ferrer-Suay M, Vorburger C. Bacterial endosymbionts protect aphids in the field and alter parasitoid community composition. Ecology. 2016;97:1712–23.

14. Costopoulos K, Kovacs JL, Kamins A, Gerardo NM. Aphid facultative symbionts reduce survival of the predatory lady beetle *Hippodamia convergens*. BMC Ecol. 2014;14:5.

15. Soldati M, Fioretti A, Ghione M. Cytotoxicity of pederin and some of its derivatives on cultured mammalian cells. Experientia. 1966;22:176–8.

16. Tiboni O, Parisi B, Ciferri O. The mode of action of pederin, a drug inhibiting protein synthesis in eucaryotic organisms. Plant Biosyst. 1968;102:337–45.

17. Bose SK, Mehta PR. A new species of *Entomophthora* on beetles. Trans Br Mycol Soc. 1953;36:52–6.

18. Kellner RLL, Dettner K. Differential efficacy of toxic pederin in deterring potential arthropod predators of *Paederus* (Coleoptera: Staphylinidae) offspring. Oecologia. 1996;107:293–300.

19. McLean AHC, Godfray HCJ. Evidence for specificity in symbiont-conferred protection against parasitoids. Proc R Soc B. 2015;282:20150977.

20. Łukasik P, Dawid MA, Ferrari J, Godfray HCJ. The diversity and fitness effects of infection with facultative endosymbionts in the grain aphid, *Sitobion avenae*. Oecologia. 2013;173:985–96.

21. Cayetano L, Rothacher L, Simon J-C, Vorburger C. Cheaper is not always worse: strongly protective isolates of a defensive symbiont are less costly to the aphid host. Proc Biol Sci. 2015;282:20142333.

22. Martinez AJ, Kim KL, Harmon JP, Oliver KM. Specificity of multi-modal aphid defenses against two rival parasitoids. PLoS ONE. 2016;11:e0154670.

23. Cayetano L, Vorburger C. Symbiont-conferred protection against Hymenopteran parasitoids in aphids: how general is it? Ecol Entomol. 2015;40:85–93.

24. Degnan PH, Moran NA. Diverse phage-encoded toxins in a protective insect endosymbiont. Appl Environ Microbiol. 2008;74:6782–91.

25. Oliver KM, Noge K, Huang EM, Campos JM, Becerra JX, Hunter MS. Parasitic wasp responses to symbiont-based defense in aphids. BMC Biol. 2012;10:11.

26. Oliver KM, Martinez AJ. How resident microbes modulate ecologically-important traits of insects. Curr Opin Insect Sci. 2014;4:1–7.

27. Loxdale HD, Lushai G. Maintenance of aphid clonal lineages: images of immortality? Infect Genet Evol. 2003;3:259–69.

28. Wu G-M, Boivin G, Brodeur J, Giraldeau L-A, Outreman Y. Altruistic defence behaviours in aphids. BMC Evol Biol. 2010;10:19.

29. McLean AHC, Parker BJ, Hrček J, Henry LM, Godfray HCJ. Insect symbionts in food webs. Philos Trans R Soc Lond B Biol Sci. 2016;371:20150325.

30. Clay K. Defensive symbiosis: a microbial perspective. Funct Ecol. 2014;28:293–8.

31. Oliver KM, Russell JA, Moran NA, Hunter MS. Facultative bacterial symbionts in aphids confer resistance to parasitic wasps. Proc Natl Acad Sci USA. 2003;100:1803–7.

32. Laughton AM, Garcia JR, Gerardo NM. Condition-dependent alteration of cellular immunity by secondary symbionts in the pea aphid, *Acyrthosiphon pisum*. J Insect Physiol. 2016;86:17–24.

33. Edition S. Cox proportional-hazards regression for survival data in R. http://socserv.socsci.mcmaster.ca/jfox/Books/Companion/appendix/Appendix-Cox-Regression.pdf.

34. Hodek I, Honek A, van Emden HF. Ecology and behaviour of the ladybird beetles (Coccinellidae). New York: Wiley; 2012.

35. Dunbar HE, Wilson ACC, Ferguson NR, Moran NA. Aphid thermal tolerance is governed by a point mutation in bacterial symbionts. PLoS Biol. 2007;5:e96.

36. Loxdale HD. The nature and reality of the aphid clone: genetic variation, adaptation and evolution. Agric For Entomol. 2008;10:81–90.

37. Hironori Y, Katsuhiro S. Cannibalism and interspecific predation in two predatory ladybirds in relation to prey abundance in the field. Entomophaga. 1997;42:153–63.

38. Abbot P. On the evolution of dispersal and altruism in aphids. Evolution. 2009;63:2687–96.

39. Muratori FB, Rouyar A, Hance T. Clonal variation in aggregation and defensive behavior in pea aphids. Behav Ecol. 2014;25:901–8.

40. Vorburger C, Sandrock C, Gouskov A, Castañeda LE, Ferrari J. Genotypic variation and the role of defensive endosymbionts in an all-parthenogenetic host-parasitoid interaction. Evolution. 2009;63:1439–50.

41. Schmid M, Sieber R, Zimmermann Y-S, Vorburger C. Development, specificity and sublethal effects of symbiont-conferred resistance to parasitoids in aphids. Funct Ecol. 2012;26:207–15.

42. Vorburger C. The evolutionary ecology of symbiont conferred resistance to parasitoids in aphids. Insect Sci. 2014;21(3):251–64.

43. Martinez AJ, Weldon SR, Oliver KM. Effects of parasitism on aphid nutritional and protective symbioses. Mol Ecol. 2014;23:1594–607.

44. Asplen MK, Bano N, Brady CM, Desneux N, Hopper KR, Malouines C, et al. Specialisation of bacterial endosymbionts that protect aphids from parasitoids. Ecol Entomol. 2014;39:736–9.

45. Martinez AJ, Ritter SG, Doremus MR, Russell JA, Oliver KM. Aphid-encoded variability in susceptibility to a parasitoid. BMC Evol Biol. 2014;14:127.

46. Weldon SR, Oliver KM. Diverse bacteriophage roles in an aphid-bacterial defensive mutualism. In: Hurst CJ, editor. The mechanistic benefits of microbial symbionts. Berlin: Springer International Publishing; 2016. p. 173–206.

47. Oliver KM, Campos J, Moran NA, Hunter MS. Population dynamics of defensive symbionts in aphids. Proc Biol Sci. 2008;275:293–9.

48. Smith AH, Łukasik P, O'Connor MP, Lee A, Mayo G, Drott MT, et al. Patterns, causes and consequences of defensive microbiome dynamics across multiple scales. Mol Ecol. 2015;24:1135–49.

49. Łukasik P, Guo H, Asch M, Ferrari J. Protection against a fungal pathogen conferred by the aphid facultative endosymbionts *Rickettsia* and *Spiroplasma* is expressed in multiple host genotypes. J Evol Biol. 2013;26(12):2654–61.

50. Łukasik P, van Asch M, Guo H, Ferrari J, Godfray HCJ. Unrelated facultative endosymbionts protect aphids against a fungal pathogen. Ecol Lett. 2013;16:214–8.

13

Differentiation and description of aromatic short grain rice landraces of eastern Indian state of Odisha based on qualitative phenotypic descriptors

Pritesh Sundar Roy[1], Rashmita Samal[1], Gundimeda Jwala Narasimha Rao[1], Sasank Sekhar Chyau Patnaik[1], Nitiprasad Namdeorao Jambhulkar[1], Ashok Patnaik[1] and Trilochan Mohapatra[1,2*]

Abstract

Background: Speciality rice, in general, and aromatic rice in particular, possess enormous market potential for enhancing farm profits. However, systematic characterization of the diversity present in this natural wealth is a major pre requisite for using it in the breeding programs. This study reports qualitative phenotypic trait based characterization of 126 short grain aromatic rice genotypes, collected from different areas of the state of Odisha, India.

Results: Out of the 24 descriptors employed, highest variability (8 different types) was observed for lemma-palea colour with a genetic diversity index (He) of 0.696. The principal component analysis reveals that the tip colour of lemma, colour of awn and colour of stigma, cumulatively explain 74 % of the total variation. The Population STRUCTURE analysis classified the population into two subpopulations which were subdivided further into four distinct groups. The western and southern districts of Odisha are endowed with maximum diversity in comparison to eastern and northern districts and at district level comparisons, Koraput and Puri districts are rich with a genetic diversity values of 0.324 and 0.303 respectively. With this set of morphological qualitative traits, based on 'phenoprinting', a newly proposed bar coding system, unique fingerprints of each genotype can be effectively generated that can help in easy identification of these genotypes.

Conclusion: Though aromatic rices represent a tiny fraction of the total rice germplasm, a small collection of 126 land races did exhibit rich diversity for all the qualitative traits. For lemma-palea colour, eight different types were detected while for tip colour of lemma, six different types were recorded, suggesting the presence of rich variability in short grain aromatic rices that are conserved in this region. The proposed 'phenoprinting' can be an effective descriptor with the unique finger prints generated for each genotype and coupled with molecular (DNA) finger printing, we can discriminate and identify each and every aromatic short grain rice genotype. The proposed system not only help in conservation but also can confer IPR protection to these specialty rices.

Keywords: Rice, Landraces, Aromatic short grain, Phenotypic, Trait, Characterization, Diversity

Background

A large and diverse set of aromatic rice landraces are being maintained in different gene banks of India, which have been assembled by various explorations spread over several decades. Many accessions of these accessions have not yet been characterized [1]. Since landraces are thought to be the intermediate group between the wild ancestors and cultivated varieties [2], characterization of this wealth is of relevance for isolation of desirable genotypes as donors of useful traits. India has rich genetic diversity of aromatic rice landraces, majority of them having small to medium grains [3–6]. Aromatic rices

*Correspondence: tmnrcpb@gmail.com
[2] Present Address: ICAR, DARE, New Delhi 110001, India
Full list of author information is available at the end of the article

constitute a small but an important sub-group (Group V) identified by Glaszmann [7] with a set of 15 isozyme markers. Other than Basmati, the indigenous short grain aromatic rices are grown in localized pockets in almost all states of the country [8, 9]. These rices possessing unique aroma, cooking and eating qualities are consumed locally as a delicacy [10, 11]. The aromatic rice varieties have evolved over thousands of years in nature and in the small farms of local farmers who select different types to suit their local cultivation practices and needs [12].

Odisha, an ancient state of India, is one of the major producer and consumer of aromatic rices. The state has its own set of aromatic short grain rices, which are being cultivated in almost all districts in different agro-climatic zones [6, 13]. These specialty rices are treated as sacred among people. Many of them like Deulabhog, Durgabhog, Krishnabhog, Prasadbhog etc. are being used in several temple duties [14]. The aromatic short grain rice landraces enriched in cooking and eating quality features hold enormous potential to be utilized as value added products, thereby providing higher economic return to small and marginal farmers [15]. Further, these native short grain aromatic rices hold significance over traditional Basmati rice for their varied and intense aroma with long retention in relatively warmer climate [3]. Characterization of these landraces is a prerequisite for understanding the extent of diversity, identification of valuable traits required for aromatic rice improvement and defining the conservation needs. Jeypore tract of Odisha is known for its biodiversity as secondary centre for origin of rice [16, 17]. Characterization of aromatic rice genotypes from Jeypore and its nearby districts can provide insights into the origin and spread of these rices. Since, rice is a self-pollinated crop and morphological description of rice plant is well established, its phenotypic descriptors could serve as potential marker for varietal identification and differentiation. Phenotypic characterization of the aromatic landraces would be a primary but essential measure leading to cataloguing of these unique high quality and diverse group, thereby strengthening their conservation and utilization strategies.

Despite that Odisha holds significance from the view point of origin and spread of rice, the status of aromatic short grain rice genotypes of this Indian state has not been studied extensively. The present study investigates structure and diversity among the aromatic short grain rice genotypes (ASGs) collected from different districts of Odisha.

Methods
Plant material collection and characterization
One hundred and twenty-six short grain aromatic rice lines (grains/panicles) were collected by germplasm experts and plant breeders of Central Rice Research Institute (CRRI), Cuttack from different parts of Odisha, India (Additional file 1: Table S1) during an exploration and collection programme between 2005 and 2009 conducted by the Institute. Further, the genotypes were grouped under 19 geographical districts based on their area of collection (Additional file 2: Figure S1). The genotypes were transplanted in randomised complete block design in different experimental plots of CRRI and the genetic purity of each landrace was maintained by periodic removal of 'off types' in four purification cycles. The experiment was conducted under controlled environmental condition with recommended dosage of fertilizers to minimize any environmental error. The purified rice genotypes were grown in 2 m × 2 m plot, from which 10 hills were sampled for characterization. Twenty-four phenotypic characters were measured at different growth stages following the guidelines of International Rice Research Institute for three consecutive years i.e. 2013–2015 [18].

Data analysis
Data for 24 qualitative phenotypic traits were recorded on 10 random plants per accession for 3 years and their means were calculated for analysis (Table 1). The qualitative phenotypic traits were coded for the presence (1) and absence (0) of each alternate form. Diversity parameters viz., number of alleles/variables (Na), effective number of alleles/variables (Ne), Shannon Index (I) and unbiased Nei's genetic diversity index (He) [19] were calculated using POPGENE v 1.32 [20] with 1000 permutations. The neighbour joining dendrogram based on Nei's unbiased pairwise genetic distances among genotypes was constructed in MEGA 6 [21]. District wise genetic diversity measures were calculated using POPGENE v 1.32 and the neighbour joining tree was constructed with pairwise genetic distance measures between districts by MEGA 6. The molecular variance was analysed by hierarchical AMOVA implemented in the software GeneAlEx6 [22] based on unbiased Fst estimator as described by Weir and Cockerham [23] to estimate genetic variation among and within districts. Significance of the F-statistics was tested by non-differentiation of probability for 10,000 randomizations. The Bayesian model-based clustering analysis of the genotypes was used for determining the optimal number of genetic clusters found among rice varieties by the software STRUCTURE [24] using admix approach with 100,000 burn-in (iteration) periods followed by 100,000 Markov Chain Monte Carlo (MCMC) replicates with ten independent runs (K) ranging from 1 to 10. The ΔK based on the change in the log probability of the data between successive K values was estimated using the parameters described by Evanno and colleagues [25] with Structure Harvester v6.0 [26] and the inferred

Table 1 Phenotypic traits used in the present study for diversity analysis of 126 aromatic short grain rice genotypes

Trait name	Variable	Trait name	Variable
Basal leaf sheath colour BLS	Green (1)	Leaf: shape of ligule LSL	Truncate (36)
	Light purple (2)		Acute (37)
	Purple lines (3)		Split (38)
	Purple (4)	Leaf: auricles LAR	Absent (39)
Flag leaf attitude of blade FLA	Erect (5)		Present (40)
	Semi erect (6)	Spikelet: colour of stigma CSTG	White (41)
	Horizontal (7)		Yellow (42)
	Deflexed (8)		Light purple (43)
Culm angle CA	Erect (9)		Purple (44)
	Semi erect (10)	Panicle: attitude of branches PAB	Erect (45)
Stem thickness ST	Week (11)		Semierect (46)
	Medium (12)		Semierect to spreading (47)
	Strong (13)		Spreading (48)
Panicle: awns PA	Absent (14)	Collar colour CC	Pale green (49)
	Present (15)	Internode colour INC	Green (50)
Panicle: colour of awns PCA	Yellowish white (16)		Light gold (51)
	Light red (17)		Purple line (52)
	Purple (18)	Panicle secondary branching PSB	Weak (53)
	Black (19)		Strong (54)
Panicle: curvature of main axis PMA	Straight (20)	Leaf angle LAG	Erect (55)
	Dropping (21)		Semi erect (56)
Spikelet: tip colour of lemma SCL	White (22)		Horizontal (57)
	Yellowish (23)		Droopy (58)
	Brown (24)	Lemma palea colour LPC	Straw (59)
	Red (25)		Brown Furrow (60)
	Purple (26)		Brown (61)
	Black (27)		Red (62)
Panicle exertion PE	Partly exserted (28)		Purple (63)
	Exserted (29)		Purple Furrow (64)
	Well exserted (30)		Black (65)
Leaf intensity of blade colour LI	Medium (31)		Brown Spot (66)
Leaf: anthocyanin colour LA	Absent (32)	Grain type GT	Short bold (67)
Leaf: pubescence of blade surface LPB	Medium (33)		Medium bold (68)
Leaf: ligule LL	Present (34)		Medium slender (69)
Leaf: colour of ligule CL	Green (35)		Medium long (70)

Variables only recorded in the present set of germplasm are given. Numbers in parenthesis in the last column represents order of variable for individual trait in the phenoprint of Fig. 3b

population clusters were produced by Structure Plot [27] (http://www.btismysore.in/strplot). Individuals with probability of membership >80 % were assigned to the same group, while those with <80 % probability membership in any single group were assigned to a admixed group [28]. We used a graphic presentation method based on alternate forms for the morphological descriptors, which we describe as 'Phenoprint'. The Phenoprint was developed using Microsoft Excel 2013, in which the binary data were represented as bar(s); black bar for the presence of the variable and grey bar for the absence to provide uniqueness for each of the genotype. Different morphological characters that contributed most to the observed phenotypic variance were identified by principal component analysis (PCA) in the software SAS 9.2 [29]. The subset regression was used to identify different principal components that contributed to the total variance in the dataset.

Results

Phenotypic trait diversity

Eighteen (75 %) out of 24 phenotypic traits showed variation among the 126 indigenous short grain aromatic rice genotypes, whereas no variation was observed for the traits like intensity of blade colour (LI), leaf anthocyanin colour (LA), pubescence of blade surface (LPB), leaf ligule (LL), colour of ligule (CL) and collar colour (CC) (Table 1). Phenotypic variation for some of the traits are given in Fig. 1. The number of variables per individual trait ranged from 1 (traits having no variation) to a maximum of 8 for lemma-palea colour with a total number of 70 variables with an average of 2.92 per trait. The genetic diversity index (*He*) for different traits ranged from 0.016 (lowest) to 0.696 (highest) for culm angle and lemma-palea colour, respectively with an average of 0.286. The number of effective variables (*Ne*) ranged from 1 to 3.292 with an average of 2.958 (Table 2). Further, frequency distribution of the phenotypic traits against the studied genotypes indicated unequal distribution of variables under each trait (Additional file 2: Figure S2). Among the 126 genotypes, presence of auricles (99 %), semi erect culm angle (99 %), split ligule (95 %), absence of awn (91 %), drooping type panicles (91 %), semi erect attitude of panicle branches (86 %), well exserted panicle (76 %), white colour stigma (74 %), light gold colour internodes (71 %), yellowish white colour of awns (64 %), branching of panicles weak (63 %), medium stem thickness (55 %), green basal leaf sheath colour (51 %), lemma having purple tip (45 %) and erect flag leaf attitude of blade (44 %) were predominant. A comparatively equal frequency of distribution for short bold (47 %) and medium bold (51 %) grain type were recorded for the genotypes. In case of lemma-palea colour that showed maximum variation with eight alternative variables, straw colour was most predominant (50 %) followed by purple furrow (14 %), brown (11 %), purple and brown furrow (10 % each), black and red (2 % each) and brown spot (1 %).

The principal component analysis (PCA) showed that 3 of the 24 morphological traits were the most important components for explaining the grouping of genotypes (Fig. 2). Principal component analysis showed that, first, second, third, fourth, and fifth principal components accounted for 46, 20, 7, 5 and 4 % of the total variance, respectively. The first three principal components together explained a cumulative 74 % of the total variance. Further, the subset regression analysis could identify tip colour of lemma, colour of awn and colour of stigma as the three most important traits. Grain type, flag leaf attitude of blade and lemma-palea colour were identified to be secondary traits to explain the maximum available phenotypic variability within the set of studied genotypes. However, ten (tip colour of lemma, colour of awn, colour of stigma, grain type, flag leaf attitude, lemma palea colour, attitude of panicle branches, basal leaf sheath colour, leaf angle and internode colour) out of 24 morphological traits could adequately explain the total variation existing in these short grain aromatic rice genotypes (Table 2).

Genetic relationship and population structure

The pairwise Nei's genetic distance was calculated to understand the genetic diversity and relatedness among the 126 short grain aromatic rice genotypes, which ranged from 0.426 (lowest) to 0.875 (highest) with an average of 0.347. The neighbour-joining dendrogram constructed based on Nei's unbiased pairwise genetic distance grouped the rice genotypes into 2 major clusters (Fig. 3a), of which, Cluster I consisted of 115 genotypes while the rest 11 genotypes grouped in Cluster II. Since, the genotypes differed from each other for at least one or more morphological characters, no duplicates with identical features were detected in the entire set of germplasm.

Bayesian analysis of population structure using the model-based approach with the admixture model proposed by Falush and colleagues [30] provided support for the existence of population structure. The maximum log likelihood (84.76) was observed for K = 2 (Additional file 2: Figure S3). The genotypes were differentiated into two subpopulations i.e. SP 1 with a membership percentage of 54.8 % while SP 2 with 45.2 %. The fixation index (Fst) values of subpopulations were 0.1415 and 0.2794 for SP 1 and 2, respectively, while the pairwise allele frequency divergence value between the subpopulations was 0.0489 (Additional file 1: Table S2), indicating the existence of moderate population structure in these genotypes. Further, 8 and 7 genotypes were identified to be admix type having a membership value of <80 % in the subpopulations SP1 and SP2, respectively. Moreover, at K = 4, the 126 short grain aromatic rice genotypes could be further differentiated into four groups. (Fig. 3c, d). Forty-four genotypes were clustered in Group II followed by 38 in Group III, 23 in Group IV and 21 genotypes in Group I. Relationship among the genotypes obtained from the STRUCTURE based clustering was in line with the NJ dendrogram except a few deviations.

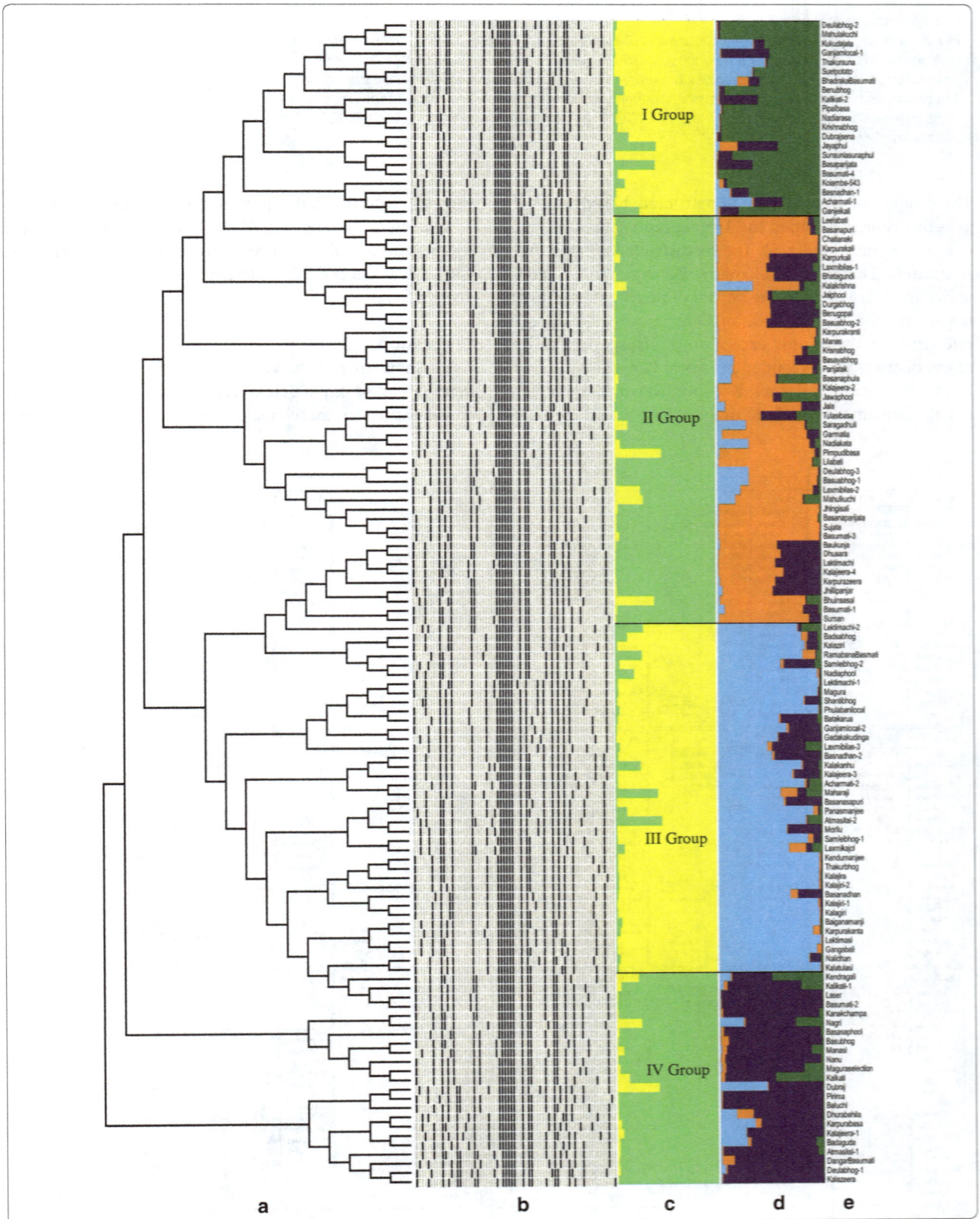

Deulabhog-2
Mahulakuchi
Kukudejata
Ganjamlocal-1
Thakursuna
Suerpotato
BhadrakaBasumati
Benubhog
Kalikati-2
Pipalbasa
Nadiarasa
Krishnabhog
Dubrajsena
Jayaphul
Sunsunnisunaphul
Basanjata
Basumati-4
Kolamba-543
Basnadhan-1
Acharnati-1
Ganjakali
Leelabati
Basanapuri
Challanaki
Karpurakali
Karpurkali
Laxmibbas-1
Bhatagundi
Kalakrishna
Jajphool
Durgabhog
Belungmati
Basuabhog-2
Karpurakranti
Manas
Krtanabhog
Bauyabhog
Parijatati
Basanaphula
Kalajeera-2
Jawaphool
Jala
Tulasibasa
Saragadhuli
Garmatla
Nadiakata
Pimpudibasa
Lilabati
Deulabhog-3
Basuabhog-1
Laxmibilas-2
Mahulkuchi
Jhingisali
Basanparijata
Sujata
Basumati-3
Baukuiya
Dhusara
Lektimachi
Kalajeera-4
Karpurazeera
Jhillipanjar
Bhunisesal
Basuati-1
Suman
Lektimachi-2
Kalazori
Badsabhog
RamabanaBasmati
Samlebhog-2
Nadiaphool
Lektimachi-1
Magura
Shantibhog
Phulabaniocal
Batakarua
Ganjamlocal-2
Godiukudinga
Laxmibilas-3
Basnadhan-2
Kalakanhu
Kalajeera-3
Acharnati-2
Maharaji
Basanasapuri
Panamanjee
Atmasallat-2
Morltu
Samlebhog-1
Laxmikajol
Kendumanjee
Thakurbhog
Kalajira
Kalajiri-2
Basanadhan
Kalajiri-1
Kalagiri
Baiganamanji
Karpurakanta
Lektimasi
Gangabati
Nalidhan
Kalatulasi
Kendragali
Kalkali-1
Laser
Basumati-2
Kanakchampa
Nagri
Basanaphool
Basubhog
Manasi
Nanu
Maguraselection
Kalikali
Dubraj
Prleeu
Baluchi
Dhunabehila
Karpurabesa
Kalajeera-1
Badiaguda
Atmasilisal-1
DangarBasumati
Deulabhog-1
Kalazeera

a   b   c   d   e

(See figure on previous page.)

Fig. 3 Genetic diversity, population structure and phenoprint representation of 126 indigenous aromatic short grain rice genotypes based on 71 variables for 24 phenotypic traits. **a** Neighbour-joining dendrogram based on Nei's genetic distance showing genetic relationship among indigenous aromatic short grain rice genotypes; **b** phenoprint illustration of indigenous aromatic short grain rice genotypes (*Each row* corresponds to a genotype and *each column* represents phenoprinting pattern of a variable across the genotype); **c** population structure for indigenous aromatic short grain rice genotypes estimated by the STRUCTURE programme at K = 2; **d** population STRUCTURE at K = 4; **e** names of the corresponding genotypes

The neighbour-joining tree constructed based on Nei's genetic distance grouped the 19 geographical regions into two major clusters (Fig. 5). The six districts i.e. Sambalpur, Kalahandi, Deogarh, Malkangiri, Bolangir and Sundargarh, which are close to the border of nearby state, Chhattisgarh on the western side of Odisha, were grouped in one cluster (I) with only exception of Nayagarh. The rest of the districts were grouped in cluster II. Genotypes from Puri, Kendrapara, Balasore, Koraput, Mayurbhanj, Dhenkanal, Ganjam, Anugul, Keonjhar, Jajpur and Phulbani showed close relationship among them. Further, analysis of molecular variance (AMOVA) based on geographical districts indicated that on an average, 92 % variation exists within districts and 8 % variation was observed among the district (Table 4).

Discussion

Assessment of genetic variation in the short grain aromatic rices of a particular region has implications for conservation of these neglected heritage resource and

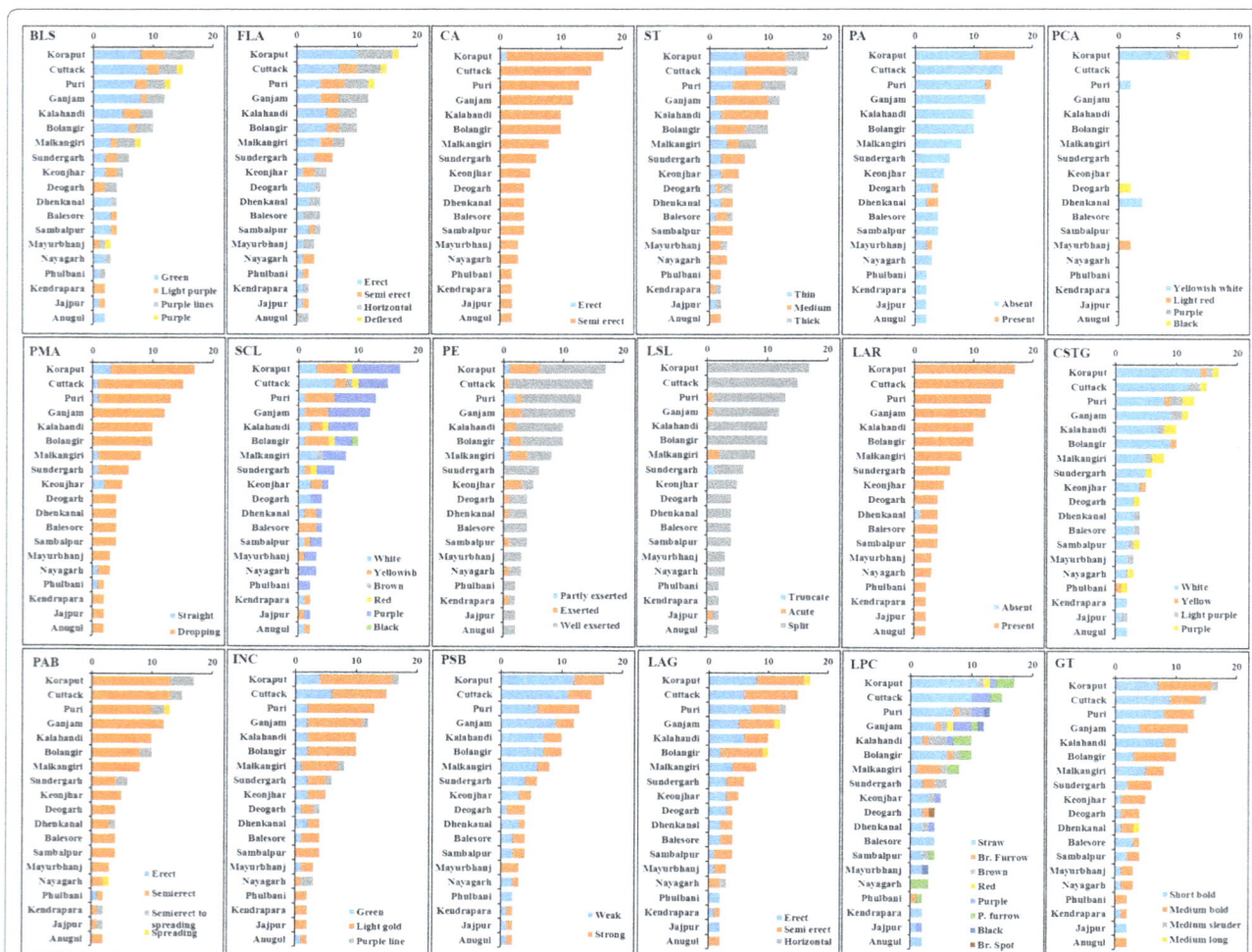

Fig. 4 Distribution of qualitative phenotypic traits in 19 geographical districts of collection

Table 3 Diversity parameters of 126 aromatic short grain rice genotypes based on 19 geographical districts of collection

District	Co-ordinates	No. of genotypes	Ne	He	I	% P
Anugul	20°47′50″N 85°1′26″E	2	1.167	0.083	0.116	8.45
Balasore	21.49°N 86.93°E	4	1.411	0.219	0.323	26.76
Bolangir	20.72°N 83.48°E	10	1.564	0.251	0.420	47.89
Cuttack	20.27°N 85.52°E	15	1.568	0.255	0.423	52.11
Deogarh	21.53°N 84.73°E	4	1.356	0.182	0.280	40.85
Dhenkanal	20.67°N 85.6°E	4	1.569	0.266	0.406	46.48
Garjam	19.38°N 85.07°E	12	1.542	0.234	0.401	50.70
Jajpur	20.85°N 86.333°E	2	1.208	0.104	0.144	25.35
Kalahandi	20.083°N 83.2°E	10	1.492	0.225	0.373	46.48
Kendrapara	20.525°N 86.475°E	2	1.292	0.146	0.202	22.54
Keonjhar	21.63°N 85.58°E	5	1.542	0.230	0.362	39.44
Koraput	18.8083°N 82.7083°E	17	1.780	0.324	0.546	67.61
Malkangiri	18.35°N 81.90°E	8	1.521	0.225	0.379	50.70
Mayurbhanj	21.933°N 86.733°E	3	1.500	0.241	0.357	32.39
Nayagarh	20.116°N 85.01°E	3	1.383	0.194	0.284	28.17
Phulbani	20.47°N 84.23°E	2	1.208	0.104	0.144	16.90
Puri	20.47°N 84.23°E	13	1.703	0.303	0.517	64.79
Sambalpur	19.48°N 85.49°E	4	1.464	0.240	0.361	30.99
Sundargarh	21.47°N 83.97°E	6	1.530	0.241	0.384	43.66

Ne average effective number of variables, *He* average Nei's genetic diversity, *I* average Shannon's information index, *% P* percent polymorphism

unlocking genes controlling valuable traits for their utilization in rice improvement [31]. Further, the state of Odisha being a major producer of rice in India and having a large assembly of indigenous rice genotypes [13], there was an imperative need to characterize them systematically.

Overall phenotypic variation

In this study, we could identify eighteen, out of 24 phenotypic traits which showed significant variation in the ASG rice collection. A total of 70 variables were detected by 24 phenotypic traits in our study, which is significantly higher than that in aromatic rice germplasm of Madhya Pradesh and Chhattisgarh of India [32], in upland rice collection of Japan [33], in aromatic rice collection from Maharashtra and Karnataka of India [34], in aromatic short grain rice genotypes of eastern India [35] and in indigenous rice landraces of Chhattisgarh state of India [36]. Highest genetic diversity value of 0.696 was contributed by lemma-palea colour, where 8 different forms were identified which is higher than the reports of Sarawgi and colleagues [37], who reported 5 types in 782 rice germplasm collection. Since lemma and palea are sepal equivalent [38], genetic base of lemma-palea colour differentiation could be utilized as a potential tool to understand the evolutionary aspects of these native genotypes. Lemma-palea pigmentation is the character of wild rice and straw hull is similar to *indica* and black hull

is similar to that in *aus* groups [39]. *Black husk4 (Bh4)* gene, encoding an amino acid transporter, fine mapped on chromosome 4, controls the black husk in *O. rufipogon* and *O. nivara*. Zhu and colleagues [40] reported that transition from black husk in wild rice to straw husk in cultivated rice is due to a functional deletion of 22 *bp* in exon 3 of *bh4* variant. The intermediate types, which were observed for this trait, could be useful to establish a lineage between aromatic rice landraces and their wild progenitors to understand their domestication process. Further, red and black husked rices having nutritive and medicinal properties, are more resistant to storage insect and pests as compared to brown husked rices [41]. Ray and colleagues [42] reported that aromatic short grain rices vary for their grain characters, which was observed in our study also, where 4 different grain types were identified. Similar to our results, high level of variation was reported for grain shape in Brazilian and Pakistan rice collections [33, 43].

In our study, genotypes having same name collected from different districts (both nearby and far apart) showed variation for one or more traits. This observation is interesting. This could be due to the naming practice being followed by traditional communities for the ASGs. Since they are offered to Gods and Goddesses in the temples, the different material in different locations might be given similar/identical names. For example Deula Bhog (Deula meaning temple; Bhog meaning offering)

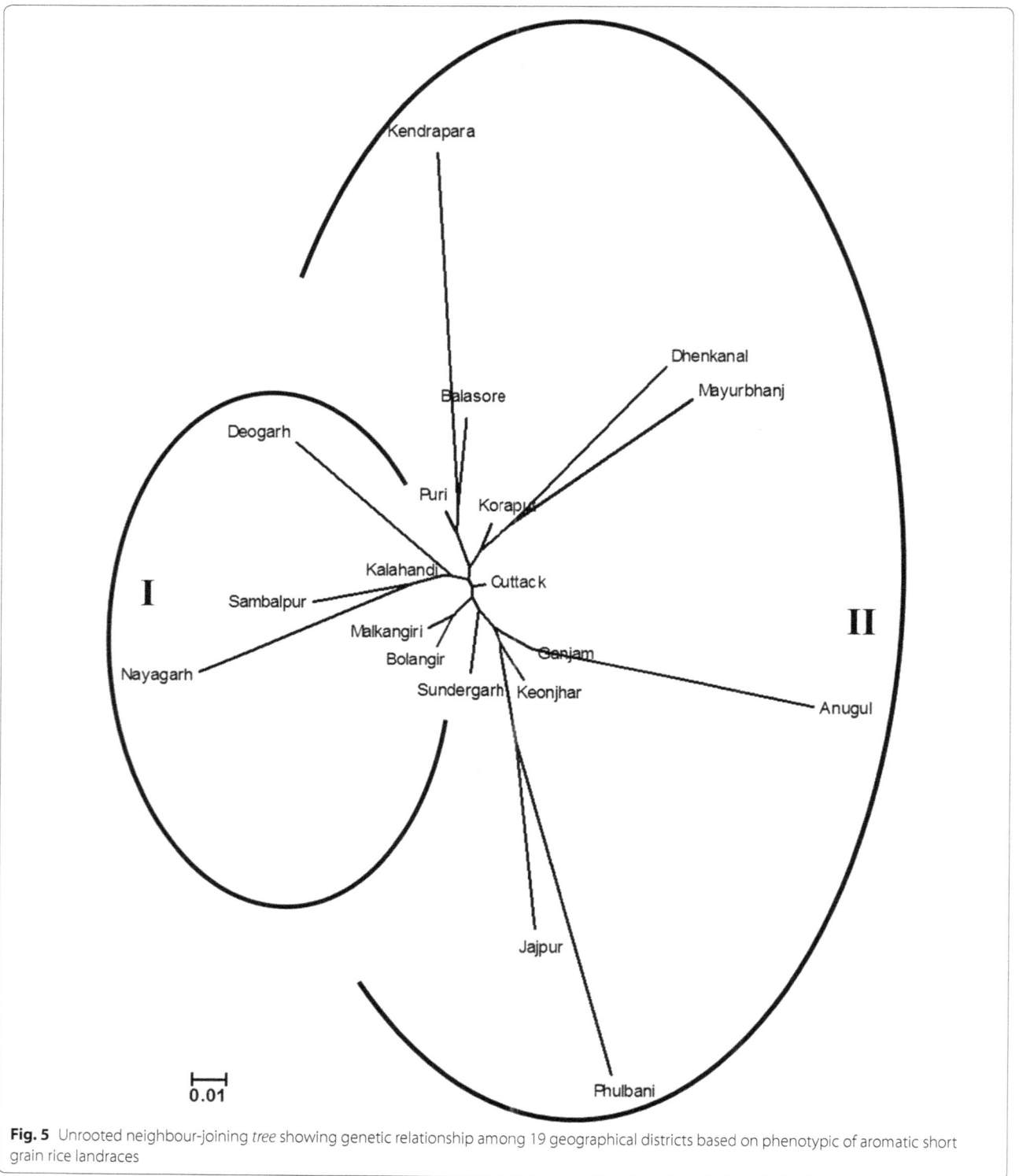

Fig. 5 Unrooted neighbour-joining *tree* showing genetic relationship among 19 geographical districts based on phenotypic of aromatic short grain rice landraces

samples collected from Puri and Koraput districts differ for basal leaf sheath colour, flag leaf attitude of blade stem thickness, panicles characteristics, tip colour of lemma, panicle exertion, shape of ligule, colour of stigma, internode colour, leaf angle and grain type. Similarly, Laxmibilas collected from Sambalpur and Deogarh districts differ

Table 4 Analysis of molecular variance for 126 short grain aromatic rice genotypes from 19 geographical districts

Source	df	SS	MS	F_{ST}	EV	%
Among population	18	120.880	6.716	0.079	0.274	8
Within population	233	742.723	3.188		3.188	92
Total	251	863.603	–		3.461	100

df degree of freedom, *SS* sum of squares, *MS* mean squares, *EV* estimated variance (p > 0.001)

for basal leaf sheath colour, flag leaf attitude of blade, tip colour of lemma, panicle exertion, colour of stigma, internode colour, panicle secondary branching, leaf angle, lemma-palea colour and grain type. The other reason of this situation arising could be introduction of the same landrace in different locations followed by selection by local growers or farmers leading to fixation of different genotypic constitution at different locations. It has been demonstrated by Kohli and colleagues [44] using molecular marker and other studies that the landraces do possess composite genetic structure and selection exercised on the base landrace is expected to result in genetically different purelines. However, what remains common in the developed lines is the target trait i.e. short grain and aroma. In our earlier report also we have detected different grades of aroma (mild/moderate/strong) in these genotypes having similar name [45]. Similarly, Bisne and Sarawgi [46] have found variation for leaf blade colour, lemma-palea colour, apiculus colour and lemma-palea pubescence in Badshah Bhog lines. Genetic variability in Indian scented rice accessions of Hansraj were also reported by Raghunathachari and colleagues [47] by using RAPD markers.

In 6 traits (leaf intensity of blade colour, leaf anthocyanin colour, leaf pubescence of blade surface, leaf ligule, leaf colour of ligule and collar colour), we could not detect any variation for the set of aromatic rice genotypes used in this study. Of the rest 6 traits, 4 related to colour of blade, leaf, collar and ligule which was green in all the genotypes. Since anthocyanin pigmentation in different parts of rice plant is due to allelic variation and complex organization of gene(s) [48], this uniformity for colour of the four traits could be due to pleiotropic effect which has remained unchanged during the process of domestication and selection by farmers [49].

The average genetic diversity value recorded in this study was relatively low (*He* = 0.286). Similar type of result was also observed in the all India collection of aromatic rice landraces [50]. However, the genetic diversity value was comparatively high than the rice landraces of Nepal [51]. The Shannon diversity index of 0.515 detected in our germplasm set is comparatively high than the rice landraces of Santhal Paraganas of Jharkhand state, India [52]. We could identify six traits i.e. tip colour

of lemma, colour of awn, grain type, flag leaf attitude of blade, lemma-palea colour and colour of stigma as the most important traits to explain the variation and these traits in combination could be utilized as key for differentiation of short grain rice genotypes at field level.

Bayesian analysis of STRUCTURE using the model-based approach suggested the presence of two optimum populations i.e. SP1 and SP2 (with a K value of 2) in these natural heritages. Choudhury and colleagues [53] also inferred 2 subpopulations in the Eastern Himalayan and Northeast Indian indigenous rice with a set of seven microsatellite markers. Similarly, 3 subpopulations were detected in Indian rice core set with molecular markers, which reveals presence of moderate population structure in the rice genotypes of India [54]. A total of 15 genotypes were determined to be admixture type, which might be due to intraspecific crosses between genotypes of SP1 and SP2 [55]. Further, the two identified populations were divided into four different groups similar to the results of Laido and colleagues [56], where similar type of groups were identified in two STRUCTURE populations in the tetraploid wheat genotypes with SSR and DArT markers.

The aromatic rice collection was found to be a reserve of many useful agronomic traits. Erect flag leaf is one of the important features and related to high yielding ability [57]. In our collection, 44 % of genotypes had erect flag leaf and these could be potential donors of the trait in high yielding aromatic rice breeding programmes. Tall plant type is prone to lodging, unresponsive to fertilizer application and hence has low yielding ability [3]. In our study, 25 genotypes were identified with thick/strong stem, which could be deployed to achieve lodging resistance under higher dose of fertilizers [58]. Similarly, 74 % of the genotypes identified with well exserted panicle type could be utilized for introgression of this trait into high yielding backgrounds. Further, 37 % of the genotypes having strong panicle secondary branching, which is another important trait related to yield, could be utilized in combination with other yield traits to overcome the low yield potential of aromatic rices [59].

Trait specific characterization of these lines would help in identifying novel genes/alleles to breed improved disease resistant quality rices. This is highly relevant in the context that in the past few decades, significant decline in cultivation and huge erosion of short grain aromatic rice germplasm in Odisha has been a major concern [60]. Artificial selection pressure and promotion of high yielding varieties and hybrids over years has led to reduced genetic base and increased genetic vulnerability in traditional rice germplasm, emphasizing the conservation, characterization and documentation of native aromatic genotypes before extinction [61–63]. The present study that describes variation in ASGs is a relevant step in this direction.

Phenoprint of aromatic short grain rice genotypes

Bar coding tools have proven to be significantly useful for varietal differentiation and their identification, almost all of which are based on DNA markers. After the establishment of a protocol for DNA barcoding of all land plants [64], this technology has been extensively used by several researchers for barcoding of many edible plants [65–70]. However, potential application of morphology based characterization cannot be ignored, since it is having useful implication to varietal discrimination at field and farm level [71]. And hence, in this study, we have characterized 126 aromatic short grain rice genotypes of Odisha, India based on qualitative phenotypic descriptors and proposed a unique bar coding approach, named 'phenoprinting', for varietal description. Once established, phenoprint could be utilized for rice and other crops as well. Given that the probability of identity based on all the polymorphic traits was low, this approach would be of use in varietal identification, farmers' rights protection and intellectual property rights issues. Since, the phenoprint deals with morphological characters, it could be generated with minimal financial inputs. The phenoprint in our study that could clearly differentiate the genotypes will be a useful guide for visual comparison of additional aromatic short grain genotypes. In fact, the traits having no variation could be easily distinguishable and the traits with maximum variability could also be precisely identified. This approach provided uniqueness to each of the lines tested, providing robustness for trait specific varietal identification. Further, the 16 genotypes identified in our study, containing more than 98 % of the variables for phenotypic traits provided the key, which could be utilized for description of ASGs and also in the rice improvement programme to overcome the genetic bottleneck in years to come.

Phenotypic diversity in different districts

The tribal dominated Koraput is endowed with diverse and valuable flora and fauna [72]. Further, Koraput region has been recently declared as 'Globally important agricultural heritage systems' by the Food and Agricultural Organization of the United Nations (FAO) [73]. Our findings, revealing the highest polymorphism percentage of more than 67 % and greater diverse group of aromatic rices in Koraput as compared to other areas, justifies the National and International attention towards Koraput for various biodiversity projects. The genetic diversity parameters were similar between Koraput and Puri but were significantly higher than the other districts. Detection of high genetic diversity in the germplasm set of Puri was interesting. Puri district of Odisha is renowned for the temple of Lord Jagannath where the famous car festival takes place annually. Besides, it also houses several other temples, monasteries and holy ashrams where aromatic rices are used in different temple rituals. Hence, import of aromatic rices from other parts of the state resulting in an assembly of a diverse group of genotypes in this district is possible. Further, 8 % of the total variation observed among districts suggested that each district have their unique group of aromatic short grain rices.

The neighbour joining tree based on pairwise genetic distance estimates, grouped the six districts of western Odisha in one group. Since, western Odisha, Jharkhand and Chhattisgarh are recognized as the centre of origin of *aus* ecotypes of rice [74], close relationship among the districts of western Odisha reflects evolution of short grain rice landraces in this region. In fact, critical observations showed that number of aromatic short grain rice genotypes and their genetic diversity is comparatively on higher side in western and southern Odisha than eastern and northern Odisha. Fuller [75, 76] reported an independent domestication of rice—Neolithic in Ganga Valley and the western Odisha with archaeo-botanical evidences. Considering Koraput in western Odisha as the secondary centre of origin for rice, greater diversity in Koraput and adjoining region is expected.

Conclusion

Native aromatic rice landraces that are highly preferred by consumers needs to be characterized which can help in varietal diagnostics purpose and their conservation [77]. Here, we have characterised native short grain aromatic genotypes of Odisha (India), one of the major consumer and producer of these native rices, based on qualitative phenotypic descriptors. For the trait lemma-palea colour, a highest of 8 different forms were detected followed by tip colour of lemma where 6 variables were identified. Tip colour of lemma, colour of awn and colour of stigma were the primary determinants for explaining the existing variation in this group of aromatic rice genotypes. The 126 genotypes were broadly grouped into 2 sub populations and 4 distinct groups by population STRUCTURE analysis, revealing the existence of moderate population structure within them. But, we did not detect any duplicates in our set. Further, western and southern districts of Odisha had maximum diverse aromatic short grain rice genotypes as compared to eastern and northern districts. Moreover, the proposed phenoprinting approach, discriminating the aromatic short grain rices based on qualitative phenotypic descriptors, could provide unique identification and description to the genotypes. Since, genetic erosion over years is a major threat to aromatic rice improvement, genetic differentiation and phenotypic description study may help in preservation of this group of indigenous short grain aromatic rices.

Additional files

Additional file 1: Table S1. Aromatic short grain rice genotypes used in the present study and region of their collection. **Table S2.** Population statistics of the estimated clusters. *SP 1* and *SP 2* are estimated subpopulations. **Table S3.** Pair-wise Nei's unbiased genetic distance of 19 geographical districts.

Additional file 2: Figure S1. Sampling sites in nineteen districts of Odisha (India) and information on population structure. **Figure S2.** Frequency distribution of phenotypic traits among 126 short grain aromatic rice landraces. **Figure S3.** The relationship between ΔK and K showing the highest peak at K = 2.

Authors' contributions
The study was designed by PSR, GJNR and TM. AP and SSCP collected the germplasm. RS collected the data. PSR and GJNR analysed the data. NNJ has helped in statistical analysis of data. PSR, GJNR and TM has drafted the manuscript. All authors read and approved the final manuscript.

Author details
[1] Central Rice Research Institute, Cuttack 753006, India. [2] Present Address: ICAR, DARE, New Delhi 110001, India.

Acknowledgements
We sincerely thank, Dr. Sergey Hegay, Senior Scientist, Biotechnology Institute, Bishkek, Kyrgyz Republic for helping in statistical analysis of data. We also acknowledge the help of Mr. Sunil K Sinha, Sr. Technical Officer, Central Rice Research Institute, Cuttack (Odisha), India, in editing the figures.

Competing interests
The authors declare that they have no competing interests.

References
1. Rana JC, Negi KS, Wani SA, Saxena S, Pradheep K, Pareek SK, Sofi PA. Genetic resources of rice in the Western Himalayan region of India: current status. Genet Resour Crop Evol. 2009;56:963–73.
2. Pusadee T, Jamjod S, Chiang YC, Rerkasem B, Schaal BA. Genetic structure and isolation by distance in a landrace of Thai rice. Proc Nat Acad Sci USA. 2009;106(33):13880–5.
3. Singh RK, Gautam PL, Saxena S, Singh S. Scented rice germplasm: conservation, evaluation and utilization. In: Singh RK, Singh US, Khush GS, editors. Aromatic rices. Kalyani: New Delhi; 2000. p. 107–33.
4. Roy JK, De RN, Ghorai DP, Panda A. Collection and evaluation of genetic resources of rice in India. Phytobreedon. 1983;1:1–9.
5. Dikshit N, Malik SS, Mohapatra P. Seed protein variability in scented rice. Oryza. 1992;29:65–6.
6. Malik SS, Dikshit N, Das AB, Lodh SB. Studies on morphological and physio- chemical properties of local scented rice. Oryza. 1994;31:106–10.
7. Glaszmann JC. Isozymes and classification of Asian rice varieties. Theor Appl Genet. 1987;74:21–30.
8. Shobha Rani N, Singh RK. Efforts on aromatic rice improvement in India. In: Singh RK, Singh US, editors. A treatise on the scented rices of India. New Delhi: Kalyani Publishers; 2003. p. 23–72.
9. Shobha Rani N, Krishnaiah K. Current status and future prospects for improvement of aromatic rices in India, In: Specialty rices of the world: Breeding, production and marketing, New York: Science Publishers, Inc.; 2001. p. 49–78.
10. Bhattacharjee P, Singhal RS, Kulkarni PR. Basmati rice: a review. Int J Food Sci Tech. 2002;37:1–12.
11. Ahuja SC, Pawar DVS, Ahuja U, Gupta KR. Basmati rice—the scented pearl. Hissar: CCS Haryana Agricultural University; 1995. p. 1–63.
12. Pachauri V, Singh MK, Singh AK, Singh S, Shakeel NA, Singh VP, Singh NK. Origin and genetic diversity of aromatic rice varieties, molecular breeding and chemical and genetic basis of rice aroma. J Plant Biochem Biotechnol. 2010;19(2):127–43.
13. Das SR. Rice in Odisha. IRRI Technical Bulletin No. 16. Los Baños: International Rice Research Institute; 2012. p. 1–31.
14. Gangadharan C. Breeding. In: Jaiswal PL, editor. Rice research in India. New Delhi: Indian Council of Agricultural Research; 1985. p. 73–109.
15. DelCruz N, Khush GS. Rice grain quality evaluation procedures. In: Singh RK, Singh US, Khush GS, editors. Aromatic rices. New Delhi: Oxford & IBH Publishing Co Pvt Ltd; 2000. p. 16–28.
16. Ramaiah K, Ghose RLM. Origin and distribution of cultivated plants of South Asian rice. Indian J Genet Plant Breed. 1951;11:7–13.
17. Arunachalam VA, Chaudhury SS, Sarangi SK, Ray T, Mohanty BP, Nambi VA, Mishra S. Rising on rice: The story of Jeypore. Chennai: MS Swaminathan Research Foundation; 2006. p. 1–39.
18. IBPGR-IRRI. Descriptors for rice (Oryza sativa L.). IBPGR-IRRI Rice Advisory Committee. Manila: IRRI; 1980.
19. Nei M. Analysis of gene diversity in subdivided populations. Proc Nat Acad Sci USA. 1973;70:3321–3.
20. Yeh FC, Yang RC, Boyle TBJ. POPGENE, version 1.32 Microsoft windowbased freeware for population genetic analysis. Edmonton: University of Alberta; 1999.
21. Tamura K, Stecher G, Peterson D, Filipski A, Kumar S. MEGA6: molecular evolutionary genetics analysis version 6.0. Mol Biol Evol. 2013;30(12):2725–9.
22. Peakall R, Smouse PE. Genalex 6: genetic analysis in Excel. Population genetic software for teaching and research. Mol Ecol Notes. 2006;6:288–95.
23. Weir BS, Cockerham CC. Estimating F-statistics for the analysis of population structure. Evolution. 1984;38:1358–70. doi:10.2307/2408641.
24. Pritchard JK, Stephens M, Donnelly P. Inference of population structure using multilocus genotype data. Genetics. 2000;155:945–59.
25. Evanno G, Regnaut S, Goudet J. Detecting the number of clusters of individuals using the software STRUCTURE: a simulation study. Mol Ecol. 2005;14:2611–20.
26. Earl DA, von Holdt BM. STRUCTURE harvester: a website and program for visualizing STRUCTURE output and implementing the Evanno method. Conserv Genet Resour. 2012;4(2):359–61.
27. Ramasamy RK, Ramasamy S, Bindroo BB, Naik VG. STRUCTURE plot: a program for drawing elegant STRUCTURE bar plots in user friendly interface. Springer Plus. 2014;3(1):1–3. doi:10.1186/2193-1801-3-431.
28. Zhang P, Li J, Li X, Liu X, Zhao X, Lu Y. Population structure and genetic diversity in a rice core collection (Oryza sativa L.) investigated with SSR markers. PLoS ONE. 2011;6(12):1–13.
29. SAS Institute. SAS/STAT Version 9.2. Cary: SAS Institute; 2010.
30. Falush D, Stephens M, Pritchard JK. Inference of population structure using multilocus genotype data: linked loci and correlated allele frequencies. Genetics. 2003;164:1567–87.
31. Rabbani MA, Pervaiz ZH, Masood MS. Genetic diversity analysis of traditional and improved cultivars of Pakistani rice (Oryza sativa L.) using RAPD markers. Electron. J Biotechnol. 2008;11:1–10.

32. Parikh M, Motiramani NK, Rastogi NK, Sharma B. Agro-morphological characterization and assessment of variability in aromatic rice germplasm. Bangladesh J Agric Res. 2012;37(1):1–8.

33. Nascimento WF, Silva EF, Veasey EA. Agro-morphological characterization of upland rice accessions. Sci Agric. 2011;68(6):652–60.

34. Mathure S, Shaikh A, Renuka N, Wakte K, Jawali N, Thengane R, Nadaf A. Characterisation of aromatic rice (Oryza sativa L.) germplasm and correlation between their agronomic and quality traits. Euphytica. 2011;179:237–46.

35. Subudhi HN, Swain D, Samantaray S, Singh ON. Collection and agromorphological characterization of aromatic short grain rice in eastern India. Afr J Agric Res. 2012;7(36):5060–8.

36. Tandekar K, Koshta N. To Study the agro morphological variation and genetic variability in rice germplasm. Middle East J Sci Res. 2014;20(2):218–24.

37. Sarawgi AK, Subba Rao LV, Parikh M, Sharma B, Ojha GC. Assessment of variability of Rice (Oryza sativa L.) germplasm using agro-morphological characterization. J Rice Res. 2013;6(1):15–28.

38. Lombardo F, Yoshida H. Interpreting lemma and palea homologies: a point of view from rice floral mutants. Front Plant Sci. 2015;6:1–6.

39. Vigueira CC, Li W, Olsen KMJ. The role of Bh4 in parallel evolution of hull colour in domesticated and weedy rice. Evol Biol. 2013;26(8):1738–49.

40. Zhu BF, Si L, Wang Z, Zhou Y, Zhu J, Shangguan Y, et al. Genetic control of a transition from black to straw-white seed hull in rice domestication. Plant Physiol. 2011;155:1301–11.

41. Ahuja U, Ahuja SC, Chaudhary N, Thakrar R. Red rices—past, present and future. Asian Agri Hist. 2007;11:291–304.

42. Ray A, Deb D, Ray R, Chattopadhayay B. Phenotypic characters of rice landraces reveal independent lineages of short-grain aromatic indica rice. AoB Plants. 2013;5:1–9. doi:10.1093/aobpla/plt032.

43. Siddiqui SU, Kumamaru T, Satoh H. Pakistan rice genetic resources. I. Grain morphological diversity and its distribution. Pakistan J Bot. 2007;39:841–8.

44. Kohli S, Mohapatra T, Das SR, Singh AK, Tandon V, Sharma RP. Composite genetic structure of rice land races revealed by STMS markers. Curr Sci. 2004;86:850–4.

45. Roy PS, Jena S, Maharana A, Rao GJN, Patnaik SSC. Molecular characterization of short grain aromatic rice landraces of Odisha for detection of aroma. Oryza. 2014;51(2):116–20.

46. Bisne R, Sarawgi AK. Agro-morphological and quality characterization of Badshahbhog group from aromatic rice germplasm of Chhattisgarh. Bangladesh J Agril Res. 2008;33(3):479–92.

47. Raghunathachari P, Khanna VK, Singh US. RAPD analysis of genetic variability in Indian scented rice germplasm (Oryza sativa L.). Curr Sci. 2000;79:994–8.

48. Reddy VS, Dash S, Reddy AR. Anthocyanin pathway in rice (Oryza sativa L.): identification of a mutant showing dominant inhibition of anthocyanins in leaf and accumulation of proanthocyanidins in pericarp. Theor Appl Genet. 1995;91:301–12.

49. Ramiah K. Rice breeding and genetics. Science monograph. 19. New Delhi: ICAR; 1953.

50. Varaprasad GS, Shobha Rani N, Padmavati G, Sesu Madhav M, Bentur JS, Laksmi VJ, et al. Catalogue on aromatic short grain rices of India. DRR Technical Bulletin No. 69. Hyderabad: Directorate of Rice Research Rajendranagar; 2013. p. 1–112.

51. Bajracharya J, Steele KA, Jarvis DI, Sthapit BR, Witcombe JR. Rice landrace diversity in Nepal: variability of agro-morphological traits and SSR markers in landraces from a high-altitude site. Fields Crop Res. 2006;95:327–35.

52. Dikshit N, Das AB, Sivaraj N, Kar MK. Phenotypic diversity for agromorphological traits in 105 landraces of rice (Oryza sativa L.) from Santhal Parganas, Jharkhand, India. Proc Natl Acad Sci India Sect B. 2013;83(3):291–304.

53. Choudhury B, Khan ML, Dayanandan S. Genetic structure and diversity of indigenous rice (Oryza sativa) varieties in the Eastern Himalayan region of Northeast India. Springer Plus. 2013;2(228):1–10. doi:10.1186/2193-1801-2-228.

54. Tiwari KK, Singh A, Pattnaik S, Sandhu M, Kaur S, Jain S, Tiwari S, et al. Identification of a diverse mini-core panel of Indian rice germplasm based on genotyping using microsatellite markers. Plant Breed. 2015;134(2):164–71.

55. Kim HJ, Jeong EG, Ahn SN, Doyle J, Singh N, Greenberg AJ, Won YJ, McCouch SR. Nuclear and chloroplast diversity and phenotypic distribution of rice (Oryza sativa L.) germplasm from the democratic people's republic of Korea. Rice. 2014;7(7):1–15.

56. Laido G, Mangini G, Taranto F, Gadaleta A, Blanco A, Cattivelli L, et al. Genetic diversity and population structure of tetraploid wheats (Triticum turgidum L.) estimated by SSR, DArT and pedigree data. PLoS ONE. 2013;8(6):1–17. doi:10.1371/journal.pone.0067280.

57. Yan W, Hu B, Zhang QJ, Jia L, Jackson A, Pan X, Huang B, Yan Z, Deren C. Short and erect rice (ser) mutant from Khao Dawk Mali 105 improves plant architecture. Plant Breed. 2012;131:282–5. doi:10.1111/j.1439-0523.2011.01943.x.

58. Ookawa T, Yasuda K, Kato H, Sakai M, Seto M, Sunaga K, Motobayashi T, Tojo S, Hirasawa T. Biomass production and lodging resistance in 'Leaf Star', a new long-culm rice forage cultivar. Plant Prod Sci. 2010;13:58–66.

59. Ogunbayo SA, Si M, Ojo DK, Sanni KA, Akinwale MG, Toulou B, Shittu A, Idehen EO, Popoola AR, Daniel IO, Gregorio GB. Genetic variation and heritability of yield and related traits in promising rice genotypes (Oryza sativa L.). J Plant Breed Crop Sci. 2014;6(11):153–9.

60. Deb D. Seeds of tradition, seeds of future, folk rice varieties of Eastern India. New Delhi: Research Foundation for Science Technology & Ecology; 2005.

61. Rabara RC, Ferrer MC, Diaz CL, Newingham MCV, Romero GO. Phenotypic diversity of farmers' traditional rice varieties in the Philippines. Agronomy. 2014;4:217–41.

62. Yamasaki M, Tenaillon MI, Bi IV, Schroeder SG, Sanchez-Villeda H, Doebley JF, Gaut BS, McMullen MD. A large-scale screen for artificial selection in maize identifies candidate agronomic loci for domestication and crop improvement. Plant Cell. 2005;17:2859–72.

63. Samal KC, Rout GR, Das SR. Study of genetic divergence of Indigenous Aromatic Rice (Oryza Sativa L.): potentials and consequences of on-farm management in traditional farming. J Agric Sci. 2014;4(4):176–89.

64. Chase MW, Cowan RS, Hollingsworth PM, vanden Berg C, Madrinan S, Petersen G, et al. A proposal for a standardised protocol to barcode all land plants. Taxon. 2007;56:295–9.

65. Pasqualone A, Lotti C, Blanco A. Identification of durum wheat cultivars and monovarietal semolinas by analysis of DNA microsatellites. Eur Food Res Technol. 1999;210:144–7.

66. Ren X, Zhu X, Warndorff M, Bucheli P, Shu Q. DNA extraction and fingerprinting of commercial rice cereal products. Food Res Int. 2006;39:433–9.

67. Salem HH, Ali BA, Huang TH, Qin DN, Wang XM, Xie QD. Use of random amplified polymorphic DNA analysis for economically important food crops. J Integr PIANT Biol. 2007;49:1670–80.

68. Terzi V, Morcia C, Gorrini A, Stanca AM, Shewry PR, Faccioli P. DNA-based methods for identification and quantification of small grain cereal mixtures and fingerprinting of varieties. J Cereal Sci. 2005;41:213–20.

69. De Mattia F, Bruni I, Galimberti A, Cattaneo F, Casiraghi M, Labra M. A comparative study of different DNA barcoding markers for the identification of some members of Lamiacaea. Food Res Int. 2011;44:693–702.

70. Nicole S, Erickson DL, Ambrosi D, Bellucci E, Lucchin M, Papa R, Kress WJ, Barcaccia G, Donini P. Biodiversity studies in Phaseolus species by DNA barcoding. Genome. 2011;54:529–45.

71. Rao LVS, Prasad GS, Chiranjivi M, Chaitanya U, Surendhar R. DUS characterization for farmer varieties of rice. IOSR J Agric Vet Sci. 2013;4(5):35–43.

72. Panda D, Bisoi SS, Palita SK. Floral diversity conservation through sacred groves in Koraput district, Odisha, India: a case study. Int Res J Environ Sci. 2014;3(9):80–6.

73. Singh AK. Probable agricultural biodiversity heritage sites in India: XVI. The Koraput region. Asian Agri Hist. 2013;17(2):97–122.

74. Sharma SD, Tripathy S, Biswal J. Origin of O. sativa and its ecotypes. In: Nanda JS, editor. Rice breeding and genetics: research priorities and challenges. New Delhi: Science Publishers, Enfield and Oxford and IBH; 2000. p. 349–69.

75. Fuller DQ. Agricultural origins and frontiers in South Asia: a working synthesis. J World Prehist. 2006;20:1–86.

76. Fuller DQ. Finding plant domestication in the Indian subcontinent. Curr Anthropol. 2011;52:S347–62.

77. Choudhury PR, Kohli S, Srinivasan K, Mohapatra T, Sharma RP. Identification and classification of aromatic rices based on DNA fingerprinting. Euphytica. 2001;118(3):243–51.

14

Convergence of soil microbial properties after plant colonization of an experimental plant diversity gradient

Katja Steinauer[1,2*], Britta Jensen[3], Tanja Strecker[3], Enrica de Luca[4], Stefan Scheu[3] and Nico Eisenhauer[1,2]

Abstract

Background: Several studies have examined the effects of plant colonization on aboveground communities and processes. However, the effects of plant colonization on soil microbial communities are less known. We addressed this gap by studying effects of plant colonization within an experimental plant diversity gradient in subplots that had not been weeded for 2 and 5 years. This study was part of a long-term grassland biodiversity experiment (Jena Experiment) with a gradient in plant species richness (1, 2, 4, 8, 16, and 60 sown species per plot). We measured plant species richness and productivity (aboveground cover and biomass) as well as soil microbial basal respiration and biomass in non-weeded subplots and compared the results with those of weeded subplots of the same plots.

Results: After 2 and 5 years of plant colonization, the number of colonizing plant species decreased with increasing plant diversity, i.e., low-diversity plant communities were most vulnerable to colonization. Plant colonization offset the significant relationship between sown plant diversity and plant biomass production. In line with plant community responses, soil basal respiration and microbial biomass increased with increasing sown plant diversity in weeded subplots, but soil microbial properties converged in non-weeded subplots and were not significantly affected by the initial plant species richness gradient.

Conclusion: Colonizing plant species change the quantity and quality of inputs to the soil, thereby altering soil microbial properties. Thus, plant community convergence is likely to be rapidly followed by the convergence of microbial properties in the soil.

Keywords: Jena Experiment, Plant colonization, Microbial biomass, Plant diversity, Plant coverage

Background

Human-induced global change is leading to worldwide changes in plant community assembly resulting in profound impacts on ecosystem functions [1, 2]. Gaining more knowledge about the mechanisms that influence biodiversity, compositional stability of plant communities, and resistance against plant colonization provide essential information to evaluate the consequences of biodiversity loss and the subsequent changes in ecosystem functions.

Generally, plant diversity increases the stability of community biomass in biodiversity experiments [3–5]. Presumably, this is due to more complete exploitation of resources with increasing plant diversity [6]. Therefore, more diverse plant communities are more resistant to the colonization of species than less diverse plant communities [7–9] and/or are more likely to contain better competitors for available resources [10]. Newly colonizing plant species must survive and grow on resources not consumed by resident plant species [7]. The prerequisite for successful plant colonization thus might be complementary resource requirements compared to the resident plant species [11]. Therefore, colonization success of a plant species is higher when its functional traits are most different from the functional traits of the resident species

*Correspondence: katja.steinauer@web.de
[1] German Centre for Integrative Biodiversity Research (iDiv) Halle-Jena-Leipzig, Deutscher Platz 5e, 04103 Leipzig, Germany
Full list of author information is available at the end of the article

[12–14]. Consistent with this expectation, higher species richness and functional complementarity were shown to increase plant biomass production in experimental studies [15, 16] due to a more complete use of available resources [17] leaving fewer vacant niches for colonizers [7, 13]. The composition and functioning of plant communities are closely linked to belowground communities and processes [18]. Colonizing plant species entering a resident community affect the biogeochemistry of ecosystems [19], alter the rate of nutrient cycling [20], and induce a shift in the structure of rhizosphere microbial communities [21], e.g., by accumulating specific pathogens in their rhizosphere [22].

Previous studies showed that the biomass and the activity of soil microorganisms increase significantly with increasing plant diversity [23–25]. The underlying mechanisms are enhanced net primary productivity, soil carbon inputs via rhizodeposition, and decomposition of plant biomass at high plant diversity [26, 27]. Different plant species, including colonizer species, release different organic compounds that change the rhizosphere conditions affecting the microbial community structure, abundance, and activity [28]. Some newly colonizing plant species might produce chemical compounds that are novel to the resident plant species, thereby having unique effects on soil microbial properties. Therefore, shifts in plant community composition and diversity due to the colonization of plant species may affect soil microbial community composition, biomass, and functions [21].

Given that there is still a need to advance knowledge about the consequences of plant colonization for soil microbial communities, we studied the effect of plant colonization of an experimental plant diversity gradient. In this split-plot experiment, one set of subplots were not weeded, while the other set of subplots were weeded. The present study was part of an established long-term grassland biodiversity experiment (Jena Experiment) with a gradient in plant diversity (1, 2, 4, 8, 16, and 60). We hypothesized that species-rich plant communities are more resistant to colonizing plant species than species-poor plant communities (hypothesis 1; Fig. 1a) [29]. Due to the colonization of functionally dissimilar plant species, we expected that plant diversity and productivity will become similar across all initial experimental plant diversity levels after plant colonization [7, 30]. Further, we expected soil respiration and microbial biomass to increase with higher plant diversity in weeded subplots (hypothesis 2, Fig. 1b, solid line) [25, 31]. As a result of plant colonization in non-weeded subplots, we expected the effects of the initial plant diversity gradient to disappear due to a homogenization of the quality and quantity of plant material entering the soil [32], thereby inducing a convergence of soil microbial properties (hypothesis 3, Fig. 1b, dotted line).

Methods

The study was conducted as part of the Jena Experiment, a long-term grassland biodiversity experiment in Jena, Germany [33]. A plant diversity gradient (1, 2, 4, 8, 16, and 60 species) was established in 2002 on 82 plots (two monoculture plots had to be given up over the course of the experiment due to very low coverage of the target species, which resulted in 80 plots for the present study). The species pool of the experiment consists of 60 plant species categorized into four functional groups (grasses, legumes, tall and small herbs; see Additional file 1: Table S1 for complete species list of plant species pool). Monocultures, two-, four-, and eight-species mixtures were replicated 16 times, 16-species-mixtures were replicated 14 times and the complete species pool of 60 species was replicated four times. Sowing density amounted to 1000 viable seeds per m² divided equally among species. Plant

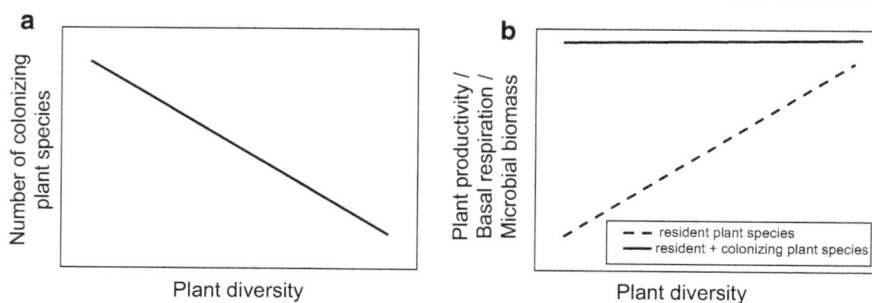

Fig. 1 Conceptual figures showing the three hypothesis. **a** The number of colonizing plant species was expected to decrease with increasing plant diversity of resident species. **b** Plant biomass production, soil basal respiration, and microbial biomass increases were expected to increase with increasing plant diversity (*dashed line*) in weeded subplots, while those variables will become similar across all initial experimental plant diversity levels after plant colonization in non-weeded subplots (*solid line*). The reader should note that the conceptual figures are graphically and qualitatively depicting the hypotheses. For simplicity, we depicted linear relationships between the variables. However, these *lines* are not based on empirical data, and we did not necessarily expect linear relationships between variables

community composition was maintained by weeding all experimental plots three times per year (May, July, and September) to remove all non-target species. Here, we use the terms "resident" for plant species initially sown in experimental plots (i.e., target plant species) and "colonizing" for plant species not sown originally in the plots (i.e., non-target plant species).

We established two independent experiments (A and B) differing in length of plant colonization (experiment A: 2 years, experiment B: 5 years) within 80 main plots resulting in 160 subplots for each experiment in a split-plot design. Therefore, the sown plant species combination and planting density was the same in both experiments. In experiment A, two subplots of 1×1 m were established in autumn 2009. In one subplot, regular weeding was continued during the entire study, whereas in the other subplot, weeding was stopped for approximately 2 years allowing for plant colonization. Using a metal corer, five soil samples (diameter 2 cm, 10 cm deep) per subplot were randomly taken in June 2011. Additionally, plant cover (%) of resident and colonizing plant species were estimated using a modified decimal Londo scale [34] and used as a proxy for plant productivity. Numerical values for species cover were coded as 1 (<1 %), 2 (1–5 %), 10 (6–15 %), 20 (16–25 %), 30 (26–35 %), 40 (36–45 %), 50 (46–55 %), 60 (56–65 %), 70 (66–75 %), 80 (76–85 %), and 90 (>85 %). Subplots of experiment A subplots were given up after this sampling campaign and therefore they were no more available for further measurements.

In experiment B, new subplots (5×3 m) were established in 2009, following the same format as in experiment A: in one subplot, regular weeding was continued, and in the other subplot, weeding was stopped. Like experiment A, five soil samples per subplot were randomly taken within 1×1 m in May 2014, i.e., after 5 years of plant colonization in the non-weeded subplots. Here, plant productivity was measured in two 0.1 m^2 plots as aboveground plant biomass (g m^{-2}).

In both experiments, soil samples were pooled, homogenized, sieved (2 mm), and approximately 5 g (fresh weight) of each soil sample was used for the measurement of soil microbial biomass and respiration. Microbial respiration (μL O$_2$ h^{-1} g^{-1} soil dry mass) was measured as mean of the O$_2$ consumption rates of 14–24 h after the start of the measurements using an O$_2$-microcompensation apparatus [35]. Soil microbial biomass C (μg C g^{-1} soil dry mass) was measured by substrate-induced respiration (SIR) after the addition of D-glucose [36]. Due to soil sieving fungal hyphae are broken up and therefore, this method mainly measures the respiration and biomass of soil bacteria. Gravimetric soil water content (%) was determined as the difference in percentages of fresh vs. dry soil (dried at 70 °C for 24 h).

First, we used General Linear Models (GLM) for a split-plot design (t-values with Satterthwaite approximation) to test effects of plant diversity (PD; manipulated at the plot level), plant colonization (COL; weeded vs. non-weeded subplots), and the interaction of plant diversity and plant colonization (PD × COL) on microbial respiration, microbial biomass (both experiments), plant cover (experiment A), plant biomass (experiment B), and number of colonizing plant species (both experiments). Second, we used a linear mixed effect model (t-values with Satterthwaite approximation) to test the effects of sown plant diversity (PD) in weeded and non-weeded plots independently on microbial respiration, microbial biomass (both experiments), plant cover (experiment A only), and plant biomass (experiment B only). Both analyses were performed using the core functions within the R statistical environment (R Development Core Team 2013) and the lme4 package [37].

Results

In both experiments, the number of newly colonizing plant species decreased with increasing plant diversity (Table 1; Fig. 2a, b; Additional file 2: Figure S2a, b; for complete species list of colonizing plant species see Additional file 1: Table S1). Two years after plant colonization (experiment A), on average 2 plant species colonized the monocultures compared to 10 plant species after 5 years of plant colonization (experiment B). Overall, in experiment B the number of colonizing plant species increased ~ fourfold per plant diversity level in comparison to the results of experiment A (after 2 years).

Experiment A—effects of 2 years of plant colonization

In experiment A, plant cover in weeded subplots was significantly higher in species-rich plant communities of resident plant species than in species-poor ones (Table 1; Fig. 2c; Additional file 2: Figure S2c). After 2 years of plant colonization, total plant cover was similar across all sown plant diversity levels in non-weeded subplots (Table 1), resulting in a significant interaction of plant diversity and plant colonization (Table 2). In weeded subplots, increasing plant diversity significantly increased both soil basal respiration (Table 1; Fig. 3a; Additional file 3: Figure S3a) and microbial biomass (Table 1; Fig. 3b; Additional file 3: Figure S3b). In non-weeded plots, basal respiration and soil microbial biomass slightly increased in species-poor plant communities and slightly declined in species-rich plant communities after plant colonization in comparison to weeded plots, rendering the plant diversity effect on soil microbial properties insignificant (Tables 1, 2). Gravimetric soil water content increased significantly with increasing plant diversity in both weeded and non-weeded subplots (Tables 1, 2).

Table 1 LM table of t- and P-values on the effects of plant diversity (PD: 1, 2, 4, 8, 16, and 60 plant species) in weeded or non-weeded plots on number of colonizing plant species, plant cover (2011), plant biomass (2014), soil basal respiration, soil microbial biomass and soil water content of 2011 and 2014

2011	Number of colonizing plant species			Plant cover			Soil basal respiration			Soil microbial biomass			Soil water content		
	df	t value	P value	df	t value	P value	df	t value	P value	df	t value	P value	df	t value	P value
Weeded				1,77.01	**4.13**	**<0.001**	1,72.18	**4.29**	**<0.001**	1,74.35	**3.57**	**<0.001**	1,74.08	**4.58**	**<0.001**
Non-weeded	1,77.00	*−1.93*	*0.057*	1,76.99	−0.30	0.767	1,71.18	0.80	0.428	1,72.33	1.14	0.256	1,73.10	3.07	**0.003**

2014	Number of colonizing plant species			Plant biomass			Soil basal respiration			Soil microbial biomass			Soil water content		
	df	t value	P value	df	t value	P value	df	t value	P value	df	t value	P value	df	t value	P value
Weeded				1,74.34	**5.57**	**<0.001**	1,74.09	**4.26**	**<0.001**	1,74.01	**2.74**	**0.008**	1,74.10	**5.30**	**<0.001**
Non-weeded	1,74.05	**−2.14**	**0.036**	1,76.04	1.09	0.281	1,71	0.23	0.815	1,68.48	0.08	0.938	1,70.09	1.29	0.202

Significant results (P < 0.05) are highlighted in bold and marginally significant results (P < 0.10) are given in italics

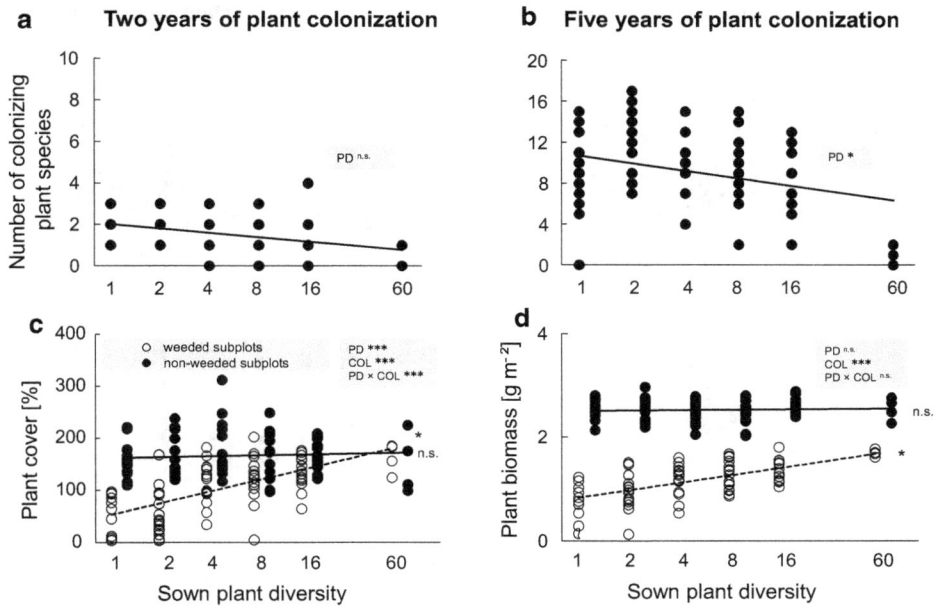

Fig. 2 Plant colonization effects on plant cover and plant biomass. Number of colonizing plant species after **a** 2 years and **b** 5 years. **c** Plant cover [%] after 2 years and **d** plant biomass [g m^{-2}] after 5 years. In **c** and **d** *dashed lines* and *open circles* display plant cover and biomass with resident plant species of weeded subplots, respectively, and *solid lines* and *circles display* plant cover and biomass with resident plant species plus colonizing plant species of non-weeded subplots, respectively. *** P < 0.001, ** P < 0.01, * P < 0.05, (*) P < 0.1, ns P > 0.1

Table 2 GLM table of t- and P-values on the effects of plant diversity (PD: 1, 2, 4, 8, 16, and 60 plant species) and plant colonization (COL) on plant cover (2011), plant biomass (2014), soil basal respiration, soil microbial biomass and soil water content of 2011 and 2014

	Plant cover			Soil basal respiration			Soil microbial biomass			Soil water content		
	df	t value	P value	df	t value	P value	df	t value	P value	df	t value	P value
2011												
PD	1, 145.61	**4.46**	**<0.001**	1, 140.86	**4.15**	**<0.001**	1, 132	**3.67**	**<0.001**	1, 145.30	**4.79**	**<0.001**
COL	1, 77	**9.42**	**<0.001**	1, 74.25	**2.06**	**0.043**	1, 76.33	**2.06**	**0.043**	1, 75.18	0.37	0.714
PD × COL	1, 77	**−3.84**	**<0.001**	1, 74.69	**−2.64**	**0.010**	1, 76.42	**−2.25**	**0.028**	1, 75.29	−1.49	0.139
	Plant Biomass			Soil basal respiration			Soil microbial biomass			Soil water content		
	df	t value	P value	df	t value	P value	df	t value	P value	df	t value	P value
2014												
PD	1, 152.79	0.61	0.543	1, 147.77	*1.90*	*0.060*	1, 143.35	1.11	0.270	1, 137.59	**5.85**	**<0.001**
COL	1, 77.94	**11.45**	**<0.001**	1, 78.53	**7.50**	**<0.001**	1, 77.23	**8.78**	**<0.001**	1, 73.92	**−6.37**	**<0.001**
PD × COL	1, 77.45	0.67	0.505	1, 77.27	−1.12	0.268	1, 76.00	−0.72	0.474	1, 73.72	**−3.79**	**<0.001**

Significant results (P < 0.05) are highlighted in bold and marginally significant results (P < 0.10) are given in italics

Experiment B—effects of 5 years of plant colonization

In experiment B, plant biomass increased with increasing plant diversity in weeded subplots, however this positive relationship disappeared in non-weeded subplots (Table 1; Fig. 2d; Additional file 2: Figure S2d).

Despite these different trends, the interaction effect of plant diversity and plant colonization was not significant (Fig. 2d; Additional file 2: Figure S2d). Generally, plant biomass was significantly higher in non-weeded subplots than in weeded subplots (+120 %). In weeded

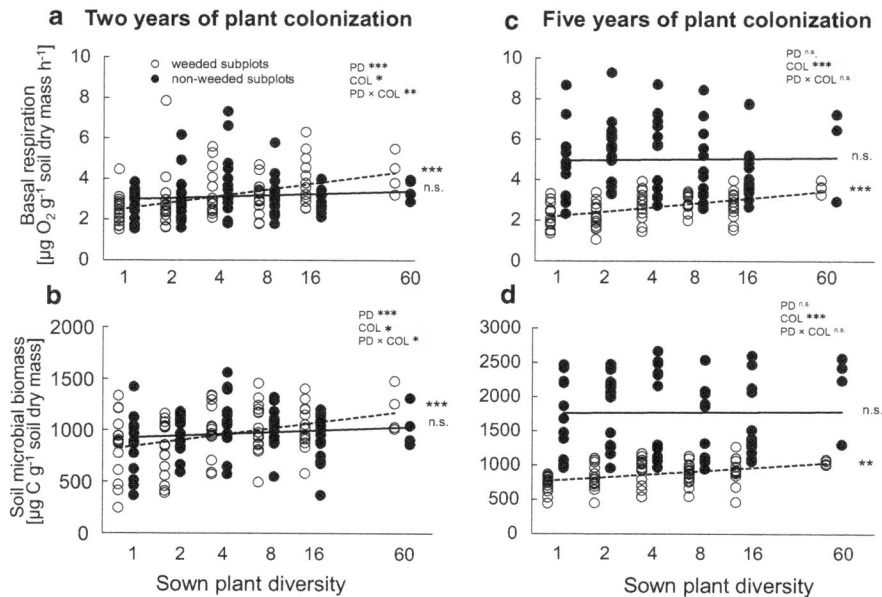

Fig. 3 Plant colonization effects on soil microbial properties. Basal respiration [μg O_2 g^{-1} soil dry mass h^{-1}] **a** 2 and **c** 5 years, and soil microbial biomass [μg C g^{-1} soil dry mass] **b** 2 and **d** 5 years after colonization by plant species. *Dashed lines* and *open circles* display basal respiration and soil microbial biomass with resident plant species of weeded subplots, respectively, and *solid lines* and *circles display* basal respiration and soil microbial biomass with resident plant species plus colonizing plant species of non-weeded subplots, respectively. ***P < 0.001, **P < 0.01, *P < 0.05, (*) P < 0.1, ns P > 0.1

subplots, soil basal respiration (Table 1; Fig. 3c; Additional file 3: Figure S3c) and microbial biomass (Table 1; Fig. 3d; Additional file 3: Figure S3d) increased significantly with increasing sown plant diversity. However, basal respiration and soil microbial biomass were not affected by sown plant diversity in non-weeded subplots, although the interaction effect of plant diversity and plant colonization was not significant (Tables 1, 2; Fig. 3c, d; Additional file 3: Figure S3c, d). Both basal respiration (+90 %) and soil microbial biomass (+104 %) were significantly higher in non-weeded subplots than in weeded subplots. Soil water content increased significantly with increasing plant diversity in weeded subplots but was not significantly affected by plant diversity after plant colonization in non-weeded subplots (Tables 1, 2).

Discussion

Total productivity (plant coverage and biomass, respectively) of resident plant communities increased with increasing plant diversity in both experiments, confirming hypothesis 1 [30]. In addition, numbers of colonizing plant species typically were high in species-poor plant communities and decreased with increasing plant diversity. In line with our findings, previous studies suggested that diverse plant communities better resist plant colonization than less diverse communities [29, 38] due

to lower levels of available resources [39]. Consequently, less resources are available for potential new colonizer species [40, 41]. Furthermore, there is evidence that large niche overlap between resident and colonizer species increases resistance against colonization [42, 43]. Generally, empty niche space is assumed to decline with increasing species richness [7, 13]. Thus, a diverse plant community should be more resistant to colonizer plant species when depending on similar resources [40, 44].

Importantly, colonizing plant species may change the quantity and quality of inputs to soil [45, 46], which has the potential to alter soil microbial functions and processes. In line with hypothesis 2, soil basal respiration and microbial biomass increased with increasing plant diversity in weeded subplots of both experiments [31]. Type and number of plant species present have considerable influence on the functions and diversity of soil microorganisms [27, 47]. Soil microorganisms are involved in processes like decomposition and nutrient mineralization and their community composition and abundances have been shown to vary with plant species [48]. Since increasing plant biomass production and release of rhizodeposits is associated with higher availability and diversity of plant-derived resources, we suggest that plant colonization positively influenced soil microbial properties in both experiments [26, 46]. Moreover, root morphological characteristics, litter types, and plant tissue qualities may affect the biomass of soil microorganisms [49]. Therefore, an increase of plant species richness, presumably resulting in a convergence of plant community composition [38] in non-weeded subplots in both experiments, may have induced a shift in soil microbial properties equalizing the effects of the initially sown plant diversity gradient, confirming hypothesis 3. After 2 and 5 years of plant colonization, basal respiration and soil microbial biomass increased in species-poor plant communities in non-weeded subplots in comparison to weeded subplots. However, in species-rich plant communities soil microbial respiration and biomass decreased after 2 years of plant colonization, which is hard to explain. In contrast, after 5 years of plant colonization, basal respiration and soil microbial biomass were considerably higher across all plant diversity levels in non-weeded subplots, but particularly at low plant diversity. This indicates that time could play a crucial role in the establishment of plant diversity effects on soil microbial properties. Previous studies [31, 50] showed that several years are required to display significant plant diversity effects on soil microbial biomass due to the slow accumulation of plant-derived resources in the soil over time [51, 52]. Although such a temporal effect could explain the differences between the two experiments in the present study, please note that we only have two sampling dates, which does not allow us to infer temporal trends.

Conclusion

Our study highlights the consequences of plant colonization for resident plant communities and soil microbial properties. The results confirmed previous findings that experimental communities with higher numbers of resident plant species are more resistant to colonization than species-poor ones [30, 38]. Further, the present results show that plant community convergence induces the convergence of microbial properties in the soil. Colonizing plant species are likely to change the quantity and quality of inputs to the soil, thereby altering soil microbial functions and processes. Future studies should investigate the potential convergence of soil microbial community composition and multiple microbial functions. Further, it remains to explore specific plant traits effects on particular microbial taxa and functions in the soil, and if novel plant traits in a colonized plant community and convergence of the functional composition of plant communities are the underlying mechanisms of the observed convergence of soil microbial properties.

Additional files

Additional file 1: Table S1. Plant species of weeded subplots (resident plant species) and non-weeded subplots (colonizing plant species). Colonizing plant species are divided into internal (belonging to the species pool of the resident plant species) and external (not belonging to the species pool of the resident plant species) plant species.

Additional file 2: Figure S1. Plant colonization effects on plant cover and plant biomass. Mean values with confidence intervals of colonizing plant species after (a) two years and (b) five years. (c) Mean values with confidence intervals of plant cover [%] after two years and (d) plant biomass [g m^{-2}] after five years. In c) and d) circles display plant cover and biomass of resident plant species of weeded subplots, respectively, and open circles display plant cover and biomass with resident plant species plus colonizing plant species of non-weeded subplots, respectively.

Additional file 3: Figure S2. Plant colonization effects on soil microbial properties. Mean values with confidence intervals of basal respiration [µg O$_2$ g^{-1} soil dry mass h^{-1}] (a) two and (c) five years, and soil microbial biomass [µg C g^{-1} soil dry mass] (b) two and (d) five years after colonization by plant species. Circles display basal respiration and soil microbial biomass with resident plant species of weeded subplots, respectively, and open circles display basal respiration and soil microbial biomass with resident plant species plus colonizing plant species of non-weeded subplots, respectively.

Authors' contributions

KS, TS, BJ, SS, and NE designed the study, KS, TS, BJ, and EL collected the data, KS analyzed the data and wrote the first draft of the manuscript, and all authors contributed to the subsequent version of the manuscript. All authors read and approved the final manuscript.

Author details

[1] German Centre for Integrative Biodiversity Research (iDiv) Halle-Jena-Leipzig, Deutscher Platz 5e, 04103 Leipzig, Germany. [2] Institute of Biology, Leipzig University, Johannisallee 21, 04103 Leipzig, Germany. [3] J. F. Blumenbach Institute of Zoology and Anthropology, Georg-August-University Göttingen, Berliner Straße 28, 37073 Göttingen, Germany. [4] Institute of Evolutionary Biology and Environmental Studies, University of Zurich, Zurich 8057, Switzerland.

Acknowledgements

We would like to thank Madhav Prakash Thakur for his suggestions in data analysis. The Jena Experiment was funded by the Deutsche Forschungsgemeinschaft (German Research Foundation; FOR 1451). NE acknowledged funding by the German Research Foundation (Ei 862/3-2). We thank the gardeners, technicians, and managers for their work in maintaining the field site and also many student helpers for weeding of the experimental plots. We would like to thank two anonymous reviewers for their helpful comments, which considerably improved the manuscript.

Competing interests

The authors declare that they have no competing interests.

Funding

Deutsche Forschungsgemeinschaft (German Research Foundation): FOR 1451 and Ei 862/3-2.

References

1. Wardle DA, Bardgett RD, Callaway RM, Van der Putten WH. Terrestrial ecosystem responses to species gains and losses. Science. 2011;332:1273–7.
2. Strayer DL. Eight questions about invasions and ecosystem functioning. Ecol Lett. 2012;15:1199–210.
3. Yachi S, Loreau M. Biodiversity and ecosystem productivity in a fluctuating environment : the insurance hypothesis. Proc Natl Acad Sci USA. 1999;96:1463–8.
4. Tilman D, Downing JA. Biodiversity and stability in grasslands. In: Samson FB et al, editors. Ecosystem management SE-1. New York: Springer; 1996. p. 3–7.
5. Isbell F, Craven D, Connolly J, Loreau M, Schmid B, Beierkuhnlein C, Bezemer TM, Bonin C, Bruelheide H, de Luca E, Ebeling A, Griffin JN, Guo Q, Hautier Y, Hector A, Jentsch A, Kreyling J, Lanta V, Manning P, Meyer ST, Mori AS, Naeem S, Niklaus PA, Polley HW, Reich PB, Roscher C, Seabloom EW, Smith MD, Thakur MP, Tilman D, et al. Biodiversity increases the resistance of ecosystem productivity to climate extremes. Nature. 2015;526:574–7.
6. Lehman CL, Tilman D. Biodiversity, Stability, and Productivity in Competitive Communities. Am Nat. 2000;156:534–52.
7. Tilman D. Niche tradeoffs, neutrality, and community structure : a stochastic theory of resource competition, invasion, and community assembly. PNAS. 2004;101:10854–61.
8. Elton CS. The ecology of invasions by animals and plants. London: English Language Book Society; 1958.
9. Kennedy TA, Naeem S, Howe KM, Knops JMH, Tilman D, Reich P. Biodiversity as a barrier to ecological invasion. Nature. 2002;417:636–8.
10. Tilman D. Resource competition and community structure. Princeton: Monographes in Population Biology (Princeton University Press); 1982.
11. Harpole WS, Tilman D. Grassland species loss resulting from reduced niche dimension. Nature. 2007;446:791–3.
12. Emery SM. Limiting similarity between invaders and dominant species in herbaceous plant communities? J Ecol. 2007;95:1027–35.
13. Eisenhauer N, Dobies T, Cesarz S, Hobbie SE, Meyer RJ, Worm K, Reich PB. Plant diversity effects on soil food webs are stronger than those of elevated CO_2 and N deposition in a long-term grassland experiment. Proc Natl Acad Sci. 2013;110:6889–94.
14. Hooper DU, Dukes JS. Functional composition controls invasion success in a California serpentine grassland. J Ecol. 2010;98:764–77.
15. Tilman D, Reich PB, Knops JMH. Biodiversity and ecosystem stability in a decade-long grassland experiment. Nature. 2006;441:629–32.
16. Cardinale BJ, Wright JP, Cadotte MW, Carroll IT, Hector A, Srivastava DS, Loreau M, Weis JJ. Impacts of plant diversity on biomass production increase through time because of species complementarity. Proc Natl Acad Sci USA. 2007;104:18123–8.
17. Wardle DA. Experimental demonstration that plant diversity reduces invasibility—evidence of a biological mechanism or a consequence of sampling effect ? Oikos. 2001;95:161–70.
18. Wardle DA, Bardgett RD, Klironomos JN, Setälä H, van der Putten WH, Wall DH, Setala H. Ecological linkages between aboveground and belowground biota. Science. 2004;304:1629–33.
19. Weidenhamer JD, Callaway RM. Direct and indirect effects of invasive plants on soil chemistry and ecosystem function. J Chem Ecol. 2010;36:59–69.
20. Ehrenfeld JG. Ecosystem consequences of biological invasions. Annu Rev Ecol Evol Syst. 2010;41:59–80.
21. Batten KM, Scow KM, Davies KF, Harrison SP. Two invasive plants alter soil microbial community composition in serpentine grasslands. Biol Invasions. 2006;8:217–30.
22. Inderjit, van der Putten WH. Impacts of soil microbial communities on exotic plant invasions. Trends Ecol Evol. 2010;25:512–9.
23. Stephan A, Meyer AH, Schmid B. Plant diversity affects culturable soil bacteria in experimental grassland communities. J Ecol Ecol. 2000;88:988–98.
24. Chung H, Zak DR, Reich PB, Ellsworth DS. Plant species richness, elevated CO2, and atmospheric nitrogen deposition alter soil microbial community composition and function. Glob Chang Biol. 2007;13:980–9.
25. Steinauer K, Tilman GD, Wragg PD, Cesarz S, Cowles JM, Pritsch K, Reich PB, Weisser WW, Eisenhauer N. Plant diversity effects on soil microbial functions and enzymes are stronger than warming in a grassland experiment. Ecology. 2015;96:99–112.
26. Zak DR, Holmes WE, White DC, Peacock AD, Tilman D. Plant diversity, soil microbial communities, and ecosystem function: are there any links? Ecology. 2003;84:2042–50.
27. Lange M, Eisenhauer N, Sierra CA, Bessler H, Engels C, Griffiths RI, Mellado-Vazquez PG, Malik AA, Roy J, Scheu S, Steinbeiss S, Thomson BC, Trumbore SE, Gleixner G, Mellado-va PG, Thomson BC, Trumbore SE, Gleixner G. Plant diversity increases soil microbial activity and soil carbon storage. Nat Commun. 2015;6(6707):1–8.
28. Hooper DU, Bignell DE, Brown VK, Brussaard L, Dangerfiled Mark, Wall DH, Wardle DA, Coleman DC. Interactions between aboveground and belowground biodiversity in terrestrial ecosystems: patterns, mechanisms, and feedbacks. Bioscience. 2000;50:1049–61.
29. Roscher C, Beßler H, Oelmann Y, Engels C, Wilcke W, Schulze ED. Resources, recruitment limitation and invader species identity determine pattern of spontaneous invasion in experimental grasslands. J Ecol. 2009;97:32–47.
30. Roscher C, Temperton VM, Buchmann N, Schulze ED. Community assembly and biomass production in regularly and never weeded experimental grasslands. Acta Oecol. 2009;35:206–17.
31. Eisenhauer N. Plant diversity effects on soil microorganisms support the singular hypothesis. Ecology. 2010;91:485–96.
32. Wardle D, Lavelle P. Linkages between soil biota, plant litter quality and decomposition. In: Cadisch G, Giller KE, editors. Driven by nature: plant litter quality and decomposition. Wallingford: CAB International; 1997. p. 107–25.
33. Roscher C, Schumacher J, Baade J, Wilcke W, Gleixner G, Weisser WW, Schmid B, Schulze E-D. The role of biodiversity for element cycling and trophic interactions : an experimental approach in a grassland community. Basic Appl Ecol. 2004;5:107–21.
34. Londo. The decimal scale for releves of permanent quadrats. Vegetatio. 1976;33:61–4.
35. Scheu S. Automated measurement of the respiratory response of soil microcompartments: active microbial biomass in earthworm faeces. Soil Biol Biochem. 1992;11:1113–8.
36. Anderson JM, Domsch K. A physiological method for the quantitative measurement of microbial biomass in soils. Soil Biol Biochem. 1978;10:215–21.
37. Bates D, Maechler M, Bolker B. lme4: linear mixed-effects models using Eigen and S4. R package. 2015
38. Petermann JS, Fergus AJF, Roscher C, Turnbull LA, Weigelt A, Schmid B. Biology, chance, or history? The predictable reassembly of temperate grassland communities. Ecology. 2010;91:408–21.
39. Fargione JE, Tilman D. Diversity decreases invasion via both sampling and complementarity effects. Ecol Lett. 2005;8:604–11.
40. Fargione JE, Brown CS, Tilman D. Community assembly and invasion: an experimental test of neutral versus niche processes. PNAS. 2003;100:8916–20.
41. Hector A, Dobson K, Minns A, Bazeley-White E, Hartley Lawton J. Community diversity and invasion resistance: an experimental test in a grassland ecosystem and a review of comparable studies. Ecol Res. 2001;16:819–31.
42. Mwangi PN, Schmitz M, Scherber C, Roscher C, Schumacher J, Scherer-Lorenzen M, Weisser WW, Schmid B. Niche pre-emption increases with species richness in experimental plant communities. J Ecol. 2007;95:65–78.
43. Frankow-Lindberg BE. Grassland plant species diversity decreases invasion by increasing resource use. Oecologia. 2012;169:793–802.
44. Turnbull LA, Rahm S, Baudois O, Eichenberger-Glinz S, Wacker L, Schmid B. Experimental invasion by legumes reveals non-random assembly rules in grassland communities. J Ecol. 2005;93:1062–70.
45. Holly DC, Ervin GN, Jackson CR, Diehl SV, Kirker GT. Effect of an invasive grass on ambient rates of decomposition and microbial community structure: a search for causality. Biol Invasions. 2008;11:1855–68.
46. Ehrenfeld JG. Effects of exotic plant invasions on soil nutrient cycling processes. Ecosystems. 2003;6:503–23.
47. Spehn EM, Joshi J, Schmid B, Alphei J, Körner C. Plant diversity effects on soil heterotrophic activity in experimental grassland ecosystems. Plant Soil. 2000;224:217–30.

48. Ehrenfeld JG, Ravit B, Elgersma K. Feedback in the Plant-Soil System. Annu Rev Environ Resour. 2005;30:75–115.

49. Porazinska D, Bardgett R. Relationships at the aboveground-belowground interface: plants, soil biota, and soil processes. Ecol Monogr. 2003;73:377–95.

50. Thakur MP, Milcu A, Manning P. Plant diversity drives soil microbial biomass carbon in grasslands irrespective of global environmental change factors. Glob Chang Biol. 2015;21:4076–85.

51. Eisenhauer N, Reich PB. Above- and below-ground plant inputs both fuel soil food webs. Soil Biol Biochem. 2012;45:156–60.

52. Kuzyakov Y, Xu X. Tansley review Competition between roots and microorganisms for nitrogen : mechanisms and ecological relevance. New Phytol. 2013;198:656–69.

15

Sequential above- and belowground herbivory modifies plant responses depending on herbivore identity

Dinesh Kafle[1*], Anne Hänel[1], Tobias Lortzing[2], Anke Steppuhn[2] and Susanne Wurst[1]

Abstract

Background: Herbivore-induced changes in plant traits can cause indirect interactions between spatially and/or temporally separated herbivores that share the same host plant. Feeding modes of the herbivores is one of the major factors that influence the outcome of such interactions. Here, we tested whether the effects of transient aboveground herbivory for seven days by herbivores of different feeding guilds on tomato plants (*Solanum lycopersicum*) alters their interaction with spatially as well as temporally separated belowground herbivores.

Results: The transient aboveground herbivory by both chewing caterpillars (*Spodoptera exigua*) and sucking aphids (*Myzus persicae*) had significant impacts on plant traits such as plant growth, resource allocation and phytohormone contents. While the changes in plant traits did not affect the overall performance of the root-knot nematodes (*Meloidogyne incognita*) in terms of total number of galls, we found that the consequences of aboveground herbivory for the plants can be altered by the subsequent nematode herbivory. For example, plants that had hosted aphids showed compensatory growth when they were later challenged by nematodes, which was not apparent in plants that had hosted only aphids. In contrast, plants that had been fed by *S. exigua* larvae did not show such compensatory growth even when challenged by nematodes.

Conclusion: The results suggest that the earlier aboveground herbivory can modify plant responses to subsequent herbivores, and such modifications may depend upon identity and/or feeding modes of the aboveground herbivores.

Keywords: Above- and belowground interaction, Induced plant defense, Priming, Feeding guilds, Resistance, Tolerance

Background

Plants respond with morphological, physiological and biochemical changes in their resistance and tolerance traits to deal with herbivores and herbivory stress [1–4]. Besides responses in local tissues which are being attacked, herbivory induces numerous changes in more distant systemic tissues, which can cause indirect interactions between spatially, and in some cases, temporally separated herbivores. Thereby, plants can even mediate indirect interactions between phytophagous organisms living above- and belowground [5–10].

Although defensive quality of roots has been analyzed less than that of aboveground plant parts, several plant species are known to systemically induce defensive compounds in roots following aboveground herbivory which may protect them from belowground herbivores [7, 8, 11–13]. Along with chemical defense, plants may also employ tolerance strategies to deal with herbivory, such as altered photosynthetic rates, compensatory growth, increased tillering, and reallocation of primary metabolites and minerals [14, 15]. Plants fine-tune their resistance and tolerance ability in order to optimize plant fitness; therefore, they may or may not employ both strategies simultaneously [12]. Any of the systemic

*Correspondence: dinesh.kafle@fu-berlin.de
[1] Functional Biodiversity, Dahlem Centre of Plant Sciences, Institute of Biology, Freie Universität Berlin, Königin-Luise-Str. 1-3, 14195 Berlin, Germany
Full list of author information is available at the end of the article

changes in root tissue due to aboveground herbivory, either in resistance or tolerance traits, may significantly impact the performance of subsequent belowground herbivores [16–18]. Recent studies also suggest that sequential herbivory events may result in the priming of plant responses which is a preconditioning by earlier herbivory that enables plants to deal with future herbivores more efficiently [19–21]. Overall, the aboveground herbivore-induced changes in root tissue can be detrimental, neutral or facilitative to the belowground herbivores depending upon several factors such as herbivore species, their feeding guild, plant species, genotypes and defense strategies [7, 16, 22–28].

One of the significant determinants of the outcome of above- and belowground herbivore interactions is the feeding mode of the herbivores. Chewing and sucking are two major feeding modes of herbivores. Coleopteran and lepidopteran insects are equipped with chewing or tearing-type mouthparts causing severe wounding injury whereas hemipteran insects such as aphids and whiteflies are equipped with piercing and sucking mouthparts to ingest the phloem-sap causing minimal injury on plant tissue [29–31]. Wound trauma inflicted by the feeding damage and type of elicitors present in oral secretion of herbivores are two major cues that regulate the induction of specific resistance or tolerance responses of the plant [4, 32]. Therefore, the feeding mode and the identity of the herbivore are key factors in plant–insect interactions as they determine specific activation patterns of plant signaling pathways that regulate a plant's response. Plant responses upon herbivory are mainly regulated by three phytohormones, jasmonic acid (JA), salicylic acid (SA) and ethylene, which are also known to play essential roles for the growth and development of the plant. A large body of evidences suggests that chewing herbivores primarily activate JA-dependent defense pathways whilst sucking herbivores induce predominantly SA- along with JA- and ethylene-dependent pathways similar to the responses induced by plant pathogenic microbes [29, 33]. But, it is important to note that their activation is highly species-specific and not limited to particular feeding guilds. Several studies have shown the activation of SA-dependent responses upon chewing herbivores and activation of JA-dependent responses upon sucking herbivores; and the phytohormones may interact antagonistically or synergistically with each other [33–35].

Here, we aimed to compare the effects of aboveground herbivory by insects from two feeding guilds (chewing caterpillars and sucking aphids) on plant traits and the plant's interaction with spatially and temporally separated belowground root-knot nematode. Root-knot nematodes (*Meloidogyne*) are endoparasites which, with the help of special gland secretions, stimulate the root cells to grow into 'giant cells' (root-knots or galls) that serve as a feeding site [36]. By inducing galls and feeding on the root tissue, root-knot nematodes weaken the ability of the root to take up water and nutrients which impairs plant performance and fitness [37]. Although nematodes do not feed by sucking up phloem sap like aphids, their feeding strategies and salivary composition have noticeable similarities [38, 39] and both are sensitive to plant resistance traits mediated by the same gene, *Mi-1* [40] which is found in tomato (*Solanum lycopersicum*). Commercial tomato cultivars are known to contain the *Mi* locus with two highly homologous genes, *Mi-1.1* and *Mi-1.2* [37] which confer resistance against aphids [40, 41], whiteflies [42] and root-knot nematodes including *Meloidogyne incognita* [37, 43]. Furthermore, subsequent studies found that the SA signaling pathway is essential for *Mi-1*-mediated defense responses, suggesting its inducibility [44, 45]. Therefore, aphids, nematodes and tomato plants are an interesting model system to study plant-mediated impacts of aboveground herbivores on belowground herbivores. In our study, we hypothesized that earlier transient aboveground herbivory by aphids would have a more pronounced impact on nematodes because of activation of the same defense pathway than transient chewing herbivory by caterpillars.

To differentiate between the effects of plant responses to herbivores of different feeding modes on the plant's interaction with root-knot nematodes (*M. incognita*, Heteroderidae), we used the sap-feeding green peach aphid (*Myzus persicae*, Aphididae) and the chewing beet armyworm (*Spodoptera exigua*, Noctuidae). Using tomato (*S. lycopersicum*, Solanaceae var. MicroTom) as a model plant, we aimed to investigate: (1) if transient aboveground herbivory has any effect on plant traits and affects spatially and temporally separated belowground herbivores; (2) if transient aboveground herbivory affects the plant's response to the subsequent belowground herbivory; (3) if those effects differ between the two aboveground herbivore species exhibiting different feeding modes. To answer these questions, we carried out a greenhouse experiment in which tomato plants were exposed to transient herbivory by either aphids, caterpillars or no aboveground herbivores, followed by nematode infestation or not. We separated the events of above- and belowground herbivory by a lag phase (a period without any herbivory) to assess the effect of transient aboveground herbivory on temporally separated belowground herbivores.

Methods

Plant material

Before germination, the seeds of tomato (*S. lycopersicum*) were surface-sterilized with 70% ethanol followed by mixture of 5.25% (w/v) sodium hypochlorite and 0.1% Polysorbate 20 (Tween 20). Then, the seeds were rinsed with deionized water and sown on paper towels in plastic boxes and kept in the greenhouse at 26 °C for a week to germinate. The seedlings of about 2 cm height were transplanted to seedling trays for a month before finally being transferred to 1 l (13 × 11 × 9 cm³) plastic pots (Pöppelmann GmbH & Co. KG, Lohne, Germany) containing 850 ml of steamed soil. The soil was collected from a research site of Freie Universität Berlin (Albrecht-Thaer-Weg) and sieved to remove the remains of roots and pebbles. The sieved soil was steamed for three hours at 90 °C using a Sterilo steamer (Harter Elektrotechnik, Schenkenzell, Germany) to exclude root herbivores. Pots were placed on individual plastic plates and the top layer of the soil was covered with sand grit to prevent the growth of green algae and infestation by fungus gnats (Sciaridae). The plants were assigned to different treatments after 3 weeks of growth in pots. During the experiment, plants were watered three times a week with 150 ml of water and randomized weekly to homogenize for variances due to abiotic factors such as light conditions.

Study insects

The green peach aphid (*M. persicae*) individuals used in this experiment were obtained from the aphid rearing of the Julius Kühn-Institut, Berlin. The larvae of beet armyworm *S. exigua* were obtained from the laboratory cultures maintained at the Freie Universität Berlin. They were reared on artificial diet (wheat germ based basic diet with a vitamin mix) in a climate chamber at 24 °C and 70% humidity under 16/8 h day/night light cycle. Second-stage juveniles (J2s) of root-knot nematodes *M. incognita* were obtained in aqueous suspension from a biological supply company, HZPC Holland B. V. (Hettema Zaaizaad en Pootgoed Coöperatie, Metslawier, The Netherlands).

Herbivory treatments

For the herbivory treatments, a total of 90 healthy and homogeneous plants were selected. Plants were subjected to six different treatments with 15 replicates each: control with no herbivory (C), aboveground herbivory with *M. persicae* aphids (Aph) or *S. exigua* larvae (Spo), belowground herbivory with *M. incognita* nematodes (Nem), and sequential above- and belowground herbivory treatments (Aph + Nem and Spo + Nem) where nematodes were added to the root of the aboveground herbivore-treated plants following a lag phase of seven days. For the aboveground herbivory treatments, the three youngest, fully expanded leaves were chosen on every plant. In the treatments with the chewing herbivore, one third instar *S. exigua* larva was added in a mesh bag and allowed to feed on the first leaf for three days starting with the oldest

among the three chosen leaves. The larva was then transferred successively to the second and the third leaf to feed for another two days on each. This way, larvae fed on three consecutive leaves for a total of seven days. In the treatments with the sucking herbivore, four individuals of *M. persicae* were added on each of the three leaves which were covered with a mesh bag. Aphids were allowed to feed on leaves for seven days and then removed carefully using a fine brush without damaging the leaves. After the removal of aboveground herbivores, the plants were kept for a lag phase of seven days without herbivory. Then, about 1875s stage juveniles (J2's) of root-knot nematodes *M. incognita* were added per pot as belowground herbivore to the roots of half of the aboveground herbivore-treated and half of the control plants. The nematodes were applied in an aqueous suspension in three holes (depth 5 cm) perforated into the soil at a distance of 3 cm from the stem. These plants were treated for 14 days with the nematodes allowing them to infest the roots and induce root galls before harvest. Upon harvest, leaf and root subsamples were collected for the phytohormone analysis. The numbers of galls induced by the nematodes were counted in three different size classes (<1, 1–2 and >2 mm) manually after keeping them submerged in water to facilitate the counting. The root (including galls) and shoot materials were then dried in an oven at 55 °C for three days before measuring the dry mass.

Sampling and measurement of phytohormone

For the phytohormone measurement, the roots of the harvested plants were washed immediately after harvest and 150–180 mg of representative fine root samples were separated and weighed. A similar amount was also collected of leaf samples from the youngest fully expanded leaf by cutting it transversely into small pieces. The leaf and root samples were kept in 2 mL screw-cap tubes, flash frozen in liquid nitrogen and stored at −80 °C until extraction. Extraction and quantification of ABA, SA, JA and JA-isoleucine (JA-Ile) were done following the procedure explained in [46]. In brief, root and leaf samples were homogenized within the tubes using FastPrep homogenizer (FastPrep®-24, MP Biomedicals, Santa Ana, CA, USA) along with 1 ml extraction solution, containing ethyl-acetate and internal deuterated standard mix: 20 ng of D4-SA, D6-ABA (OlChemIm Ltd., Olomouc, Czech Republic) and D6-JA-Ile and 60.4 ng of D6-JA (HPC Standards GmbH, Cunnersdorf, Germany). Supernatant was collected after centrifuging the homogenized samples for 5 min at high speed (18,000×g). Samples were extracted a second time with 1 mL pure ethyl-acetate, then supernatants were combined and dried in a Vacufuge (Eppendorf, Hamburg, Germany). The dried samples were re-eluted in 400 μL of 70% (v/v) methanol (MeOH) and 0.1% acetic acid by shaking 10 min at room temperature. The re-eluted extracts were subjected to a UPLC-ESI–MS/MS Synapt G2-S HDMS (Waters, Milford, Massachusetts, USA) for identification and quantification of phytohormones as described in [46]. The peak area integration was performed using MassLynx Software v. 4.1 (Waters, Milford, Massachusetts, USA). The amount of hormone per g of sample fresh weight was calculated by comparing the peak area of the plant derived hormone in a given sample with the corresponding peak area of the deuterated internal standard in the same sample. From the pool of 15 replicates, eight replicates from each treatment were chosen randomly for hormonal measurement.

Carbon and nitrogen concentration measurement

Dried leaf and root materials were ground in Eppendorf tubes by using a mixer mill (Mixer Mill MM 400, Retsch GmbH, Haan, Germany) and dried again for at least 24 h. Then, their carbon and nitrogen concentration were determined by using an elemental analyzer (Euro EA, HEKAtech GmbH, Wegberg, Germany).

Statistical analysis

All the statistical analyses were performed in 'R', version 3.2.2 [47]. One-way and two-way factorial ANOVAs were performed to test the significance of the treatments; aboveground herbivory (AGH), belowground herbivory (BGH) and their interactions (AGH*BGH). Statistical significance was set at $P < 0.05$. All the data were checked for normality and homogeneity of variances using Shapiro–Wilk test and Bartlett test, respectively, to make sure that they meet the assumptions of ANOVA. The data of number of galls and root C concentration were transformed using log and square transformation, respectively, while the data of shoot biomass and root C/N ratio were transformed using inverse transformation before being checked for assumptions of ANOVA. The phytohormone data were analyzed with Generalized Linear Models (GLM) assuming gamma distribution of errors as the data were not normally distributed. Means and standard errors (SE) are reported in the result section. To determine the effects of the particular aboveground herbivores, the means were additionally compared with Tukey HSD test as post hoc analysis.

Results

Plant biomass

Shoot biomass: Both above- and belowground herbivory had significant main and interaction effects on shoot biomass (AGH: $F_{[2, 84]} = 16.58$, $p = 0.001$; BGH: $F_{[1, 84]} = 5.05$; $p = 0.027$; AGH*BGH: $F_{[2, 84]} = 9.08$; $p < 0.001$). When applied alone, both aphid and S. exigua herbivory reduced the shoot biomass. The negative effect

of S. exigua remained stable under single or sequential herbivory exposure; the negative effect of aphid herbivory was abolished when followed by nematode infestation although nematode infestation alone did not significantly affect shoot biomass (Fig. 1a).

Root biomass: Aboveground herbivory had no significant main effect on root biomass, while belowground herbivory significantly reduced root biomass, which was significantly affected by the interaction with aboveground herbivory (AGH: $F_{[2, 84]} = 1.13$, $p = 0.33$; BGH: $F_{[1, 84]} = 7.07$; $p = 0.001$; AGH*BGH: $F_{[2, 84]} = 6.20$; $p = 0.003$). Earlier S. exigua herbivory followed by the nematode treatment reduced the root biomass by about 25% as compared to S. exigua alone and control plants; aphids and nematodes alone and in combination did not significantly differ from the control (Fig. 1b).

Carbon and nitrogen concentration

We measured the changes in C and N concentration in leaf and root tissue following herbivory to estimate changes in allocation of these major constituents of plant metabolites and because plant as well as herbivore performance parameters are known to depend on C/N contents.

Leaf C and N concentration: None of the herbivory treatments had any significant effect on the foliar C concentration. Aboveground herbivory had a significant main effect on leaf N concentration and a significant interaction effect with belowground herbivory as S. exigua feeding increased foliar N which was stronger and only significant when its herbivory was followed by nematode infestation. Belowground herbivory alone had no effect on foliar N concentration (Table 1).

Root C and N concentration: Both above- and belowground herbivores had main and interaction effect in root C and N concentration (Table 1). The S. exigua herbivory reduced the C concentrations in the root tissues, while the nematode treatment after S. exigua herbivory abolished this effect. Nematodes alone increased root N concentration compared to control plants. This effect was still present in plants previously damaged by S. exigua, but the nematodes had no effect on root N concentration if plant were fed by aphids earlier. Aphids or S. exigua alone had no effect on root N concentration.

C/N ratio: As the C concentration was similar in all treatments; the change in leaf C/N ratio was dependent on changes in leaf N concentration and therefore had similar patterns as leaf N concentration (Table 1). Plants treated with S. exigua followed by nematode decreased the C/N ratio of the leaves but these herbivores alone had no effects. Similarly, there were significant main and interactive effects of both above- and belowground herbivores on the C/N ratio of the roots. Single herbivory by

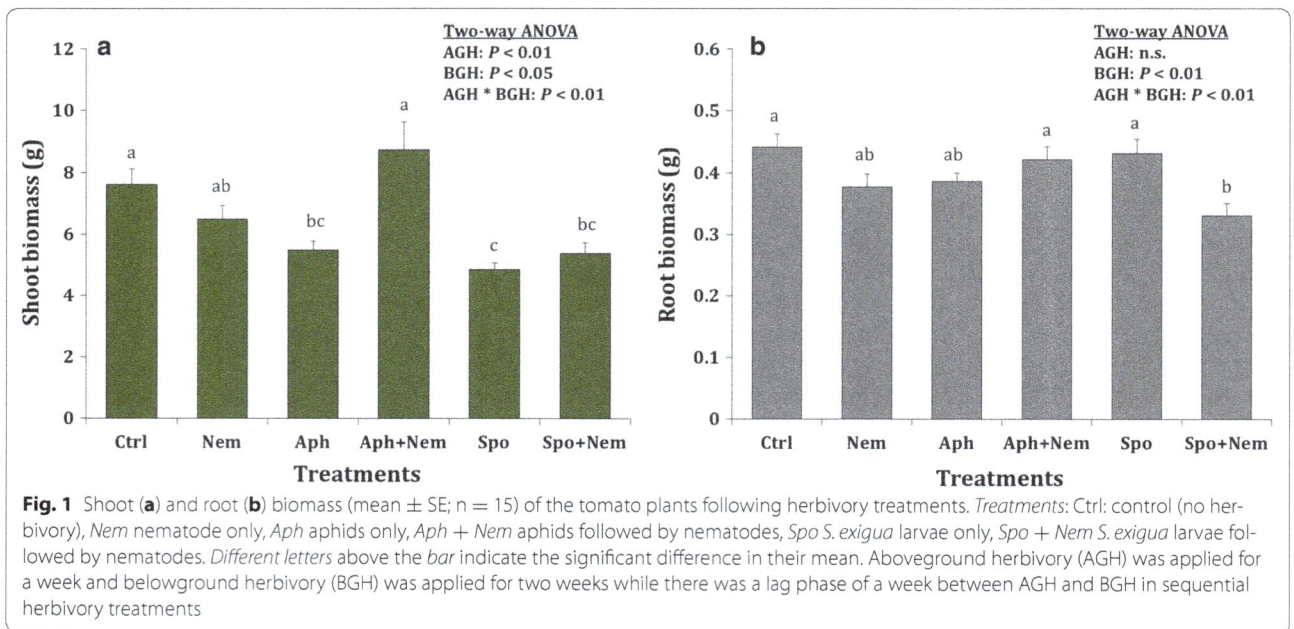

Fig. 1 Shoot (**a**) and root (**b**) biomass (mean ± SE; n = 15) of the tomato plants following herbivory treatments. *Treatments*: Ctrl: control (no herbivory), *Nem* nematode only, *Aph* aphids only, *Aph* + *Nem* aphids followed by nematodes, *Spo* S. *exigua* larvae only, *Spo* + *Nem* S. *exigua* larvae followed by nematodes. *Different letters* above the *bar* indicate the significant difference in their mean. Aboveground herbivory (AGH) was applied for a week and belowground herbivory (BGH) was applied for two weeks while there was a lag phase of a week between AGH and BGH in sequential herbivory treatments

Table 1 The effect of above- and belowground herbivory treatments on C and N concentration (percentage) and their ratios in leaves and roots of the tomato plants

Tissue	Concentration (Mean ± SE; n = 15)					
	Ctrl	Nem	Aph	Aph + Nem	Spo	Spo + Nem
Leaf						
C	38.99 ± 0.36[a]	38.60 ± 0.38[a]	38.66 ± 0.43[a]	38.37 ± 0.5[a]	38.61 ± 0.38[a]	38.56 ± 0.35[a]
N	2.89 ± 0.08[b]	2.92 ± 0.10[b]	2.98 ± 0.08[ab]	2.75 ± 0.04[b]	3.03 ± 0.09[ab]	3.27 ± 0.08[a]
C/N	13.63 ± 0.41[a]	13.41 ± 0.41[a]	13.11 ± 0.38[ab]	14.00 ± 0.25[a]	12.94 ± 0.41[ab]	11.88 ± 0.29[b]
Root						
C	41.86 ± 0.56[ab]	43.37 ± 0.42[a]	41.06 ± 0.69[b]	43.05 ± 0.43[ab]	37.94 ± 0.51[c]	42.64 ± 0.56[ab]
N	2.46 ± 0.08[b]	3.02 ± 0.04[a]	2.61 ± 0.07[b]	2.68 ± 0.06[b]	2.53 ± 0.06[b]	3.08 ± 0.05[a]
C/N	17.18 ± 0.48[a]	14.38 ± 0.21[cd]	15.81 ± 0.36[abc]	16.15 ± 0.37[ab]	15.1 ± 0.39[bcd]	13.88 ± 0.21[d]

	ANOVA results					
	AGH		BGH		AGH:BGH	
	F	P	F	P	F	P
Leaf						
C	0.267	0.766	0.542	0.464	0.094	0.911
N	7.127	*0.001*	0.047	0.8281	4.279	*0.017*
C/N	6.418	*0.003*	0.194	0.6606	3.56	*0.033*
Root						
C	9.706	*<0.001*	38.31	*<0.001*	4.751	*0.011*
N	3.28	*0.042*	61.39	*<0.001*	10.45	*<0.001*
C/N	12.58	*<0.001*	20.04	*<0.001*	9.95	*<0.001*

Treatments: Ctrl: control (no herbivory), *Nem* nematode only, *Aph* aphids only, *Aph* + *Nem* aphids followed by nematodes, *Spo* S. *exigua* larvae only, *Spo* + *Nem* S. *exigua* larvae followed by nematodes. AGH and BGH stand for above- and belowground herbivory respectively. (AGH. *df*: 2, 84; BGH. *df*: 1, 84; AGH:BGH. *df*: 2, 84). Italic fonts indicate the significant effects (P < 0.05) of the treatments. Mean ± SE followed by different letters are significantly different from each other (Tukey HSD test: P < 0.05)

S. exigua and nematode, and sequential herbivory by *S. exigua* followed by nematodes decreased the C/N ratio in the roots as compared to control plants.

Phytohormone induction

There were significant main effects of the above- and belowground herbivores on both salicylic acid (SA) and jasmonic acid (JA) content of the leaf tissues at a time point that was three weeks after the aboveground herbivory and after two weeks of exposure to nematodes. Nematodes had a significant negative main effect on leaf SA content (Fig. 2a), while nematodes either alone or following *S. exigua* herbivory increased the leaf JA content, which did not occur on plants previously infested with aphids (Fig. 2b). Above- and belowground herbivores had significant main effects and interaction effects on root SA content, while the root JA content was affected by the aboveground herbivores only. Whereas nematodes and *S. exigua* alone and in combination significantly reduced SA contents in the roots, previous aphid herbivory abolished this effect of nematodes on root SA (Fig. 2c). On the other hand, *S. exigua* larvae alone or followed by nematodes decreased root JA content compared to control plants (Fig. 2d).

Number of galls induced by nematodes

The total number of galls and number of galls per mg of root tissue induced by nematodes did not differ between the treatments. There was a significant reduction of the number of small galls per mg of root tissue (<1 mm) in the plants previously treated with aphids compared to plants treated with nematodes only ($p = 0.01$) while total number of small galls (not corrected for root mass) tended to be reduced ($p = 0.07$) (Additional file 1).

Discussion

Our study demonstrated that the transient aboveground herbivory by both chewing and sucking herbivores had significant impact on root and shoot parameters, nutrient allocation and the activation of signaling components (phytohormones). The consequences of transient aboveground herbivory on plant traits had no major effect on overall nematode performance (in terms of total number of galls), but plants previously exposed to aphids showed

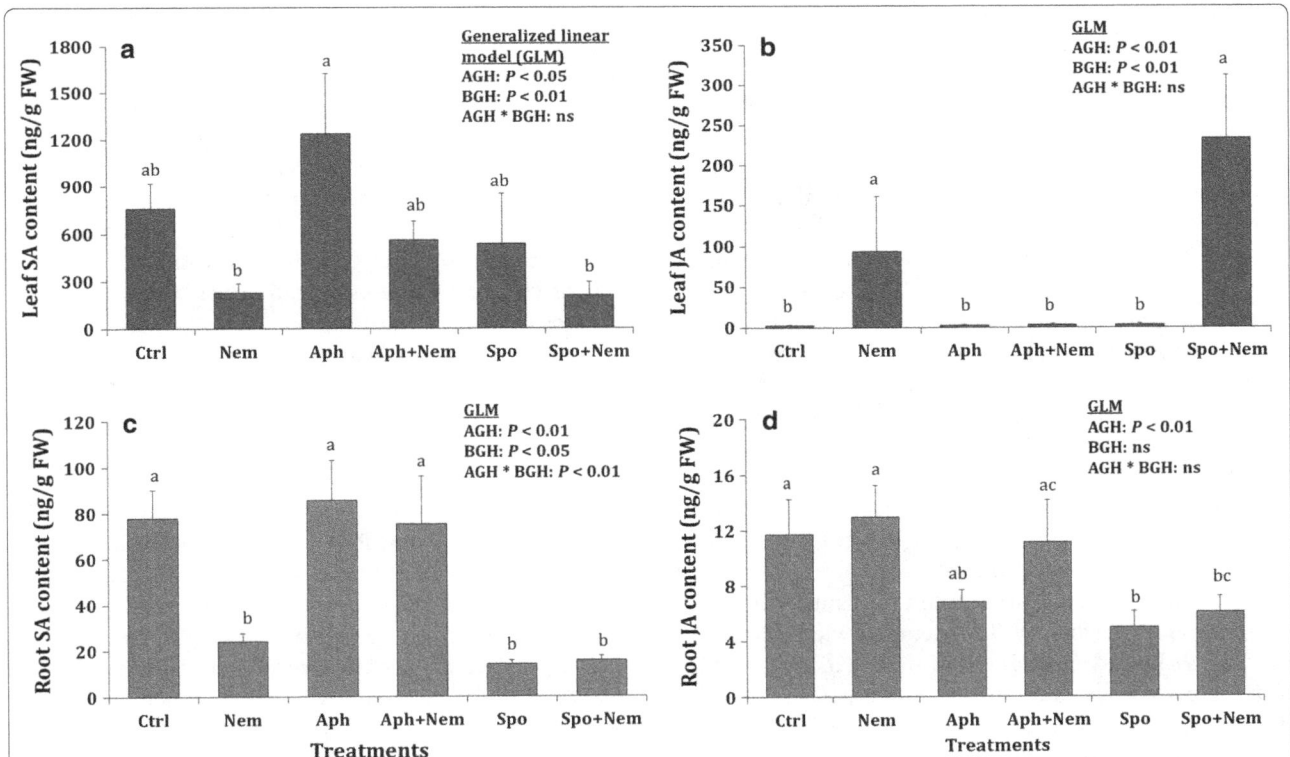

Fig. 2 Leaf (**a**, **b**) and root (**c**, **d**) SA and JA content of the tomato plants (mean ± SE; n = 8) following herbivory treatments. *Treatments*: Ctrl: control (no herbivory), *Nem* nematode only, *Aph* aphids only, *Aph + Nem* aphids followed by nematodes, *Spo S. exigua* larvae only, *Spo + Nem S. exigua* larvae followed by nematodes. *Different letters* above the *bar* indicate the significant difference in their mean. Aboveground herbivory (AGH) was applied for a week and belowground herbivory (BGH) was applied for two weeks while there was a lag phase of a week between AGH and BGH in sequential herbivory treatments

a reduced number of small galls per unit root mass. Transient aboveground herbivory changed the plant response to the later root infestation by nematodes. The way in which the plant response was altered by the sequential herbivory, was different for the two aboveground herbivores highlighting the significance of the herbivores' identities. As the two herbivores used in this experiment exhibit different feeding modes, the plants' distinct response to them could in part have resulted from the different feeding modes [29, 33].

Effect of transient aboveground herbivory on belowground herbivores

A recent meta-analysis suggests that the aboveground herbivore, if it arrives first on the plant, is expected to have negative effects on the performance of belowground herbivores [26]; however we found no negative effects of aboveground herbivores on the overall performance of nematodes in terms of total number of galls. For example, an experiment [24] with cultivated and wild maize plants (*Zea mays mays* and *Z. mays mexicana*) showed that feeding by the aboveground chewing herbivore *Spodoptera frugiperda* had a significant negative effect on the root chewing herbivore *Diabrotica virgifera* in terms of root colonization and weight gain, but only if *S. frugiperda* was added first to the plant. Thus, we expected an overall negative effect of aboveground herbivores, which were added first on the plant, on the belowground herbivore. Additionally, we expected even stronger responses of the plants treated first with aphids, as tomato plants are known to respond with a similar arsenal of defenses against aphids and nematodes, namely the *Mi-1* gene dependent defense responses which require SA signaling [44, 45]. However, although *Mi-1* gene was found in commercial tomatoes, some tomato varieties lack it [44] and it remains unclear whether the MicroTom cultivar contains it. While we did not find elevated SA levels in roots or shoots of plants that had been exposed to aphids three weeks earlier, we found reduced levels of root SA in nematode-infested plants and in plants that had been attacked by *S. exigua* and nematodes. The negative effect of nematodes and *S. exigua* on root SA which likely resulted in a reduced root resistance due to a lack of SA-mediated defense was absent in plants with earlier transient aphid herbivory. The plants previously attacked by aphids showed no reduction in root SA after nematode herbivory, and such plants had a reduced number of smaller galls. The finding of a negative effect of nematodes on root SA that was negated on plants with earlier aphid herbivory highlights the effect of earlier aboveground herbivory and their identity on plant response to subsequent belowground herbivores.

Plant response upon above- and belowground herbivory

Both above- and belowground herbivores had significant effects on plant growth that differed in direction and magnitude. While *S. exigua* herbivory reduced the shoot biomass independent of a later nematode infestation, the negative effect of aphid herbivory on shoot biomass was abolished, when subsequently nematodes fed on the same plants. This suggests that aphid-treated plants showed a compensatory growth of shoots upon nematode herbivory, while *S. exigua* larvae-treated plants did not compensate for the loss in biomass upon nematode herbivory. Nematode addition may have facilitated a tolerance response of the tomato plants such as a compensatory growth to replenish the biomass loss due to aphid herbivory. On the other hand, nematode infestation or *S. exigua* herbivory alone had no significant effects on root biomass, whereas these herbivores in sequential combination reduced the root biomass. This result further demonstrates the altered plant response to subsequent belowground herbivores due to their earlier exposure to aboveground herbivores. Such reduction in root biomass was not evident in the plants treated with aphids followed by nematodes suggesting the significance of aboveground herbivore identity.

Regarding the allocation pattern of C and N in leaf and root tissues, most noticeable effects were found in the N concentration of the plants subjected to sequential above- and belowground herbivory: plants previously exposed to *S. exigua* and followed by nematode infestation contained higher N concentration in both leaf and root tissue. In root tissue, nematodes also increased the N concentration but not when the plants had been previously exposed to aphid feeding. Interestingly, the direction of change in N concentration was opposite to the changes in biomass of the plants infested by *S. exigua* followed by nematode. These results suggest that the nutritional quality may be improved in the shoot and root tissues of the plants whose biomass was decreased in the *S. exigua* followed by the nematode treatment. Systemic nutrient translocation to and away from the site of herbivory is another well-known tolerance response of the plants upon herbivory. Plants allocate carbon and nitrogen in specific cells and tissues to be used for compensatory growth or defense of valuable plant parts which are critical for survival and reproduction [48]. In addition, such diversion of nutrients results in poor nutritional quality of the feeding site with possible negative effects on growth of herbivores [3, 16, 49–51]. Further, increased N in both shoot and root tissues in plants treated with *S. exigua* and later with nematodes may indicate the acquisition of additional N from the soil pool to meet the increased demand of N for either compensatory growth or for biosynthesis of N-based defense compounds such as protease inhibitors (PIs). However, such potential increase in N compounds did not contribute to resistance against nematodes. On the contrary, increased N may also promote herbivore performance by enhancing the nutritional quality of plant tissue.

We measured the defense-regulatory phytohormones JA and SA which may allow some estimation on the level of induced defense in the leaf and root tissue upon sequential above- and belowground herbivory. The defensive functions of SA and JA in tomato against herbivores has been studied in detail in previous studies. For example, SA was found to be an essential component of the *Mi-1* mediated resistance against both aphid and nematode in tomato plants [44, 45]. An earlier study [52] has demonstrated that the JA is also an essential and dominant regulatory component for the induction of not only direct plant defense compounds such as polyphenol oxidases (PPOs) but also indirect plant defense compounds such as volatiles. In addition, defense signaling pathways mediated by these phytohormones are known to coordinate with several other pathways in a complex regulatory network that governs growth and defense physiology of plants and understanding the role of each of such pathways is still a challenge in ecological studies. We found herbivore-specific alterations of phytohormone levels in both leaf and root tissues. Nematode herbivory increased the leaf JA content but not on plants that had been previously exposed to aphids, whereas prior *S. exigua* herbivory did not alter this JA-induction by nematodes. The roots of plants previously attacked by caterpillars had lower JA levels independent of a later nematode infestation. On the other hand, both *S. exigua* and nematode herbivory either alone or in combination decreased the root SA content, while previous aphid herbivory reversed the negative effect of nematodes on root SA, which might be related to the lower number of small nematode galls per root mass in previously aphid infested plants. Although speculative, this finding may indicate an increased nematode resistance of plants upon aphid exposure due to stronger SA-mediated defenses, which would be in line with the concept of defense priming [19, 53]. However, whether a priming of plant defense is involved in the interactions between above- and belowground herbivory that we determined would require further investigation. In general, plant defense is considered to be costly for example in terms of resources that are required for the production of defense compounds [54]. And if the costs of defense outweigh the cost of herbivory, plants may employ other strategies such as tolerance which is an alternative plant strategy to cope with herbivory stress [15]. In our study, tomato plants were able to compensate for the loss of shoot biomass due to

aphid herbivory when they were later exposed to nematodes indicating a tolerance response that is only triggered by the sequential herbivory.

Role of herbivores' identity and feeding mode in plant–insect interaction

As we hypothesized, herbivore identity was a key factor to bring specific changes in plant traits. All the changes in measured parameters such as biomass, C and N distribution and phytohormone content in both leaf and root tissue upon nematode herbivory were dependent on the identity of the shoot herbivores. For example, plants previously treated with *S. exigua* herbivory contained higher N concentration in both leaf and root tissue upon nematode infestation, while previous aphid feeding had no such effect. There is some evidence that the induced response of tomato differs upon herbivory by insects of different feeding guilds; for example, aphid (*Macrosiphum euphorbiae*) feeding was found to induce peroxidase and lipoxygenase, but not PPO and PIs, while noctuid insect *Helicoverpa zea* feeding induced PPO, PIs, and lipoxygenase, but not peroxidase [55]. Similarly, another study [56] showed that herbivory by *S. exigua* increased the PI activity by three times as compared to control plants, whereas aphid (*M. euphorbiae*) herbivory did not induce such effects in tomato plants. For the efficient use of limited resources, plants respond to herbivores by activating a specific array of resistance and tolerance to deter herbivores which share similar characteristics such as feeding mode. Therefore, such specific defense strategies targeted at herbivores with different feeding modes might explain the differences we find in the plant response to sequential attack by aphids, caterpillars and nematodes.

Conclusion

In summary, our study showed for the first time that transient aboveground herbivory modified the plant response to subsequent root herbivory, and herbivores' identity and probably the feeding mode of the aboveground attacker had significant influence on such modification. Although earlier transient herbivory had no detrimental effect on the overall performance of belowground herbivore, the plant responded with compensatory shoot growth to sequential aphid and nematode herbivory. Herbivore-induced plant responses such as compensatory growth and root exudation may affect species across different trophic levels which may eventually affect species composition and diversity in terrestrial ecosystems [9, 57]. Our study provides a small glimpse on the complexity of plant-herbivore interactions and shows that it is important to study interactions between multiple organisms above- and belowground to complement our understanding of plant-herbivore ecology.

Authors' contributions

DK and SW conceived and designed the experiments. AH implemented the experiments and collected data. TL and DK carried out the phytohormone measurements. AH and DK analyzed the data. DK wrote the manuscript, while TL, AS and SW contributed and advised on data analysis and final manuscript preparation. All authors read and approved the final manuscript.

Author details

[1] Functional Biodiversity, Dahlem Centre of Plant Sciences, Institute of Biology, Freie Universität Berlin, Königin-Luise-Str. 1-3, 14195 Berlin, Germany. [2] Molecular Ecology, Dahlem Centre of Plant Sciences, Freie Universität Berlin, Haderslebener Str. 9, 12163 Berlin, Germany.

Acknowledgements

The authors are grateful to Caspar Schöning for his suggestions and help during the entire research period. We are grateful to Monika Fünning, Annegret Plank and Cynthia Kienzle for their help in practical works. We thank Inga Mewis, Julius Kühn-Institute (JKI) Berlin for providing the aphids. We thank Andreas Springer for the help with phytohormone measurements at MS-Core facility, FU Berlin.

Competing interests

The authors declare that they have no competing interests.

Funding

This research work was funded by The German Research Foundation (DFG), Collaborative Research Centre 937 "Priming and Memory of Organismic Responses to Stress".

References

1. Gatehouse JA. Plant resistance towards insect herbivores: a dynamic interaction. New Phytol. 2002;156:145–69.
2. Kessler A, Baldwin IT. Plant responses to insect herbivory: the emerging molecular analysis. Annu Rev Plant Biol. 2002;53:299–328.
3. Schwachtje J, Minchin PEH, Jahnke S, van Dongen JT, Schittko U, Baldwin IT. SNF1-related kinases allow plants to tolerate herbivory by allocating carbon to roots. Proc Natl Acad Sci USA. 2006;103:12935–40.
4. Howe GA, Jander G. Plant immunity to insect herbivores. Annu Rev Plant Biol. 2008;59:41–66.
5. Masters GJ, Brown VK, Gange AC. Plant mediated interactions between aboveground and belowground insect herbivores. Oikos. 1993;66:148–51.
6. Blossey B, Hunt-Joshi TR. Belowground herbivory by insects: influence on plants and aboveground herbivores. Annu Rev Entomol. 2003;48:521–47.
7. Bezemer TM, Wagenaar R, van Dam NM, Wackers FL. Interactions between above- and belowground insect herbivores as mediated by the plant defense system. Oikos. 2003;101:555–62.
8. Bezemer TM, van Dam NM. Linking aboveground and belowground interactions via induced plant defenses. Trends Ecol Evol. 2005;20:617–24.
9. Ohgushi T. Indirect interaction webs: herbivore-induced effects through trait change in plants. Annu Rev Ecol Evol Syst. 2005;36:81–105.
10. Wurst S, Van Dam NM, Monroy F, Biere A, Van der Putten WH. Intraspecific variation in plant defense alters effects of root herbivores on leaf chemistry and aboveground herbivore damage. J Chem Ecol. 2008;34:1360–7.
11. Rasmann S, Agrawal AA. In defense of roots: a research agenda for studying plant resistance to belowground herbivory. Plant Physiol. 2008;146:875–80.
12. van Dam NM. Belowground herbivory and plant defenses. Annu Rev Ecol Evol Syst. 2009;40:373–91.
13. Erb M, Glauser G, Robert CAM. Induced immunity against belowground insect herbivores- activation of defenses in the absence of a jasmonate burst. J Chem Ecol. 2012;38:629–40.
14. Vandermeijden E, Wijn M, Verkaar HJ. Defense and regrowth, alternative plant strategies in the struggle against herbivores. Oikos. 1988;51:355–63.
15. Strauss SY, Agrawal AA. The ecology and evolution of plant tolerance to herbivory. Trends Ecol Evol. 1999;14:179–85.
16. Kaplan I, Halitschke R, Kessler A, Rehill BJ, Sardanelli S, Denno RF. Physiological integration of roots and shoots in plant defense strategies links above- and belowground herbivory. Ecol Lett. 2008;11:841–51.
17. Johnson SN, Hawes C, Karley AJ. Reappraising the role of plant nutrients as mediators of interactions between root- and foliar-feeding insects. Funct Ecol. 2009;23:699–706.
18. Huang W, Siemann E, Yang XF, Wheeler GS, Ding JQ. Facilitation and inhibition: changes in plant nitrogen and secondary metabolites mediate interactions between above-ground and below-ground herbivores. Proc R Soc B Biol Sci. 2013;280(1767):7.
19. Conrath U, Beckers GJM, Flors V, Garcia-Agustin P, Jakab G, Mauch F, Newman M-A, Pieterse CMJ, Poinssot B, Pozo MJ, Pugin A, Schaffrath U, Ton J, Wendehenne D, Zimmerli L, Mauch-Mani B. Priming: getting ready for battle. Mol Plant Microbe Interact. 2006;19:1062–71.
20. Frost CJ, Mescher MC, Carlson JE, De Moraes CM. Plant defense priming against herbivores: getting ready for a different battle. Plant Physiol. 2008;146:818–24.
21. Karban R, Myers JH. Induced plant-responses to herbivory. Annu Rev Ecol Syst. 1989;20:331–48.
22. Wurst S, Ohgushi T. Do plant- and soil-mediated legacy effects impact future biotic interactions? Funct Ecol. 2015;29:1373–82.
23. Wurst S, van der Putten WH. Root herbivore identity matters in plant-mediated interactions between root and shoot herbivores. Basic Appl Ecol. 2007;8:491–9.
24. Erb M, Robert CAM, Hibbard BE, Turlings TCJ. Sequence of arrival determines plant-mediated interactions between herbivores. J Ecol. 2011;99:7–15.
25. Erb M, Ton J, Degenhardt J, Turlings TCJ. Interactions between arthropod-induced aboveground and belowground defenses in plants. Plant Physiol. 2008;146:867–74.
26. Johnson SN, Clark KE, Hartley SE, Jones TH, McKenzie SW, Koricheva J. Aboveground-belowground herbivore interactions: a meta-analysis. Ecology. 2012;93:2208–15.
27. Kutyniok M, Muller C. Plant-mediated interactions between shoot-feeding aphids and root-feeding nematodes depend on nitrate fertilization. Oecologia. 2013;173:1367–77.
28. Kutyniok M, Mueller C. Crosstalk between above- and belowground herbivores is mediated by minute metabolic responses of the host Arabidopsis thaliana. J Exp Bot. 2012;63:6199–210.
29. Walling LL. The myriad plant responses to herbivores. J Plant Growth Regul. 2000;19:195–216.
30. Goggin FL. Plant-aphid interactions: molecular and ecological perspectives. Curr Opin Plant Biol. 2007;10:399–408.
31. Kempema LA, Cui XP, Holzer FM, Walling LL. Arabidopsis transcriptome changes in response to phloem-feeding silverleaf whitefly nymphs. Similarities and distinctions in responses to aphids. Plant Physiol. 2007;143:849–65.
32. Bonaventure G. Perception of insect feeding by plants. Plant Biol. 2012;14:872–80.
33. Bari R, Jones J. Role of plant hormones in plant defence responses. Plant Mol Biol. 2009;69:473–88.
34. Mewis I, Appel HM, Hom A, Raina R, Schultz JC. Major signaling pathways modulate Arabidopsis glucosinolate accumulation and response to both phloem-feeding and chewing insects. Plant Physiol. 2005;138:1149–62.
35. Thaler JS, Humphrey PT, Whiteman NK. Evolution of jasmonate and salicylate signal crosstalk. Trends Plant Sci. 2012;17:260–70.
36. Williamson VM, Hussey RS. Nematode pathogenesis and resistance in plants. Plant Cell. 1996;8:1735–45.

37. Milligan SB, Bodeau J, Yaghoobi J, Kaloshian I, Zabel P, Williamson VM. The root knot nematode resistance gene *Mi* from tomato is a member of the leucine zipper, nucleotide binding, leucine-rich repeat family of plant genes. Plant Cell. 1998;10:1307–19.

38. Bird DM, Kaloshian I. Are roots special? Nematodes have their say. Physiol Mol Plant Pathol. 2003;62:115–23.

39. Carolan JC, Caragea D, Reardon KT, Mutti NS, Dittmer N, Pappan K, Cui F, Castaneto M, Poulain J, Dossat C, Tagu D, Reese JC, Reeck GR, Wilkinson TL, Edwards OR. Predicted effector molecules in the salivary secretome of the pea aphid (*Acyrthosiphon pisum*): a dual transcriptomic/proteomic approach. J Proteome Res. 2011;10:1505–18.

40. Rossi M, Goggin FL, Milligan SB, Kaloshian I, Ullman DE, Williamson VM. The nematode resistance gene *Mi* of tomato confers resistance against the potato aphid. Proc Natl Acad Sci USA. 1998;95:9750–4.

41. Vos P, Simons G, Jesse T, Wijbrandi J, Heinen L, Hogers R, Frijters A, Groenendijk J, Diergaarde P, Reijans M, Fierens-Onstenk J, de Both M, Peleman J, Liharska T, Hontelez J, Zabeau M. The tomato *Mi-1* gene confers resistance to both root-knot nematodes and potato aphids. Nat Biotechnol. 1998;16:1365–9.

42. Nombela G, Williamson VM, Muniz M. The root-knot nematode resistance gene Mi-1.2 of tomato is responsible for resistance against the whitefly *Bemisia tabaci*. Mol Plant Microbe. 2003;16:645–9.

43. Kaloshian I, Lange WH, Williamson VM. An aphid-resistance locus is tightly linked to the nematode-resistance gene, *Mi*, in tomato. Proc Natl Acad Sci USA. 1995;92:622–5.

44. Branch C, Hwang CF, Navarre DA, Williamson VM. Salicylic acid is part of the *Mi-1*-mediated defense response to root-knot nematode in tomato. Mol Plant Microbe Interact. 2004;17:351–6.

45. Li Q, Xie QG, Smith-Becker J, Navarre DA, Kaloshian I. *Mi-1*-mediated aphid resistance involves salicylic acid and mitogen-activated protein kinase signaling cascades. Mol Plant Microbe Interact. 2006;19:655–64.

46. Nguyen D, D'Agostino N, Tytgat TOG, Sun P, Lortzing T, Visser EJW, Cristescu SM, Steppuhn A, Mariani C, van Dam NM, Rieu I. Drought and flooding have distinct effects on herbivore-induced responses and resistance in *Solanum dulcamara*. Plant, Cell Environ. 2016;39:1485–99.

47. R Core Team. R: a language and environment for statistical computing. 2015. R Foundation for Statistical Computing, Vienna. https://www.r-project.org.

48. Creelman RA, Mullet JE. Biosynthesis and action of jasmonates in plants. Annu Rev Plant Phys. 1997;48:355–81.

49. Babst BA, Ferrieri RA, Gray DW, Lerdau M, Schlyer DJ, Schueller M, Thorpe MR, Orians CM. Jasmonic acid induces rapid changes in carbon transport and partitioning in Populus. New Phytol. 2005;167:63–72.

50. Gomez S, Ferrieri RA, Schueller M, Orians CM. Methyl jasmonate elicits rapid changes in carbon and nitrogen dynamics in tomato. New Phytol. 2010;188:835–44.

51. Gomez S, Steinbrenner AD, Osorio S, Schueller M, Ferrieri RA, Fernie AR, Orians CM. From shoots to roots: transport and metabolic changes in tomato after simulated feeding by a specialist lepidopteran. Entomol Exp Appl. 2012;144:101–11.

52. Thaler JS, Farag MA, Pare PW, Dicke M. Jasmonate-deficient plants have reduced direct and indirect defences against herbivores. Ecol Lett. 2002;5:764–74.

53. Hilker M, Schwachtje J, Baier M, Balazadeh S, Bäurle I, Geiselhardt S, Hincha DK, Kunze R, Mueller-Roeber B, Rillig MC, Rolff J, Romeis T, Schmülling T, Steppuhn A, van Dongen J, Whitcomb SJ, Wurst S, Zuther E, Kopka J. Priming and memory of stress responses in organisms lacking a nervous system. Biol Rev. 2015. doi:10.1111/brv.12215.

54. Karban R. The ecology and evolution of induced resistance against herbivores. Funct Ecol. 2011;25:339–47.

55. Stout MJ, Workman KV, Bostock RM, Duffey SS. Specificity of induced resistance in the tomato, *Lycopersicon esculentum*. Oecologia. 1998;113:74–81.

56. Rodriguez-Saona CR, Musser RO, Vogel H, Hum-Musser SM, Thaler JS. Molecular, biochemical, and organismal analyses of tomato plants simultaneously attacked by herbivores from two feeding guilds. J Chem Ecol. 2010;36:1043–57.

57. Bardgett RD, Wardle DA, Yeates GW. Linking above-ground and belowground interactions: how plant responses to foliar herbivory influence soil organisms. Soil Biol Biochem. 1998;30:1867–78.

PERMISSIONS

The contributors of this book come from diverse backgrounds, making this book a truly international effort. This book will bring forth new frontiers with its revolutionizing research information and detailed analysis of the nascent developments around the world.

We would like to thank all the contributing authors for lending their expertise to make the book truly unique. They have played a crucial role in the development of this book. Without their invaluable contributions this book wouldn't have been possible. They have made vital efforts to compile up to date information on the varied aspects of this subject to make this book a valuable addition to the collection of many professionals and students.

This book was conceptualized with the vision of imparting up-to-date information and advanced data in this field. To ensure the same, a matchless editorial board was set up. Every individual on the board went through rigorous rounds of assessment to prove their worth. After which they invested a large part of their time researching and compiling the most relevant data for our readers.

The editorial board has been involved in producing this book since its inception. They have spent rigorous hours researching and exploring the diverse topics which have resulted in the successful publishing of this book. They have passed on their knowledge of decades through this book. To expedite this challenging task, the publisher supported the team at every step. A small team of assistant editors was also appointed to further simplify the editing procedure and attain best results for the readers.

Apart from the editorial board, the designing team has also invested a significant amount of their time in understanding the subject and creating the most relevant covers. They scrutinized every image to scout for the most suitable representation of the subject and create an appropriate cover for the book.

The publishing team has been an ardent support to the editorial, designing and production team. Their endless efforts to recruit the best for this project, has resulted in the accomplishment of this book. They are a veteran in the field of academics and their pool of knowledge is as vast as their experience in printing. Their expertise and guidance has proved useful at every step. Their uncompromising quality standards have made this book an exceptional effort. Their encouragement from time to time has been an inspiration for everyone.

The publisher and the editorial board hope that this book will prove to be a valuable piece of knowledge for researchers, students, practitioners and scholars across the globe.

LIST OF CONTRIBUTORS

Carola Petersen, Ruben Joseph Hermann, Mike-Christoph Barg, Rebecca Schalkowski, Philipp Dirksen, Camilo Barbosa and Hinrich Schulenburg
Department of Evolutionary Ecology and Genetics, Zoological Institute Christian-Albrechts University, Am Botanischen Garten 1-9, 24118 Kiel, Germany

Wopke van der Werf
Centre for Crop Systems Analysis, Wageningen University, P.O. Box 430, 6700 AK Wageningen, The Netherlands

Yi Zou
Centre for Crop Systems Analysis, Wageningen University, P.O. Box 430, 6700 AK Wageningen, The Netherlands
Department of Environmental Science, Xi'an Jiaotong-Liverpool University, Suzhou 215123, China

Haijun Xiao
Institute of Entomology, Jiangxi Agricultural University, Nanchang 330045, China

Felix J. J. A. Bianchi
Farming Systems Ecology, Wageningen University, P.O. Box 430, 6700 AK Wageningen, The Netherlands

Frank Jauker
Department of Animal Ecology, Justus Liebig University, Heinrich-Buff-Ring 26-32, 35932 Giessen, Germany

Shudong Luo
Institute of Apicultural Research, Chinese Academy of Agricultural Sciences, Beijing 100093, China

Paulo Milet-Pinheiro
Institute of Evolutionary Ecology and Conservation Genomics, University of Ulm, Helmholtzstraße 10-1, 89081 Ulm, Germany
Present Address: Departamento de Química Fundamental, Universidade Federal de Pernambuco, Av. Prof. Moraes Rego, s/n, Recife 50670-901, Brazil

Kerstin Herz and Manfred Ayasse
Institute of Evolutionary Ecology and Conservation Genomics, University of Ulm, Helmholtzstraße 10-1, 89081 Ulm, Germany

Stefan Dötterl
Department of Ecology and Evolution, University of Salzburg, Hellbrunnerstrasse 34, 5020 Salzburg, Austria

Sinead Phelan and Ewen Mullins
Dept. Crop Science, Teagasc, Oak Park, Carlow, Ireland

Vilma Ortiz
Dept. Crop Science, Teagasc, Oak Park, Carlow, Ireland
Plant Biology and Crop Science, Rothamsted Research Station, West Common, Harpenden, Hertfordshire AL5 2JQ, UK

Oz Barazani and Yoni Waitz
Institute of Plant Sciences, Israel Plant Gene Bank, Agricultural Research Organization, 75359 Rishon LeZion, Israel

Arnon Dag
Department of Fruit Tree Sciences, Institute of Plant Sciences, Agricultural Research Organization, Gilat Research Center, 85280 M.P. Negev 2, Israel

Zohar Kerem
Institute of Biochemistry, Food Science and Nutrition, Faculty of Agricultural, Food and Environmental Quality Sciences, The Hebrew University of Jerusalem, 76100 Rehovot, Israel

Yizhar Tugendhaft
Department of Fruit Tree Sciences, Institute of Plant Sciences, Agricultural Research Organization, Gilat Research Center, 85280 M.P. Negev 2, Israel
Institute of Biochemistry, Food Science and Nutrition, Faculty of Agricultural, Food and Environmental Quality Sciences, The Hebrew University of Jerusalem, 76100 Rehovot, Israel

Michael Dorman
Department of Geography and Environmental Development, Ben-Gurion University of the Negev, 84105 Beer-Sheva, Israel

Mohammed Hamidat and Thameen Hijawi
Arab Agronomist Association, Al Nahda St., Ramallah and Al-Bireh Governorate, 4504 Al-Bireh, Palestine

Erik Westberg and Joachim W. Kadereit
Institut für Spezielle Botanik und Botanischer Garten, Johannes Gutenberg- Universität Mainz, 55099 Mainz, Germany

Jacek Bartlewicz, Olivier Honnay and Hans Jacquemyn
Biology Department, Plant Conservation and Population Biology, KU Leuven, Kasteelpark Arenberg 31, 3001 Heverlee, Belgium

Bart Lievens
Laboratory for Process Microbial Ecology and Bioinspirational Management (PME and BIM), Department of Microbial and Molecular Systems (M2S), KU Leuven, Campus De Nayer, Fortsesteenweg 30A, 2860 Sint-Katelijne Waver, Belgium

Kalle Remm and Liina Remm
Institute of Ecology and Earth Sciences, University of Tartu, Vanemuise 46, 51014 Tartu, Estonia

Nadine Ali
Plant Protection Department, Faculty of Agriculture, Tishreen University, PO Box 2233, Latakia, Syrian Arab Republic IRD, UMR CBGP, 755 Avenue du Campus Agropolis, CS30016, 34988 Montferrier-sur-Lez Cedex, France

Johannes Tavoillot, Thierry Mateille and and Odile Fossati-Gaschignard
IRD, UMR CBGP, 755 Avenue du Campus Agropolis, CS30016, 34988 Montferrier-sur-Lez Cedex, France

Guillaume Besnard
CNRS, UMR EDB, Université Toulouse III Paul Sabatier, Bâtiment 4R1, 118 Route de Narbonne, 31062 Toulouse Cedex 9, France

Bouchaib Khadari and Laila Essalouh
UMR AGAP, SUPAGRO, Campus CIRAD, TAA-108/03, Avenue Agropolis, 34398 Montpellier Cedex 5, France

Ewa Dmowska and Grażyna Winiszewska
Museum and Institute of Zoology PAS, Wilcza 64, 00-679 Warsaw, Poland

Mohammed Ater
Faculté des Sciences et Techniques, Université Abdelmalek Essaadi, BP 2062, 93030 Tétouan, Morocco

Abdelhamid El Mousadik and Mohamed Aït Hamza
Laboratoire LBVRN, Faculté des Sciences d'Agadir, Université Ibn Zohr, BP 8106, 80000 Agadir, Morocco

Aïcha El Oualkadi and Abdelmajid Moukhli
INRA, CRRA, BP 513, 40000 Marrakech, Morocco

Ahmed El Bakkali
INRA, UMR APCRPG, BP 578, 50000 Meknes, Morocco

Elodie Chapuis
IRD, UMR CBGP, 755 Avenue du Campus Agropolis, CS30016, 34988 Montferrier-sur-Lez Cedex, France IRD, UMR IPME (IRD/Université de Montpellier/CIRAD), 911 Avenue Agropolis, BP 64501, 34394 Montpellier Cedex 5, France UMR PVBMT, 3P-CIRAD, 7 chemin de l'Irat, Ligne paradis, 97410 Saint Pierre, Réunion

Sophie Groendahla
Cologne Biocenter, Workgroup Aquatic Chemical Ecology, University of Cologne, Zuelpicher Strasse 47b, 50674 Koeln, Germany

Patrick Fink
Cologne Biocenter, Workgroup Aquatic Chemical Ecology, University of Cologne, Zuelpicher Strasse 47b, 50674 Koeln, Germany Institute for Zoomorphology and Cell Biology, Heinrich-Heine University of Duesseldorf, Universitaetsstrasse 1, 40225 Duesseldorf, Germany

Shicai Shen, Gaofeng Xu, Guimei Jin, Shufang Liu, Yanxian Yang, Aidong Chen and Fudou Zhang
Agricultural Environment and Resource Research Institute, Yunnan Academy of Agricultural Sciences, Kunming 650205, Yunnan, China

David Roy Clements,
Biology Department, Trinity Western University, 7600 Glover Road, Langley, BC V2Y1Y1, Canada

Hisashi Kato-Noguchi
Department of Applied Biological Science, Faculty of Agriculture, Kagawa University, Miki, Kagawa 761-0795, Japan

Thomas Sauvage, William E. Schmidt, and Suzanne Fredericq
Department of Biology, University of Louisiana at Lafayette, Lafayette, LA 70503, USA

Shoichiro Suda
Department of Marine Science, Biology and Chemistry, University of the Ryukyus, Nishihara, Okinawa 903-0213, Japan

Jennifer L. Kovacs
Spelman College, 350 Spelman Lane, S.W., Atlanta, GA 30314, USA

Candice Wolf and Seth Wolf
University of Vermont, Larner College of Medicine, 89 Beaumont Ave, Burlington, VT 05405, USA

Dené Voisin
Neuroscience Institute, Georgia State University, Petit Science Center, Atlanta, GA 30303, USA

Pritesh Sundar Roy, Rashmita Samal, Gundimeda Jwala Narasimha Rao, Sasank Sekhar Chyau Patnaik, Nitiprasad Namdeorao Jambhulkar and Ashok Patnaik
Central Rice Research Institute, Cuttack 753006, India

Trilochan Mohapatra
Central Rice Research Institute, Cuttack 753006, India
ICAR, DARE, New Delhi 110001, India

Katja Steinauer and Nico Eisenhauer
German Centre for Integrative Biodiversity Research (iDiv) Halle-Jena-Leipzig, Deutscher Platz 5e, 04103 Leipzig, Germany
Institute of Biology, Leipzig University, Johannisallee 21, 04103 Leipzig, Germany

Britta Jensen, Stefan Scheu and Tanja Strecker
J. F. Blumenbach Institute of Zoology and Anthropology, Georg-August-University Göttingen, Berliner Straße 28, 37073 Göttingen, Germany

Enrica de Luca
Institute of Evolutionary Biology and Environmental Studies, University of Zurich, Zurich 8057, Switzerland

Dinesh Kafle, Anne Hänel and Susanne Wurst
Functional Biodiversity, Dahlem Centre of Plant Sciences, Institute of Biology, Freie Universität Berlin, Königin-Luise-Str. 1-3, 14195 Berlin, Germany

Tobias Lortzing and Anke Steppuhn
Molecular Ecology, Dahlem Centre of Plant Sciences, Freie Universität Berlin, Haderslebener Str. 9, 12163 Berlin, Germany

Index

www.ingramcontent.com/pod-product-compliance
Lightning Source LLC
Chambersburg PA
CBHW082030190326
41458CB00010B/3324